Natural Growth Inhibitors and Phytohormones in Plants and Environment

Natural Growth Inhibitors and Phytohormones in Plants and Environment

by

Valentine I. Kefeli
Biomost, Inc., Pennsylvania, U.S.A.

and

Maria V. Kalevitch
Robert Morris University, Pennsylvania, U.S.A.

Editor

Bruno Borsari
Slippery Rock University, Pennsylvania, U.S.A.

KLUWER ACADEMIC PUBLISHERS
DORDRECHT / BOSTON / LONDON

A C.I.P. Catalogue record for this book is available from the Library of Congress.

ISBN 1-4020-1069-9

Published by Kluwer Academic Publishers,
P.O. Box 17, 3300 AA Dordrecht, The Netherlands.

Sold and distributed in North, Central and South America
by Kluwer Academic Publishers,
101 Philip Drive, Norwell, MA 02061, U.S.A.

In all other countries, sold and distributed
by Kluwer Academic Publishers,
P.O. Box 322, 3300 AH Dordrecht, The Netherlands.

Printed on acid-free paper

Printed in the Netherlands.

CONTENTS

PREFACE

This book represents the authors' lifetime dedication to the study of inhibitors and phytohormones as well as its practical applications for achieving a more sustainable agriculture. Their work focuses on the functions of various groups of active molecules, their direct effect upon plant growth, but also implications for their impact upon the surrounding environment are explored. The main idea of the book evolved from the need to determine a balance among natural growth inhibitors and phytohormones. This approach was pursued through a better understanding of their biochemical pathways, their effects on plants physiological functions, and their influence upon stress factors on plant ontogenesis. Therefore, this effort proposes a more holistic approach to the study of plant physiology, in which the plant-soil interactions are discussed, with a profound description of different allelochemicals and their effects on plants growth.

A rigorous attention is also paid to discuss the role of microorganisms in ecosystems and their capability to synthesize physiologically active substances, which trigger also unique plant-microbial interactions. These synergies are leading scientists to the discovery of major breakthroughs in agriculture and pharmacology that are revolutionizing old epistemologies and thus, contributing to the emergence of a philosophy of interconnectedness for the whole biosphere.

The initial conceptions about plant hormone-inhibitor interactions were based upon the research accomplished by the authors at the Russian Academy of Sciences during the last three decades. The outcome of these efforts have been published in major American and European, scientific journals. Such an impressive record of scholarly endeavors inspired Kefeli to write the book "Natural Growth Inhibitors and Phytohormones" which was published in Holland, in 1978. Since then, many other works were completed, including a seminal paper on "Natural Inhibitors of Phenolic and Terpenoid Nature", that appeared in 1984, in Biological Review, Cambridge, UK.

Since the late eighties Dr. Maria Kalevitch (Filimonova) has been contributing to this research agenda with her investigations on growth hormones and inhibitors produced by gray molds, which infect blueberry plants. Her research substantiated the main concept of effect of these biologically active substances found in plants and other organisms as well. This theoretical framework enhanced a step further the authors' interest in directing their studies to the broader ecosystem-level and the niches played by these groups of molecules, within their respective food webs. Dr. Kalevitch's pioneering research continued through the nineties allowing her to pursue a better understanding about the role of allelopathogens produced by woody plants. The opportunities offered by the effect of these molecules as natural herbicides constituted the initial conceptual foundation that fostered the publication of this new book.

The innovative theoretical underpinnings and ideas were presented and discussed at numerous international meetings and symposia. The theory of natural allelochemicals and their possible use as environmentally friendly herbicides was introduced also in the authors' teaching agendas, during the most recent years (1998-2000), at different colleges and universities across the USA. Curricula in which the above mentioned principles were proposed consisted primarily in courses in, but not limited to, Environmental Science, Environmental Health, Ecology, and also student post-graduate research.

This book maintains a comprehensive approach to the study of plant physiology to its end. Its educational objectives are pursued also through the presentation of research that spanned within a 50-60 years time frame and didactic methods of easy implementation. The content of this work extends itself to reach many other disciplines among the environmental and agricultural sciences in particular, for its direct implications to a renovated philosophy of ecosystem management, restoration and sustainability. Its approach to the idea of plant growth stimulators-inhibitors is founded upon chemical and biological tests, as these methodologies allowed the investigators to evaluate the environmental impact of the 'regulators' under study. Appropriate laboratory techniques made ft possible to maintain the consistency necessary to validate the accuracy of the research findings.

It is impossible to foresee and to predict at this time the acceptance of all the ideas and concepts presented in this work, within a broad-spectrum range of the scientific community. However, despite its limitations, this effort should be unanimously appreciated as it demonstrates a real passion to revitalize energies for the study of plant physiology and its potential to contribute to the development of a more sustainable world. Additionally, the educational perspective of the book cannot pass unnoticed as it contributes successfully to the enhancement of teaching and learning in the plant sciences. Finally, and most importantly, the work of Kefeli and Kalevitch will serve to stimulate more debates on the issue and engage the efforts of present and future researchers in an interdisciplinary manner.

Professor Bruno Borsari
Slippery Rock University
Slippery Rock, PA
USA, 2002

ACKNOWLEDGMENT

We would like to thank the people who shared their valuable domain knowledge and inspired us in writing this book. We express cordial thanks to the Professor M. Kh. Chailakhjan - editor of the first edition of this book.

Authors express special appreciation to the members of BIOMOST, INC, USA- Mrs. Margaret Dunn and Mr. Tim Danehy for their sponsorship of our research for all these years.

Our cordial thanks go to Dr. Alex E. Kalevitch without whose research, diligent and thorough preparation of this publication, the book itself will not be possible. By his heroic effort this book was born!

Our book certainly will not be possible without the strong support of our colleagues from different European and American institutions whose contributions to this book always greatly and sincerely appreciated.

Our thanks go to our families, especially to Mrs. Galina Mzhen, without whose patience, love, support, and total devotion this book will never come to life.

Dr. Valentine Kefeli,
Member of New York Academy of Sciences

And

Dr. Maria Kalevitch,
Member of American Society of Microbiology

Slippery Rock, PA, USA. 2002

INTRODUCTION

Concept of plant growth stimulation and inhibition is tightly connected with the function of some secondary substances. They include indoles, phenols, terpenoids, adenine derivates and others.

These classes of substances contain neutral and active products. The last form is present in minute amounts, and after their function is concluded in the plant, they can be easily inactivated (phytohormones and some inhibitors). Here we are discussing the products which can regulate plant growth and development, including differentiation, tissues growth and specialization, and organ interactions. Organ interactions involve such processes as: polarity, regeneration, reaction of the whole plant on the environmental factors: gravity, light, temperature, gas regim, microorganisms, and much more. Some of these substances will act as allechemicals or allelopathogens, which are secreted from the leaves and roots of the trees and herbs, and demonstrate some effect on the other plants. Allelochemicals are relative stable in the soil, and resist the neutralizing effects of microorganisms.

In case of higher plants they express both: more or less resistance or sensitivity to microorganisms. In some cases microbes can change the growth activity of higher plants, examples are *Fusarium* that produces gibberellins which stimulate stem elongation, *Taphrina* - produces cytokinins and auxins which activate tumor formation or suppress apical dominance. Another great example is *Botrytis* , it produces abscisic acid (ABA) which is an active inhibitor in the root and shoot growth retardation. However, many aspects of hormonal biosynthesis and hormonal functions in fungus are still unclear.

But now is more or less obvious that ABA regulates stomata movement; cytokinins involve in cell multiplication and greening of the leaves; auxins induce cell elongation and root regeneration; phenolic inhibitors arrest seed germination, cell elongation, and decrease the intensity of respiration. Some growth regulators synthesize in one place, and transported over the long distance, where they produce growth regulating effect. Some, like phenolic inhibitors synthesize in one certain place, and are not so movable as their precursors- amino acids or organic acids (Kefeli, Kadyrov, 1971; Kefeli, 1978; Mok and Mok, 2001; Richards et al., 2001)

Plants in the environment can be recipients and accept the effect of other plants and microorganisms. On the other hand, plants could be donors and produce growth regulating substances which could be secreted and act as botanical herbicides which inhibit selectively the growth of some species. This might be an ecological tool for the formation of natural ecosystem (Zucconi, 1996)

With the help of auxins and cytokinins roots of legumes and some other species can form nodules, where the process of nitrogen fixation by *Rizobium* bacteria proceeds. At the same time bacteria of the same family, called *Agrobacterium* induce tumor formation and genome changes (transgenosis) via tumor formation in the other species (Lynn, Chang 1990; Chilton, 2001; Long, 2001).

Thus hormones and inhibitors can be environmental factors with positive or negative signs and effects. Biosynthesis, transport and function of growth regulating substances are under the genetic and environmental control (Kende, 2001). In general these factors regulate the process of biological rhythms (so-called "Biological clock function"). Needless to say how important it is to know the targets of hormones and inhibitors action (Golden, Stayer, 2001)

These targeted centers could be on the cellular level, where growth inhibitors and stimulators could interact with the cellular receptors. However, not much yet clear about how these receptors are connected with the system of effect multiplication. However this processes either integrate cellular receptor effects on the tissues targeting center or function on the organ level.

The investigation of mutant plants helped to understand how genes of dwarfism block the process of biosynthesis of hormone-stimulators (like gibberellin), and activate the process of inhibitor function (ABA and some phenolic biosynthesis). Some mutants are able to produce inhibitors (ABA, ethylene) as factors of resistance to UV damage. The process of effect multiplication could be a part of the hormone-inhibitor field function, which regulates such processes as dormancy, growth, and transition to flowering, fruit production (Kefeli, 1997)

No doubt not all ideas, which are presented in this book, could be accepted by researches. But we hope that even some doubtful conceptions will stimulate the plant physiologists to confirm or appose them. That was our main task when we wrote this book, to spark the research creativity and enginuity.

Professor Valentine I. Kefeli
Professor Maria V. Kalevitch
(Slippery Rock, U.S.A.)

CHAPTER 1

SYSTEM OF GROWTH AND DEVELOPMENT REGULATION IN THE PLANT

1.1 Properties of natural hormones

Since the 1960s' the knowledge of main phytohormones and natural inhibitors in green plants has remained unchanged. It consists of the group of stimulators such as auxins, gibberellins, cytokinins as well as natural inhibitors like abscisic acid, ethylene and phenolic inhibitors.

Beside the group of growth regulators there is a number of secondary substances present in both groups, and characterized by inhibitory, or stimulatory properties.

1.1.1 Auxins

Auxins belong to one of the best-known class of phytohormones. In plants, auxins are represented mainly by indol-3-acetic acid (IAA). Auxins take part in regulating various growth and formation processes such as: elongation and curvature of the coleoptile, promotion of rhizogenesis, inhibition of axillary bud growth, prevention of fruit abscission, induction of parthenocarpy in fruits. The site of auxin synthesis is the apical meristem of stems and roots. The amount of auxins formed in the tips of stems is greater than that in roots. Formed in apices, auxins are translocated down the stem and up the root, retarding the growth of lateral organs (shoots, buds and roots). Cutting off the tip of a shoot or root promotes vigorous development of lateral organs. The application of lanolin paste with auxin on the apex restores the dominant position of this region (the apex) while suppressing the growth of lateral apices and leaf rudiments. The tips of etiolated seedlings of wheat and oat containing large amounts of auxins readily release these products into agar blocks. If such auxin-rich agar blocks are applied to seedlings they are able to cause the curvature of the coleoptiles in the extension zone.

Auxins produce also pronounced effects in remote plant organs. Being synthesized in regions of high meristematic activity, in the stem apex, auxins move downward, stimulate extension of cells in coleoptiles located far from the apices, and prevent the abscission of fruits and leaves by attracting nutrients to the places in which they are accumulated. (*Figure 1.1*)

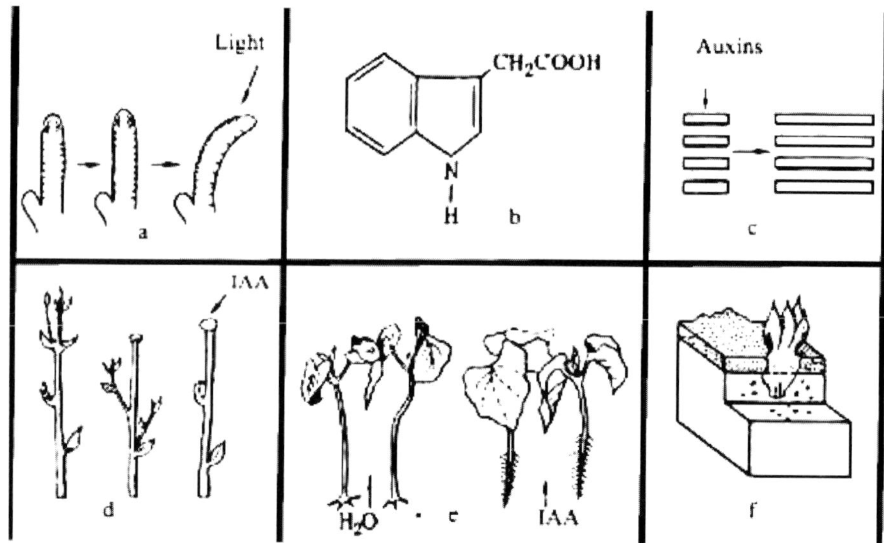

*Figure1.*1 Effects of Auxins in plants.
a. – Phototropism; b. – Structural formula of IAA; c. – cell elongation of coleoptiles; d. – Apical dominance; e. – root formation (left – root formation with water, right – treatment with IAA); f. – xylemagenesis.

However, it would be wrong to consider auxins only as factors that regulate cell extension. Auxins are involved in active regeneration processes occurring in a plant, such as restoration of missing organs, initiation of vegetative buds, reproduction of callus cells. Without auxins, calluses formed on plant tissues and isolated into a culture plate could not divide, nor start organogenesis. Auxins and their analogues are able to promote the effects of roots regeneration. This effect was well investigated in many laboratories of plant physiology and it can be easily reproduced in green house conditions (*Figure 1.2).*

1.1.2 Gibberellins

Gibberellins are formed chiefly in the leaves of a plant though there is evidence in the literature that they can also be synthesized in roots. Whereas the hysiological features of auxins were described as far back as in the 1920's the properties of gibberellins were discovered much later, by the middle fifties. The class of gibberellins is extremely large. We already know some forty of these hormones with a closely similar molecular structure. All these hormones are fluorene compounds having a structure of gibbane.

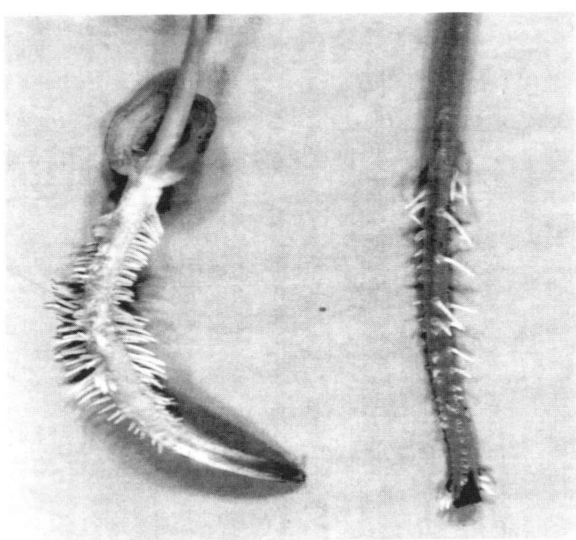

*Figure 1.2.*Rooting and Cloning of Green Bean.
Left – treatment of cuttings by auxin – indolbutiric acid (70 mg/L.)
Right – control (water). It is observed definite hormonal effect "brush of root".
(Slippery Rock University, Biology Department, 1998)

The discovery of endogenous auxins by F.Went and N. Kholodny in the 1920s' stimulated more research producing the creation of synthetic analogues, which are used both as root stimulators and herbicides. In general auxin function in plants is based on its biosynthesis (Bartel, 1997), transpost (Estelle, 1968; Jenses et al., 1998), and effect on growth processes (Keller, Volkenurgh, 1998).

The main property of gibberellins consists in their ability to stimulate elongation of plant shoots and to induce growth of stems in rosette and dwarf forms. In promoting stem elongation, gibberellins do not affect the number of internodes, but only stimulate their cell elongation (*Figure 1.3*). Similar to auxins, gibberellins are able to attract nutrients to the sites of their localization and thus, retard growth processes in other parts of the plant. Usually, plants treated with gibberellins develop intensively the above ground part and their lateral buds come to life while the growth of their roots is strongly inhibited.

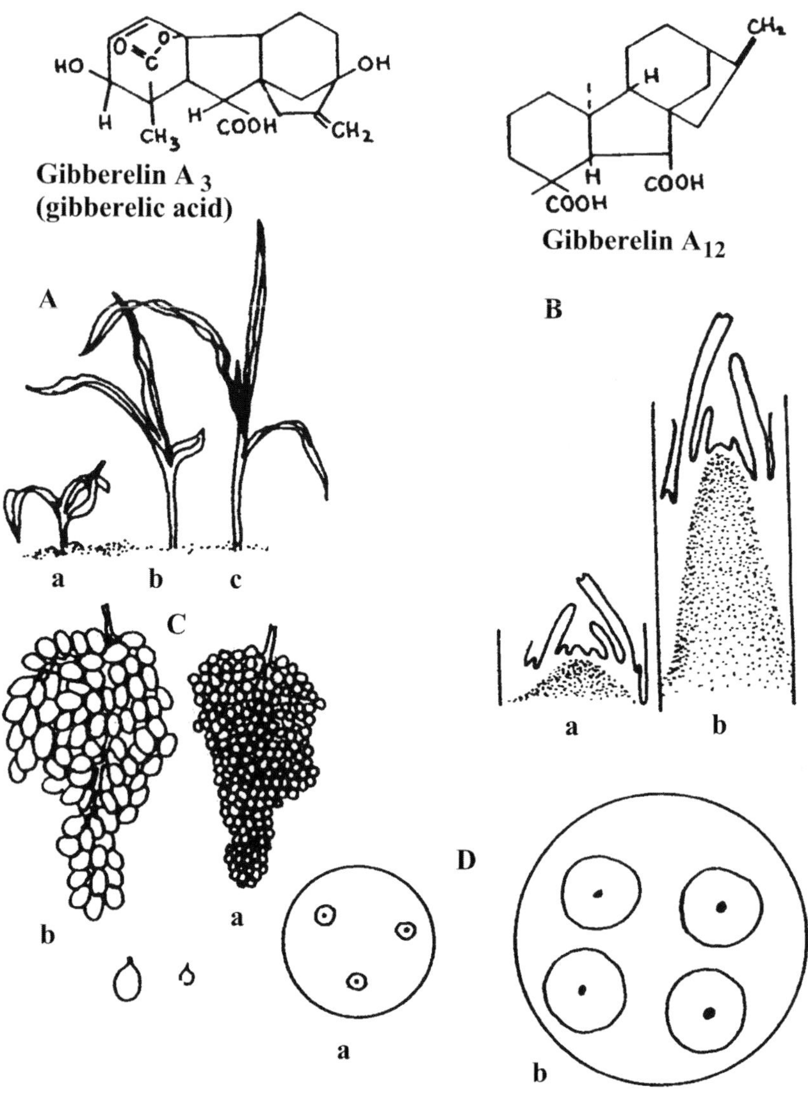

Figure 1.3 Main regulatory properties of gibberellins.
A – effect on growth of dwarf maize: a. – dwarf plant, b. – dwarf plant treated with gibberellin (100 μg), c. – normal heights variety; B – effect of incorporation of ^{14}C – thymidine in Samolus apices: a. – control, b. – gibberellin; C – effect on growth of grapes: a. – control, b. – gibberellin treatment; D – induction of α-amylase formation (spots – hydrolyzed starch): a. – control, b. – gibberellin treatment.

Figure 1.4. Rudbeckia bicolor – long day plan, GA sensitive
Left – in the garden; Right – close up plant.

Unlike auxins, the translocation of gibberellins in plants is acropetal. In order to say their regulatory action, the growth models must always preserve a stem growing point which is directly affected by gibberellin. The microscopic analysis of the tips of plants has shown that gibberellin affects predominantly the central zone responsible for forming stem cells. This zone exhibits active cell division, intensive transition of the latter into the expansion zone and then differentiation of vegetative bud or flower (Figure 1.4)

Discovery of gibberellins occurred in Japan in 1920's. Further investigations about these compounds were conducted by A. Lang (USA), M. Chailakhjan (Russia), as well as by other researchers who demonstrated the multifunctional properties of this class of hormones (see Figure 1.3)

1.1.3. Cytokinins.

The class of cytokinins includes mainly purine derivatives which are synthesized predominantly in roots. Transported to the aereal (above-ground) parts of a plant they stimulate cell division. Unlike auxins and gibberellins, endogenous cytokinins were detected in plants only after the researchers of the Skoog laboratory (USA) isolated a component, which strongly promoted cell division from a DNA hydrolizate. The compound (in an isolated tissue culture and which) was later called kinetin (Miller, 1961). A few years later, in 1964, Letham isolated endogenous cytokinin zeatin out of immature maize caryopses.

Along with the stimulation of cell division cytokinins are also involved in the formation and growth of buds in undifferentiated callus tissue. In fact they are able to extend the life span and maintain the regular metabolism in aging leaves, and even cause their secondary greening. Cytokinins also transport essential nutrients to the areas of potential growth such as fruits, seeds, and tubers.

It is a well known fact that natural cytokinins originally accumulate in the roots, and later move up along with natural juices (bleeding sap) to the above-ground parts of the plant. However, we do not exclude their presence in buds and young leaves. For example, natural cytokinins are found in significant amounts in coconut milk and in the growing and developing apple.

The synergistic effect of cytokinins takes place in combination with other phytohormones like auxins.

It is an interesting fact that natural cytokinins do not affect the whole plant, however their role in cell division in callus tissue is obvious and serves as a tool for the identification of this phytohormone in plant tissues. The greening of old yellow leaves is a classical example in this case.

Natural cytokinins belong to the purines - group of double-ringed nitrogenous bases that serve as the genetic code in the synthesis of nucleic acid molecules. Cytokinins activate cell division and seed germination. They also stimulate buds differentiation and greening of yellow leaves as we mentioned above. Cells that actively divide contain more cytokinins than non-dividing cells.

The ability of cytokinins to induce organ formation was observed in different plant tissues of tobacco and leaves of flowering plants such as: *Begonia rex, Isaatis tinctoria* and *Convolvulus arvensis.*

The cytokinin called kinetin affects the formation of pigment-containing organelles such as plastids, and regulates the synthesis of lignin.

Cytokinins promote the intense growth of cotyledons , and their intense growth leads to the immediate breakage of the seed covering. Cytokinins also promote the development_of flowering buds even in dormant plants. Research was done with grapes *(Vitis vinifera)* to demonstrate the correlation between a radioactive marker position and cytokinin.

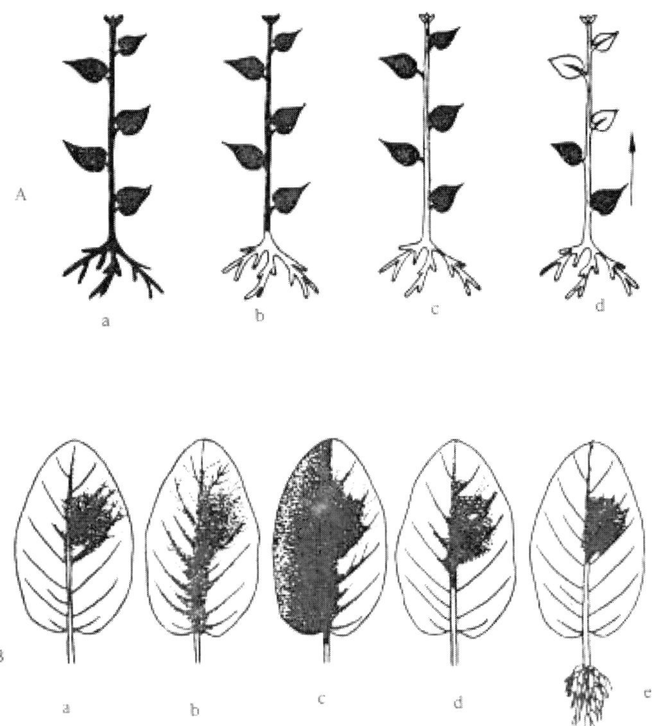

Figure 1.5. Plant senescence and cytokinins.

A- Plant senescence types: a. -senescence of the entire plant, b. -senescence only of the upper part of the plant, c. --senescence of the entire leaf canopy, d. - same, only for lower leaves;

B -Effects of biologically active substances on the aging of the plant (conception of K. Mothes and his collegues (Germany):

a. - application of labeled marker 14 C-a-aminoisobutyric acid, b. - same application, old leaf, c- cytokinin called kinetin was applied at a distance from the marker. The labeled compound moves toward the cytokinin application. d. - If the cytokinin is applied on the marker no movement is observed. e. - As the natural cytokinins move from the roots up, the marker stays in place.

Young leaves usually do not show any effect after cytokinin treatment. On the other hand, old leaves show a tremendous change, as tissues are able to get rid of their pigments, if the amount of cytokinins is low. The transport of cytokinins within the whole plant is not yet well studied however, some information on this topic now is becoming available (experiments of K. Mothes and coworkers Figure 1.5).

1.2 Properties of Natural Inhibitors

1.2.1 Ethylene

Ethylene occupies an intermediate position between phyto-hormones and growth inhibitors. On the one hand, ethylene is known to be a substance causing defoliation, on the other, it is a factor which promotes fruit ripening. The quicker its content increases the more intensive will be the fruit maturing process.

The synthesis of ethylene occurs simultaneously with growth. At the same time in surviving tissues, ethylene formation process is not affected by inhibitors, and it is not connected with the plant's genome. Another difference between growing and surviving tissues is a different response to phytohormones. Inhibition of ethylene synthesis by cycloheximide goes parallel with the depression of pigment synthesis in a flower. Phytohormones are involved in the formation of ethylene. Germination of cotton seeds is accompanied by a vigorous synthesis of ethylene. Ethylene biosynthesis is light dependent. Ethylene is considered to be a regulator that takes part in growth inhibition, causes defoliation and inactivation of auxins. Although the functions of ethylene have not yet been made absolutely clear, it appears for this compound to be related to the growth and senescence processes in plants.

The chemical characteristics of ethylene include: molecular weight 28.05, freezing point- 181°C, boiling point- 103°C. It is well dissolved in ethers, less in alcohol, almost non-dissolved in water. Ethylene is formed more often in meristems (stem, non-differentiated cells), seldom in the inner knots of seedlings. In apple tree, large amounts of ethylene are found in dormant buds and senescent leaves ready for abscission. Hormone of senescence - ethylene is involved in leaf abscission, ripening of the fruit, and it affects the dormancy of the buds (*Figure 1.6*).

Besides the aging process ethylene is also involved in stress reactions in the plant. Stress is the reaction of the organism to harsh conditions that lead to immediate hormonal and metabolic responses and change. The excretion of ethylene in plants is equivalent to that of adrenaline rush in humans and animals when these are exposed to stress. Sometimes ethylene is so-called "plant fever" and was discovered in 1901 in St.Petersburgh (Russia) by D. Nelubov.

The interactions between the above mentioned phytohormones are not only antagonistic. For example, auxins suppress the ripening of the fruit, however high dosages of auxins (as stress factor) tend to stimulate the production of ethylene. Ethylene on the other hand can suppress the biosynthesis of auxins and the effect of ethylene in this case is common in the growth pattern of superior plants.

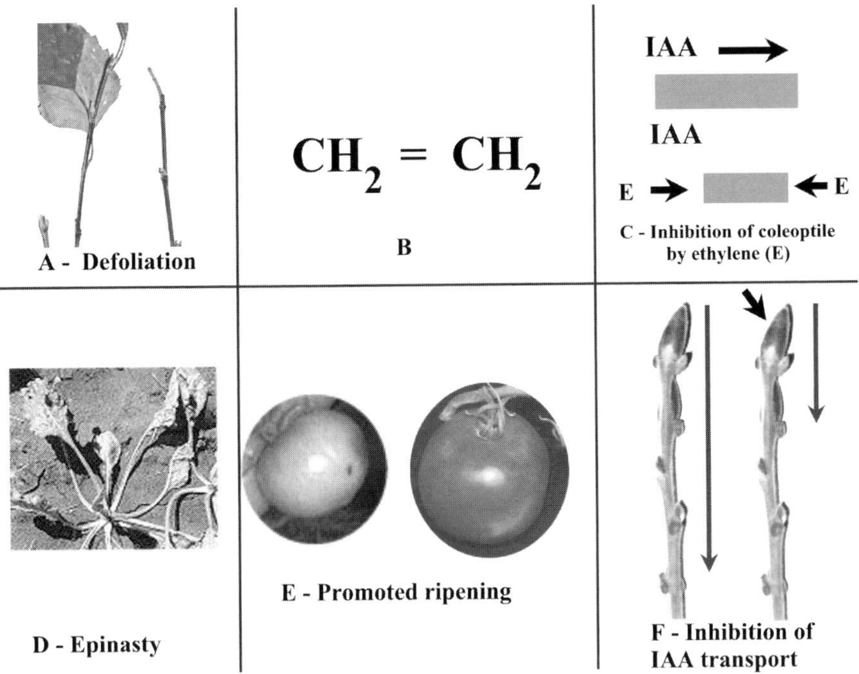

Figure 1.6. Effects of Ethylene.
a– defoliation; b- chemical structural formula; c - the inhibition of coleoptile (etiolated part of newly germinated seed) elongation previously treated with auxins – IAA; d - epinasty (the curving of the lower part of the seedling during the growth); e -the promoted ripening; f - the inhibition of IAA transport.

1.2.2 Abscisic acid and phenolic inhibitors

Beside phytohormones plants contain compounds that inhibit growth processes. These compounds classified as natural growth inhibitors are extremely diverse and they are represented mainly by phenolic and terpenoid compounds (*Figure 1.7,1.8*).

Natural growth inhibitors are found not only in dormant organs, but also in growing parts of plants leaves, stems and roots (Kefeli, 1978; Kefeli and Dashek, 1984)). A terpenoid inhibitor was detected in bleeding sap moving in spring along the stems of woody plants. Later this growth inhibiting substance was identified as abscisic acid (ABA)

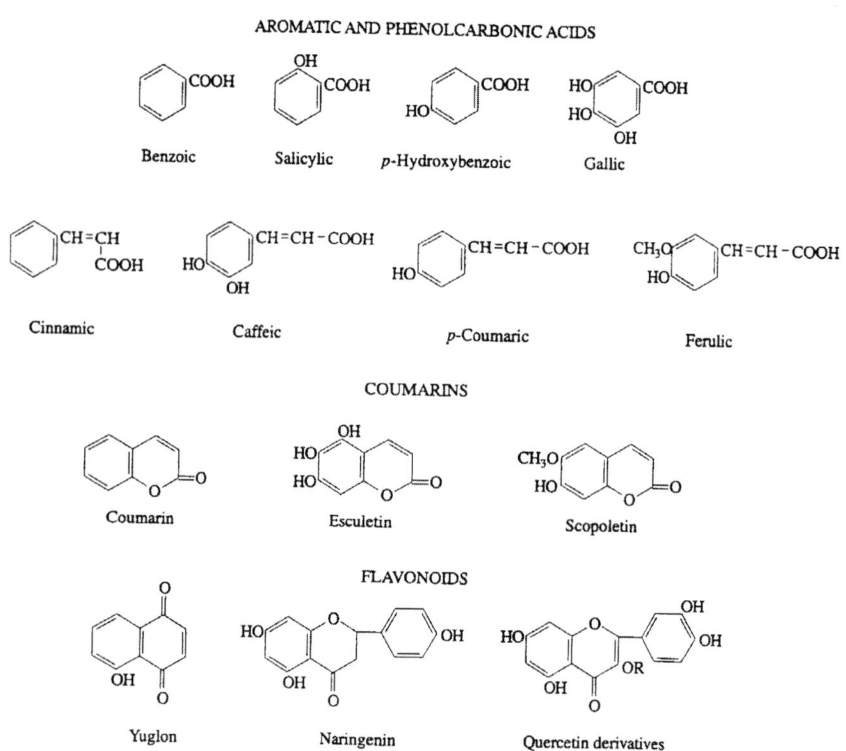

Figure 1.7. Phenolic inhibitors of various growth processes in plants – non hormonal effects.

In ripening fruits growth inhibitors are accumulated in seeds, their amount increasing with seed maturity. The maximum amount of growth inhibitors is found in the skin of dormant tubers and in the autumn buds of woody plants. In contrast to active growth, dormancy is characterized by the fact that dormant organs contain only growth inhibitors, while growing organs have both inhibitors and phytohormones. The accumulation of growth inhibiting substances in the dormant organs of plants is not associated with any definite time (for instance, with autumn), but is rather determined by the physiological rhythm of growth. Growth inhibitors isolated from dormant plants are able to suppress the growth of opening buds and germinating seeds. Natural growth inhibitors can inhibit both native growth processes (opening of buds, stem growth, germination of seeds, tubers, formation of roots) and phytohormone-induced processes.

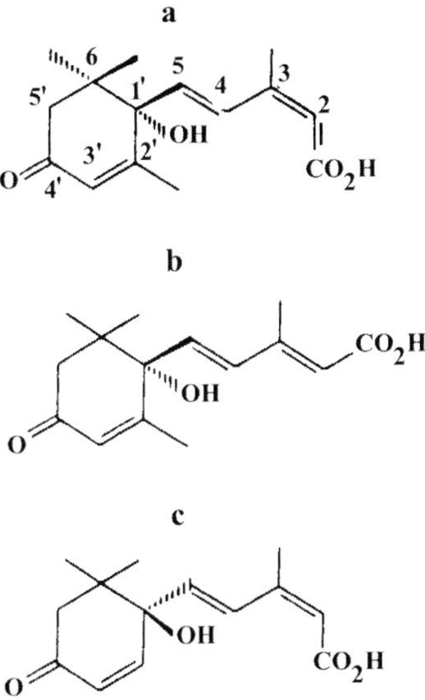

Figure 1.8. Structural formulas of (a) natural S(+) –2-cis,4-trans-ABA, (b) its trans-form, and (c) R – enantiomer.

The first assumptions about the natural substances capable of inhibiting plant growth appeared at the end of the 1920s' and the beginning of the 1930s'. By that time, the existence of growth-stimulating phytohormones had already been proven. The substances that suppressed plant tissue growth were later recognized as inhibitors. However, it was only after the application of chromatography, in combination with bioassay techniques that made it possible to separate phytohormones from their antagonists, growth inhibitors, and to study their chemical and physical properties.

It was demonstrated that the accumulation of an inhibitor named β-inhibitor complex was essential for the induction of seed and bud dormancy. By the end of the dormancy period the β-inhibitor content decreased considerably. Similar results were obtained in studies of seasonal changes in sycamore and birch trees, as they are closely related to their growth and dormancy cycles.

Further investigators have shown that the main components of the β-inhibitor complex are the derivatives of phenolcarbonic acid and more complex phenolic compounds: coumarins, flavonoids, and their glucosides, in particular (Figure 1.9). A great deal of attention over many years was focused on the identification of the nature of these substances and in describing their functions, in normal plant metabolism.

However, later, β-inhibitor complex was found also to contain abscisin or dormin, which came to be known as abscisic acid (ABA) *(Figure. 1.10)*.

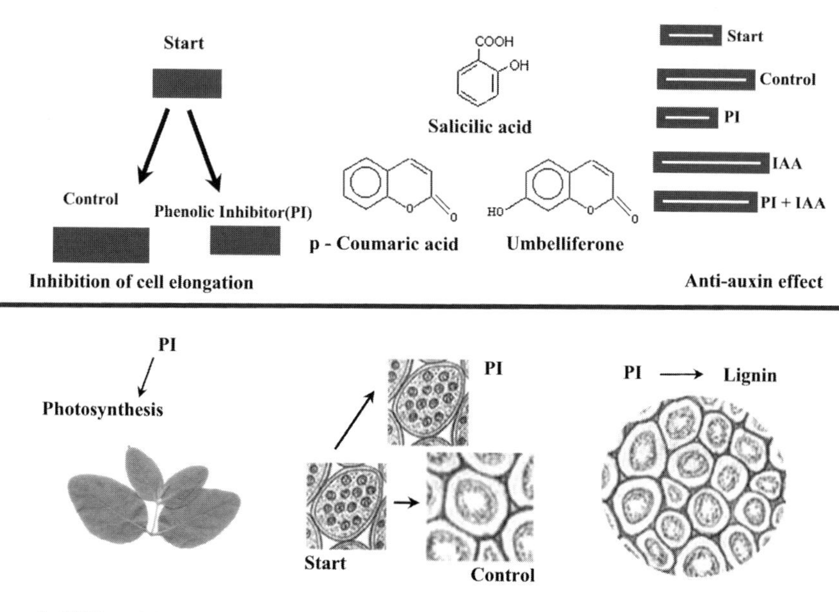

Figure1.9 Property of phernolic inhibitors (PI)

A number of publications have been devoted to the detection, identification, and location of phenolic compounds, as well as their effects on growth. It is important to note that it is the chloroplasts that are the sites of the primary biosynthesis of most phenolic substances. Phenolic substances are accumulated in the vacuoles, and in the case of lignin, they impregnate the cell wall. In some cases, a negative, and in other cases, a positive correlation was described between the content or polyphenols and indole-3-acetic acid in plant tissues. (On the one hand, and the growth processes in them, on the other.) Phenolic compounds are accumulated and the auxin level decreases prior to the beginning of the dormancy period in buds, tubers, and seeds (Figure 1.9) The opposite is observed during the withdrawal from dormancy release as it was clearly demonstrated for phloridzin in apple and isosalipurposide in willow trees (Kefeli , 1978).

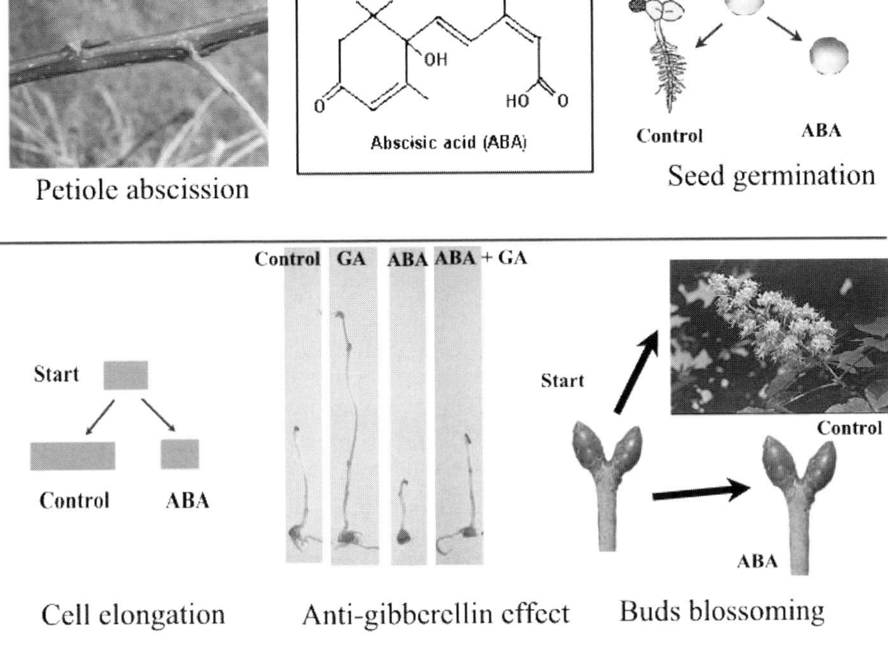

Figure 1.10 Property of Abscisic acid (ABA)

As it was mentioned, growth inhibitors are represented by some phenolic compounds and abscisic acid (ABA). Various effects of abscisic acid and phenolics on growth under stress and normal conditions are discussed, as well as ABA interaction with other phytohormones

1.2.3 Biosynthesis of Phytohormones and Natural Growth Inhibitors

Higher plants differ from microorganisms in the occurrence of a well-developed system of hormonal regulation. Plants synthesize phytohormones in small amounts. Newly formed hormones exert their effects on various growth processes in micro doses. These are readily inactivated by oxidation, conjugation, and breakdown during common metabolic pathways. Phytohormonal excess is harmful and it is often toxic to plants.

Like higher plants, microorganisms can synthesize auxins and gibberellins, but in amounts higher by several orders. Exogenous auxins and gibberellins have a weak or noll effect on the growth of the microorganisms themselves. Microbial growth is independent of phytohormone biosynthesis. It is on this basis that Lang and his colleagues (1962) chose Fusarium fungus for modeling gibberellin formation, thus producing a "pure" model of biosynthesis. Substances that act as phytohormones in higher plants are accumulated in a free form in the microbial culture media. They are neither conjugated nor oxidized and the hormone surplus (for instance, auxin) has no toxic effect on fungal growth (Filimonova, 1987)

It is safe to assume that the formation of the hormonal system in higher plants is a result of a long period of plant evolution, which is coupled with the transformation of some microbial secondary metabolites into hormones of the higher plant. We are not yet sure about the role of plant hormones in microbial existence.

However, limitations at the level of the common metabolic precursor exist. Shicimic and chorismic acids are precursors for the biosynthesis of both indolic and phenolic substances. There are many reviews on shikimiate pathway (Herrmann, Weeaver, 1999) and isoprenoid biosynthetic pathway (Chappel, 1995) which deal with biochemical characterisation of biosynthesis as well as destruction of hormones including auxins, gibberellins, and cytokinins(Bilyeu et al., 2001). On the other hand, mevalonic acid and other C^5-products are the participants of the biosynthetic pathway leading to gibberellins and abscisic acid (*Figure 1.11*).

We support the idea of a divergence in these metabolic pathways, auxins or phenolics being the final products of one path, whereas gibberellins or abscisic acid beinge produced at the end of the other path. The final products often behave as antagonists (*Figure 1.11, 1.12*). It can be assumed that the syntheses of one or the other group of phytohormones or inhibitors are mutually exclusive, being controlled at the very beginning. Production of ABA and phenolics is drastically enhanced in response to stress or transition to dormancy; at the same time, gibberellin, auxin, and cytokinin formation is inhibited (*Figure 1.12*)

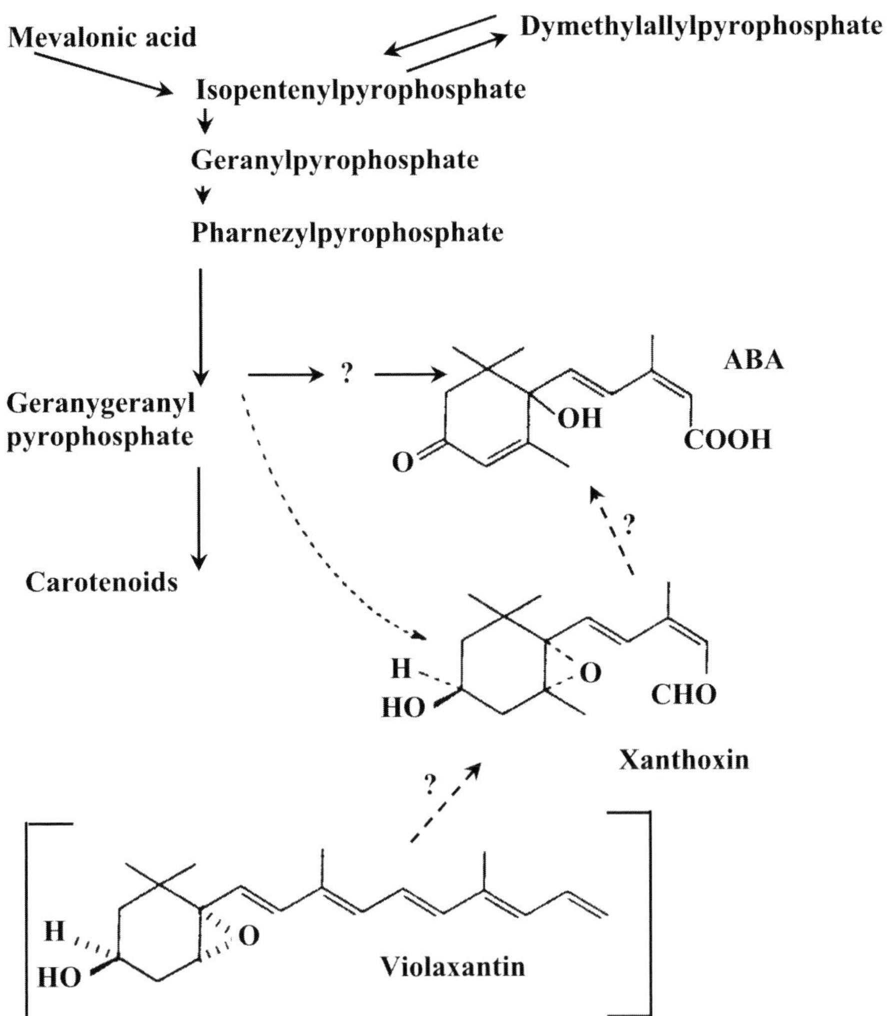

Figure 1.11. ABA biosynthesis pathways.

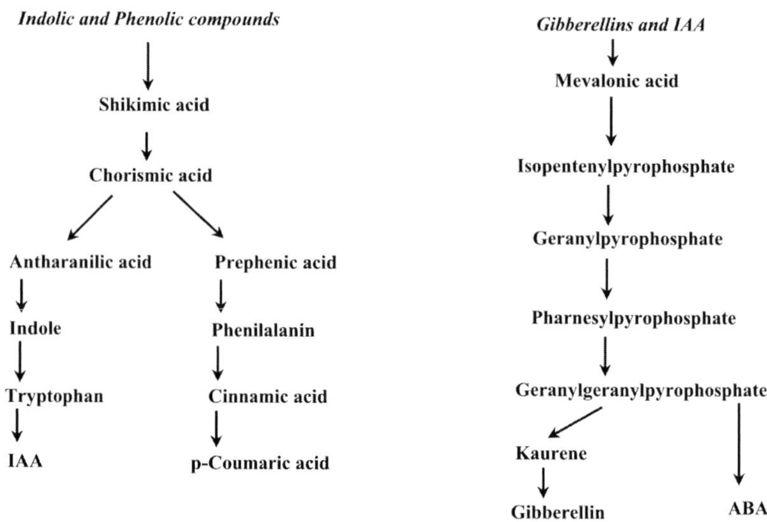

Figure1. 12 Pathway of natural growth-regulator biosynthesis from common precursors (metabolic bifurcations).

Another possibility can be suggested; i.e., every phytohormone can control its own synthesis by virtue of its inhibitory effect (by a feed-back mechanism) on the allosteric center of the enzyme synthesizing its precursor.

While analyzing phytohormone and inhibitor biosynthetic pathways, one can assume that the existence of common precursors appears to be an evolutionarily developed possibility to limit phytohormone production. This can be a tool by which the plant hormonal status is controlled in the course of plant ontogeny.

Treatment of the segments of etiolated stems of cabbage and pea plants with natural growth inhibitors extracted from the leaves of these intact plants resulted in the inhibition of auxin synthesis from L-tryptophan; i.e., ^{14}C-tryptophan incorporation into bound IAA decreased, and IAA formation was hindered. This inhibition manifested itself not only in a more slowly occurring L-tryptophan transformation into indole auxins and their analogues, but also in the active diversion of labeled precursor to the synthesis of slightly active glucobrassicin (cabbage tissues) or in its retention in the nonmetabolized state (pea tissues). In other words, natural inhibitors blocked or changed the direction of metabolic transformation of L-tryptophan, the IAA precursor, which resulted in the inhibition of IAA biosynthesis.

We have observed that p-coumaric acid and QGC (Quercetin-Glucosil-Coumarate), its derivative, were able to suppress growth and inhibit the conversion of ^{14}C-anthranilic acid and ^{14}C-tryptophan into indolic metabolites. Other phenolics and their precursors tat have no effect on growth, namely tyrosine and phenylalanine, did not suppress these transformations.

Thus, there are specific branches of metabolic pathways in the system of phytohormone biosynthesis in higher plants leading to the formation of the natural phenolic growth inhibitors and abscisic acid. Such metabolic branching was not revealed in microorganisms. It is tempting to suppose that the appearance of growth inhibitors represents an evolutionary "adaptation" that is aimed at regulating growth. (growth regulation through changes in the rates of phytohormone biosynthesis.)

Therefore, control over plant growth must be considered as the result of a balance between phytohormones and their antagonists, the relative content of which in the tissues is strictly coordinated. Hormones that control growth processes in plants appear to be supracellular control mechanisms. The complex of endogenous inhibitors controls the biosynthesis and level of endogenous phytohormones, starting from their synthesis and up to their utilization in the target tissues.

Much progress toward elucidating the mechanism by which the natural regulators exercise there controlling influence will be achieved if using a complex approach to the problem, including investigations at the cellular, tissue, organ, and the whole plant level.

CHAPTER 2

NATURAL GROWTH INHIBITORS AND PHYTOHORMONES UNDER THE CONTROL OF THE PLANT'S GENOME

2.1 Genome - growth regulators.

There are large gaps in our understanding of some of the most important aspects of the chain "hormones-genome-properties" such as:

- hormonal control of gene expression;
- genetic control of hormone biosynthesis;
- hormonal changes in mutants, transformants and haploids under changes in development (Gulfoyle et al., 1998; McCourt, 1999; Sakamoto et al., 2001; Yaxley et al., 2001)

It is clear that the main properties of growth and development should involve a certain control mechanisms. As a result of intensive studies it is now clear that hormones play a vital role in the control of growth. Now it is known that there are at least three major classes of growth promoting hormones, namely auxins, gibberellins, and cytokinins. In addition to this, other classes of plant hormones exist, particularly the "growth inhibitors" such as phenolic inhibitors, abscisic acid (ABA), and ethylene, which are apparently involved in numerous growth processes.

The fact that developmental processes are basically hormonal- and gene-controlled is self-evident since there are genetic and hormonal variations which affect almost every aspect of development, ranging from the external morphology and internal anatomy to physiological characteristics such as growth rate, blooming time and dormancy period. Growth and development are such orderly processes that must require the right genes to be expressed in the right cells at the right time. That is to say that growth and development are essential processes, involving selective gene expression, and these processes involve the activity of specific groups of genes. These, in turn control the synthesis of some enzymes including those regulating hormonal and inhibitory biosynthesis (Gulfoyle et al., 1998; Kefeli 1992; Klee, Estelle, 1991; Le thi Muoi, 1985).

This characteristic, unique among some specialized plant cells is called the variable gene expression theory, or growth and differentiation theory.

Hormonal substances and their phenolic modificators can regulate not only growth and development but also photosynthesis. The visible manifestations of growth and differentiation include variations in the development of the cell wall and organelles like the plastids. It is clear that differentiation must also extend to certain aspects of metabolism when one keeps in mind that some tissues are specially adapted to particular functions, such as photomorphogenesis, photosynthesis, secretion and storage of reserve materials.

In order to clarify the concept "genome-hormones-properties" of various plant forms and to discuss the connection between growth and photosynthesis the hormonal status of mutant and transformant plants will be discussed.(Klee, Estelle, 1999; Kefeli, 1992; Kefeli et al., 1992)

There is no doubt, that the precursors of plant growth regulators are tightly connected with the primary products of photosynthesis. Thus, photosynthesis is a source of phenolics, auxins, ABA and gibberellins. This connection is not always accomplished easily and is not obviously specific. There is some evidence that IAA affects the formation of phenolics, as well as feed-back effects, like that of cinnamic acid on phenolic biosynthesis. As Table 2.1 shows, cinnamic acids inhibit intracellular PAL-activity. Whether feed-back inhibition of phenyl-alanine-ammonilyase (PAL) by cinnamic acids has physiological significance in buckwheat, is, however, questionable, as free cinnamic acids do not accumulated in the tissue and may thus be present in sufficiently high concentrations to cause inhibition of PAL (Amrhein et al., 1976).

The end products of cinnamate metabolism in buckwheat, flavonol glycoside rutin and depside chlorogenic acid had little, if any, effect on intracellular PAL-activity. The crucial question is, however, whether these compounds penetrate *in vivo* into the intracellular site of PAL.

As phenolic compounds have been shown to interfere with indole biosynthesis, the opposite situation (inhibition of phenol synthesis by indole compounds), was considered to occur possibly at the level of PAL. The precursors of L-tryptophan, anthranilic and indolepyruvic acids, had a peculiar effect on formation certain compounds from radioactive phenylalanine firstly by repressing the formation and then, after a lag phase of about 90 min., allowing it to proceed at the rate of the control. The common precursor of the three aromatic amino acids, shikimic acid, showed, however, no effect, while the aromatic amino acids, tyrosine and tryptophan, again caused inhibition. As in this type of experiment the effect of an exogenously added compound on the rate of equilibration of the similarly exogenously supplied substrate, phenylalanine, with the endogenous substrate pool of PAL is unknown and cannot be determined. The interpretation of the results presented in Table 2.1. The results obtained with tryptophan, with indoleacetic acid and other auxins are described below.

Indoleacetic acid (IAA) was found to inhibit the formation of labeled water

(^3HOH)from labeled phenylalanine with a potency similar to that of the cinnamic acids, while the auxin-analogues, *a*-NAA and 2,4-D, were less active. Surprisingly, the inactive auxin-analogue, β-NAA, showed inhibitory activity considerably higher in the intact cell than the active auxin-analogue *a*-NAA (*a*-naphthyl-acetic acid), indicating that the effect is not completely specific for auxins. All compounds tested were found to be competitive inhibitors of buckwheat PAL, extracted from an acetone powder of illuminated hypocotyls. The K-values were in the range of the Km of the enzyme (4.5 x 10^{-5} Mol/l).

Table 2.1 Interference of various compounds with intracellular formation of ^3HOH from L-(3-^{3}H)-phenilalanine by segments of illuminated buckwheat hypocotyls (Amrhein et al., 1976)

Compounds	*Effect of 3HOH formation from L-(3-3H)-phenilalanine*
Cinnamic acid	Inhibition: I 50 0.05 mM
ρ-Coumaric acid	Inhibition: I 50 0.1 mM
Caffeic acid	Inhibition: I 50 0.1 mM
Chlorogenic acid	10-20% inhibition with 1 mM
Rutin	no effect with 1 mM
Shikimic acid	no effect with 1 mM
L-tyrosine	50% inhibition with 1 mM
Anthranilic acid	in the presence of 1 mM, rate of 3HOH formation is identical with that of control after 90 min lag-phase
Indolpyruvic acid	same as for anthranilic acid
Indole	50% inhibition with 1 mM
L-tryotophan	80% inhibition with 1 mM
D,L-p-fluorophenyl-alanine	100% inhibition with 0.5 mM

IAA and β-NAA were the most potent inhibitors, which is in agreement with the data obtained from the intact cell assays. Inhibition of PAL-activity *in vivo* should result in the reduced accumulation of metabolites derived from cinnamic acid. Light-induced anthocyanin formation was used as an indicator of the *in vivo* production of cinnamic acid. All the compounds tested reduced the production of anthocyanins in isolated hypocotyls, incubated under light, but in these experiments *a*-NAA showed a nearly tenfold higher inhibitory activity as compared to β-NAA. This result clearly indicates that the effect of auxins on anthocyanin biosynthesis is rather complex and involves more than one site of action. Therefore, the inhibition of intracellular PAL by high concentrations of exogenous auxin may, nevertheless, be partially involved in the inhibition of anthocyanin synthesis. This is made possible by the fact that the IAA-mediated inhibition can fully or partially be overcome by the application of phenylalanine or cinnamic acid. It is necessary to mention that growth inhibition sometimes is accompanied by phenolics accumulation. This situation could be observed, for example, when pea plants are exposed to high-intensity light (xenon arc lamp). The suppression of growth is accompanied by an increase in the quantities of quercetin derivatives, of which quercetin-3-glucosyl-p-coumarate (QOC) represents the greatest amount (75%). Exogenous application of p-coumaric acid (PCA) and QGC to pea stem segments depresses their growth. However, in pea leaves the rate of photosynthesis increases until the light intensity of 200.000 erg/cm^2/sec, and then remains at the same level.

The accumulation of photosynthetic products (photosynthates) during inhibited growth occurred not in the same way as during vigorous stem elongation. However, the use of these products, including phenolic compounds, for cell lignification processes and for building up a cell skeleton during stem elongation is greatly reduced (see below). Phenol derivatives and p-coumaric acid acquire a new function in growth inhibition (see below) because they are not fully used for lignification of elongated cells. This phenomenon can be demonstrated by the following scheme.

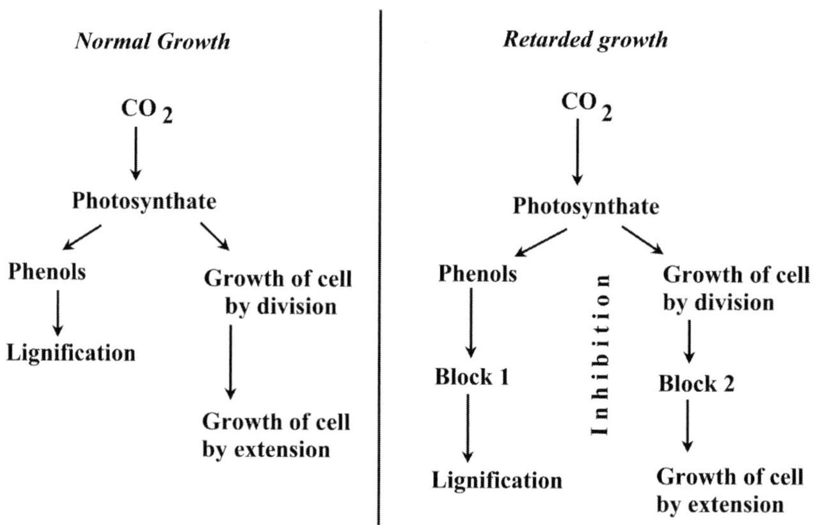

Thus, the juvenile period in plant ontogeny is a heterogeneous stage of development which involves repeated blocking of growth processes with the result that one type of growth is switched over to another. In this situation some phenolics could play a role of markers or factors regulating the growth of plants.

We compared also the behavior of dwarf mutants in the light of various intensity. Fourteen-days old pea plants (cv. Torstag-tall) and the mutants semi- dwarf K-29 and dwarf K-202 were analyzed to reveal the QGC level. In the dark-grown plants had the tallest stems and their leaf plates were reduced in size. At light (50.200, 400 W/ m^2) the stem growth of all pea forms was suppressed. Their leaf plate became larger, but in light of high intensity its area decreased. Inhibition of stem growth is enhanced with increasing light intensity. The dwarf forms appeared to be more sensitive to light even of low intensities. In light of high intensity the stem of the tall variety was shortened and light-grown dwarf plants were formed. Inhibition of stem growth as result of mutation or light action is correlated with the thickening of leaf plate, because the number of layers of mesophyll cells is increased (Table 2.2)

The changes occur in area and anatomical structure of the leaf may offer a plausible reason to explain an increase of photosynthetic activity. The content of gibberellins decreased in all plant forms with suppressed stem growth.

However, depression of stem growth induced by mutation or light action is correlated with an increase of QGC accumulation up to 20-30 $mg.g^{-1}$ of dry matter. This is probably caused by an increase of photosynthetic activity in thickened leaves of the dwarf plants. Thus, we used a xantha-mutant pea in order to answer this question.

Table 2.2 Anatomical characteristics of leaves of 40-days old original and mutant pea plants

Thickness(mk) of	*Light intensity, W,m^{-2}*		
	50	*200*	*400*
		Torstag	
Leaf	88.7± 1.00	116.9± 0.09	168.5±1.05
Palisade mesophyll	30.2± 1.00	40.1± 1.03	57.0±1.20
Spongy mesophyll	41.9± 2.30	57.6± 1.63	87.4±2.71
		Mutant – 202	
Leaf	190.5± 3.21	270.0± 4.49	303.5± 4.10
Palisade mesophyll	82.6± 1.72	106.7± 1.99	120.4± 2.30
Spongy mesophyll	91.3± 2.27	140.7± 3.90	161.2± 1.40

This mutant is characterized by a block (defective gene) in chloroplastogenesis. Plastids of this mutant differ from normal ones by smaller size, more spherical form and absence of lamellar structure. Therefore, photosynthesis was completely suppressed and for this reason, eight-days old mutants did not differ in their size from the original seedlings.

It is possible to conclude that all the morphogenetic reactions except chloroplastogenesis originate similarly in both forms of plants. The block of chloroplastogenesis in the mutant caused inhibition of chlorophyll synthesis and photosynthesis. Carotenoids were not synthesized in etiolated plants, but in light their content in xanthomutants was about (12%) in comparison to green ones. The content of phenolics calculated on fresh and dry matter shows similar levels in xanthomutants as in green plants (Absalov et al., 1985; Kof et al., 1994)

Dwarf mutations and light of high intensity increased the QGC-levels. In order to find out whether there are differences in metabolism of the QGC, precursor PCA in the different types of pea plants was studied. The cv. Torstag was chosen and treated with light 200 W/m^{-2} (tall) or light 400 W/m^{-2} ("light" dwarf) and in mutant dwarf plants K-202 at light 200 W/m^{-2}.

The mutant dwarf and the "light" dwarf plants possess similar catabolism of p-coumaric acid (PCA). Considerable incorporation of radioactivity into the methanol-insoluble fraction which includes generally cell walls was shown. Maximal ^{14}C-incorporation was observed in tall Torstag plants (28.7% of total incorporation), while it was less prominent in "light" (20.9%) and mutant (18.3%) dwarf plants. ^{14}C-incorporation into low-molecular methanol-soluble metabolites was the lowest in the tall plants (Table 2.3). Methanol extracts from "light" and mutant dwarf plants were enriched with radioactive metabolites, first of all, with glucose ester of PCA (Table 2.3).

Table 2.3 Content of phenolic in xanthomutant peas.

Forms	*Content of*			
	QGC in mg.g^{-1}		*PCA in mg.g^{-1}*	
	fresh	dry	fresh	dry
Normal at light	5.0	36.3	0.04	0.30
Normal in darkness	1.3	17.1	0.30	0.40
Mutant at light	5.7	36.9	0.04	0.20
Mutant in darkness	1.2	17.5	0.02	0.30

What is the mode of action of PCA, quercetin and QGC, and their effects on growth? QGC inhibited the growth of pea stem sections only at 8.000 $mg.l^{-1}$, while the growth of wheat coleoptile sections - at 4.000 mg/l^{-1} . The concentration of endogenous conjugated flavonoids is very high in pea tissues. Quercetin (quercetin-glucoside) did not inhibit growth of wheat coleoptile sections even at 220 mg/l^{-1} of concentration (semi- saturated solution). PCA started to suppress the growth at 175 mg/l^{-1}. QGC, which consists of inert quercetin glucoside and of active PCA suppressed the growth of wheat coleoptile sections at a concentration 20 times higher than that of PCA. However, PCA (0.41-0.0082 mg/l^{-1} or less) can activate cell division in tobacco suspension cultures.

Pea plants contain large amounts of QGC (10 mg/g^{-1} of dry matter) whereas the content of free PCA, other hydroxycinnamic acids, and their glucose esters is much lower. However, the concentration of QGC changes widely and this depends on the variety of pea plants. Dwarf forms induced by mutation or light action are accompanied by a high QGC accumulation (Table 2.4)

Table 2.4 Radioactivity of methanol-soluble of PCA from pea

Variety, light-intensity, W.m⁻²	*Radioactivity, % of methanol extract*			
	free PCA	*Glucose ester of PCA*	*QGC*	*Nonidentified metabolite*
Torstag, 2000	58.48+2.9	24.77+1.2	4.15+ 0.2	12.54 + 0.6
Torstag, 400	42.97+2.1	37.33+0.6	6.89+ 0.3	12.81 + 0.6
K-202 200	37.96 +1.9	40.50+2.0	6.61+ 0.3	14.93 + 0.7

When stem growth was blocked, striking changes were observed in the leaves. Leaf thickness increased as a result of increased number of parenchyma and mesophyll cell layers. This phenomenon is accompanied by an increase of photosynthetic activity and it may constitute a reason for QGC accumulation.

The experiments with the xanthomutant of pea have shown that the formation of phenolic compounds is not dependent on photosynthesis and the structural organization of chloroplasts. Probably, all these processes proceed under direct control of light through the activation of PAL and 4-cinnamic acid hydroxylase. However, the accumulation of QGC in peas is accompanied by a depressed stem growth. The metabolism of 2-^{14}C-trans-PCA in "light" and mutant dwarf plants is similar but distinguished from that in tall plants. Hydroxycinnamic acids are known to be lignin and flavonoid precursors. PCA incorporation in methanol-nonsoluble material can be explained as being lignin, a constituent of the cell wall.

The formation of conjugated forms of PCA (glucose ester and QGC) was higher in "light" and mutant dwarf plants. Such increase of glucose ester of PCA in comparison with free PCA was observed at blue light.

We suppose a close relation between the formation of glucose esters of PCA and stem growth depression. It is possible that the incorporation of hydroxycinnamic acid into the cell wall of dwarf plants is inhibited. In cell wall formation nonutilized precursors accumulate as methanol-soluble low-molecular conjugates, which are able to affect cell elongation and division. The latter is especially interesting because depression of stem elongation is accompanied by an activation of cell division in the mesophyll.

Thus, differences in QGC levels and in PCA metabolism in tall and dwarf pea plants, the absence of direct connection with photosynthesis, and their participation in some growth processes show an important role of different phenolic conjugates in common plant metabolism and in growth regulation. Thus, dwarfism is tightly connected to the systems "auxin-phenolics" and "gibberellin-ABA". However, the gene control of cytokinins is investigated on the other model (Kefeli, et al, 1990).

Transgenic plants have been obtained by this procedure. It was shown, that regenerated plants with transferred gene 4 are dwarfed and root with difficulty. In this research for the transfer of gene 4 was used, either deletion derivatives of the virulent Ti plasmid where the genes for hormone synthesis are inactivated, or the "disarmed" Ti plasmid pGV 3850 into which gene 4 is inserted.

The same method of gene transfer was used in our experiments. Use of Ti plasmid pGV5 as the vector and the use of Ri plasmid as assistant. This method helped to regenerate transgenic plants, which were constructed in the laboratory of Professor E. Piruzian (Russian Academy of Science). As a result the gene for synthesis of a cytokinin of the trans-ribosyl zeatin type was inserted into T-DNA of tobacco plants (*Nicotiana tabacum cv. Samsun*). Transgenic tobacco plants carrying gene 4 and rooting normally could be obtained by the use of this vector system.

The newly constructed pGK5-5n4 plasmid was introduced into *Agrobacterium rhizogenes LBA9402*, which carries an Ri plasmid, by three strain conjugation. Exconjugants were selected on (YEB) medium with antibiotics, such as rifampicin 50 ng ml^{-1}, KM (kanamycin) 12.5 ng ml^{-1} and Carbenicilin 100 ng ml^{-1}. The presence of Barn HI-Hind III fragment (received after use of restriction enzyme) was demonstrated by Southern blotting, using the fragment as a probe. The fragment was detected in 4 out of 6 clones tested and one of them was used for tobacco transformation by the leaf disc method. Shoots resistant to kanamycin were found after 3-4 weeks.

Regenerated tobacco plants carrying gene 4 (as demonstrated by Southern blotting) were cultivated in Erlenmayer flasks on solidified MS media without growth substances. Plants were sectioned and one part of the material was used for clonal propagation and the other part for establishing callus cultures. The MS with IAA and GA was used for establishment and maintenance of callus cultures. Callus cultures of untransformed controls were established in parallel with transformants.

The levels of auxin and cytokinins in leaves and roots from sterile growing plants were estimated by determining the activity of the phytohormones through bio-tests after purification of the metabolic extract of fresh material by ether of ethyl acetate and paper chromatography in the system isopropanol-ammonia-water 10:1:1. For auxin estimation segments of wheat coleoptile of cv. Albidum 43 were used; for cytokinin estimation the Amaranthus test was performed.

The ABA content was measured by gas-liquid chromatography (GLC). The extraction and determination of IAA oxidase was carried out as described earlier. The composition of the test solution was as follows: 2 ml 0.02 M phosphate buffer pH 6.1, I ml I mM $MnCl_2$, I ml I mM 2.4-dichlorophenol, 2 ml 1 mM IAA and 4 ml enzyme extract. The enzyme activity was expressed in g IAA oxidized per mg of protein per I mM.

Chlorogenic acid was isolated by two-dimensional chromatography using n-butanol-acetic acid-water 4:1:1 and 15% acetic acid as solvents. The spot corresponding to the Rf of a standard sample of chlorogenic acid was eluted and the absorbance of the eluate was measured at 280 nm with a spectrophotometer.

Table 2.5 Chloroplast pigments in cotton and mutants.

Forms	Chlorophyll (a+b)		Carotenoids	
	mg/g^{-1} (fr. mass)	mg/g^{-1} (dry. mass)	mg/g^{-1} (fr. mass)	mg/g^{-1} (dry. mass)
Cotton:				
green	7.60		2.80	
etiolated	0.02		0.01	
albino (growth in light)	0.47		1.30	
Pea:				
green	1.318	9.400	0.356	2.500
etiolated	0.002	0.026	0.002	0.026
albino (growth in light)	0.019	0.122	0.044	0.290
albino (growth in dark)	0.002	0.026	0.005	0.070

The formation of cytokinins appears to be under a distinct gene control. It is important to mention, that in a plant system phenolics also play a role of modifiers of the hormonal activity. The previous genetic modifications concern the growth or hormonal content.

Table 2.6 Flavonoids and anthocyanins in cotton mutants ($mg/\ g^{-1}$ of fresh mass).

Forms	*Flavonol glucoside*	*Anthocyanin*
Green	13.0	20.0
Etiolated	0.00	0.00
Albino(growth in light)	12.5	10.0

2.2 Photosynthetic changes during genome mutation.

A discussion to investigate albino mutants with suppressed photosynthesis will follow as the growth of green and albino seedlings was investigated during the first week since germination.

The starting material for albino plants were two cotton lines: L-73 and L-453 from the collection of *Gossipium hirsutum L.* of Tashkent University (Uzbekistan). The authors wish to thank Profs. D. Musatov and A. Almatov for the supply of these plants. In 1977 lines L-73 and L-453 were bred and hybrid seeds were obtained. These were treated by 0.25% nitrosomethylurea and grown in 1978 as F1(first generation). In the population of M1 and M2 the splitting (3:1) was observed for the absence of chlorophyll (Mendelian ratio). All plants of this family were self-pollinated and the seeds obtained were planted again in 1980 as F3. In our experiments we used seeds of F3, F4 and F5 generations.

Pea albino mutants were obtained in the same way using 0.025% solution of nitrosomethylurea. As the albino mutants possess carotenoids, they were called anti-mutants by the authors. Plants were grown under phytotronic conditions, temperature 24^0C, 70% air humidity. Luminescent lamps were used as a source of illumination. The age of the experimental plants was 7 days for pea, and 8-10 days for cotton.

The growth of cotton plants (11 days- old), green vs albino, was similar (green - 11.9 cm, albino - 10.9 cm). However, the photosynthesis of albino plants was not observed. Chloroplast pigments of cotton mutants were reduced as seen in Table 2.4 although we could not observe the photosynthetic activity of albino mutants. The flavonoid content was in these plants similar to that in the green ones (Table 2.5), and the amount of anthocyanins (pigments) were decreased only twice. No dramatic difference was observed in the content of auxins and ABA level in albino mutants, where it was even higher than in green plants.

Thus, our observations show that flavonoids, anthocyanins are synthesized in the albino plants (Table 2.6). According to our results, their level does not change remarkably under the suppression of the morphogenesis of chloroplasts; chlorophyll and carotenoids formation. Why? There are two hypothetical possibilities:

a) either an additional biosynthetic center exists, or

b) the reduced level of chloroplast pigments is sufficient for the formation of phenolic substances and ABA in protoplasts.

To the present these questions have not been answered yet. Additionally, another question arises from comparing the results from the analyses of etiolated and albino plants. Albino plants are sensitive to the light and show the following morphogenetic reactions:

a) depression of stem growth;

b) activation of cotyledon (cotton) or leaf (pea) development;

c) activation of anthocyanin and flavonoid synthesis;

d) phototropism.

All these reactions are known to be regulated by the phytochrome system. Therefore, all these reactions may not be directly connected with photosynthesis and morphogenesis of chloroplasts which occured within plant tissues.

The summary Table 2.7. presented below illustrates the processes involved in plant productivity based on the discussed results.

Table 2.7. Processes involved in plant productivity.

Genes responsible for genesis of chloroplasts	Non-active genes (albino plants)	Genes responsible for growth	Genes responsible for photomorphogenesis
Mutagens	Mutation (no photosynthesis)	Active process	Active process

In our experiments mutagens have induced the mutation of genes responsible for the genesis of chloroplasts. Nevertheless in albino plants the gene blocked actually induced growth and morphogenesis, synthesis of phytohormones and phenolics. Therefore, the system of hormones which regulate ontogeny is now developed based on the gene control of the whole hormonal system as a primary regulating block. It is important to mention that this gene controlled hormonal signals on one hand, and hormones and their analogues were able to control gene expression on the other hand. Also one gene change induces various changes in growth regulators modifications (pleiotropic effects) which should be investigated carefully in the future.

2.3 Photomorphogenesis and Light Effect on Phytohormones

The photomorphogenetic effect of light on plant growth begins with the activation of the phytochrome and then, according to the Mohr-Sitte hypothesis (1972), extends to the genome (a region of potentially active genes). This is followed by a number of metabolic changes which determine the completion of the photo-morphogenetic response. The compounds that seem to be closest to the phytochrome in the metabolic chain are phytohormones. Without going much into the details of the history of research of this problem, let us consider the modern theory of interaction of components in the plant: light – phytochrome - genome-phytohormones - general links of metabolic chain-growth and morphogenesis.

The photomorphogenetic effect of light, at least in its first stages, is not associated with the direct activity of the phytohormones, but it is rather dependent on the induction and repression of a certain set of enzymes. At the same time the interrelations between light,growth and morphogenesis, and phytohormones are of considerable importance (Table 2.8).

Table 2.8. Effects of light and phytohormones (indolylacetic acid - IAA, gibberellic acid - GA, and kinetin) on the processes of growth and morphogenesis.

Process and object	*Acting factor*	
	Red light	*Photosynthesis*
Growth *of Phaseolus* hypocotyl	Straightening of curvature	IAA: causing curvatute
Germination of lettuce seeds	Stimulation	IAA: negative response or inhibition
Growth of epicotyl and hypocotyl of legumes	Depression	IAA: promotion
Growth of stems of pea, pumpkin, cucumber, sunflow, *Phaseolus* plants	Inhibition of elongation, increase in number of internodes	GA: stimulation of total elongation of seedling, no effect on number of internodes
Germination of bur marigold (Bidens) and lettuce seeds	Stimulation	GA: promotion, ability to substitute effect of red light
Germination of lettuce seeds	Stimulation	Kinetin: stimulation, synergism with light
Reproduction of duckweed (Lemna)	Stimulation	Kinetin: stimulation, ability to substitute red light

These data provide evidence that in some stages of growth phytohormones start to participate in the chain of reactions leading from light absorption to morphogenesis. However, no specific substance has been isolated yet to act as a primary product in the induction of the phytochrome by light energy. Nevertheless, facts are known which point to the existence of some substances involved in a fast transfer of light stumuli to the tissues, these substances forming vigorously under the effect of red light.

A review of data relating to the functions of phytohormones in phytomorphogenesis shows that they all have one common drawback - phytohormones are introduced exogenously with the endogenous level of these factors being unknown. Thereffre, when phytohormones are applied, the following situations may arise:

1) an exogenous phytohormone is not needed for morphogenesis (there is a rich supply of this phytohormone in the tissue);

2) an exogenous phytohormone is required, but it fails to get where it is needed in the cell and is quickly decomposed;

3) an exogenous phytohormone is required but only in combination with another hormonal factor;

4) instead of a phytohormone, an inhibitor is needed for creating morphogenetic equilibrium (for instance, to inhibit stem growth and accelerate the growth of leaves).

What is the mechanism of action of red light on the synthesis of phytohormones?

The free auxin content in cereal coleoptiles is known to decrease under the effect of red light. The content of gibberellins in pea seedlings increased when the seedlings were exposed to the long illumination with a long photoperiod. In a short-time, illumination conditions that remained practically the same (30 sec. exposure of lettuce seeds to red light) promoted seed germination and the appearance of cytokinins only 48 hours since the start of the experiment.

Experiments with ^{14}-C-anthranilic acid were carried out in collaboration with Professor Kutachek (in 70-s) and showed (Figure 2.1) that in kohlrabi seedlings the pathway of the synthesis of indole products, precursors of auxins (anthranilic acid - tryptophan) is practically insensitive to light (light from luminiscent lamps, 2075 lx, 48 hours). At the same time, two-day illumination of etiolated seedlings caused a decrease in the rate of stem growth. The length of etiolated seedlings was 48 mm, and that of light-exposed – only 27 mm. Consequently, there must exist some systems in plants which are involved in direct regulation of stem growth, but these substances are not auxins.

So, at the present time there is no direct evidence in favor of the fact that a high level of phytohormones or their active biosynthesis may determine photomorpho-genetic processes. It can be assumed that photoinactivation systems, rather than the systems of synthesis will be of key importance in the occurrence of photomorphogenetic effects, under light exposure.

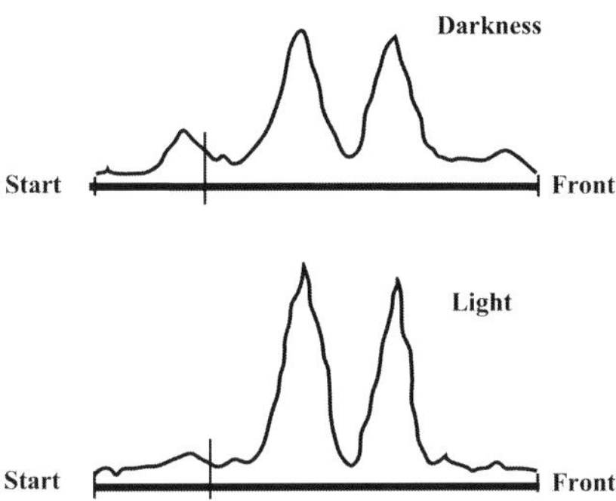

Figure 2.1 Biosynthesis of L-tryptophan from ^{14}C-anthranilic acid in kohlrabi seedlings in the dark and under light. Two large peaks from left to right – L-tryptophan and anthranyl-glycoside.

As it is known, phytohormones in plants can be inactivated through different mechanisms. First, by direct action of light on a phytohormone molecule; second, indirectly, through the formation of inactive bound forms and, in particular, glycosides; third, through the oxidation system regulated by specific cofactors and, finally, through the formation of complexes with natural growth inhibitors.

2.4 Light and Functions of Phenolic Compounds in Photomorphogenesis

The effect of light on the biosynthesis of phenolic compounds is not ambiguous since it activates a number of photoreceptors in the leaf mesophyll. Table below shows the main photoreceptors involved in the regulation of flavonoid and phenyl propanoid levels (P – phytochrome; B - blue light photoreceptor; FR - far red light photoreceptor) (Table 2.9).

Table 2.9. List of photoreceptors. From Smith (1972).

Plant	*Substance*	*Organ*	*Photoreceptor*
Mustard	Anthocyan	Hypocotyl; cotyledon	P; B
Pumpkin	Phenil-propanoid	Hypocotyl; cotyledon	P;B;FR (?)
Red cabbage	Anthocyan	Leaves	B;P
Turnip	Anthocyan	Hypocotyl; cotyledon	P;B;FR
Pea	Flavonol	Young leaves	P;B
Buckwheat	Flavonol	Shoots	P;B;
Sorghum	Anthocyan	First internode	B
Tomatoes	Flavonol	Fruit cuticle	P
Apple-tree	Anthocyan	Fruit skin	P;B
Phaseoulus	Anthocyan	Young leaves	P;B
Maize	Anthocyan	Endosperm culture	B;P (?)
Sphirodella	Anthocyan	Fronds	B;P (?)
Touch-me-not	Anthocyan	Stem	B;P
Potatoes	Chlorogenic acid	Tuber excisions	B

Legend: p=phytochrome, B=blue light photoreceptor, FR=far red light photoreceptor.

Red light caused an increase in the content of anthocyans, phenylcarbonic acids and some quercetin-derivative flavonols. At the same time, some classes of phenols are not photo dependent or, at least, are not activated by the phytochrome system, for instance, some flavones and isoflavones.

Mohr (1972) offers an interpretation of the photo dependent biosynthesis of phenolic compounds in individual cell compartments. Light activates the biosynthesis of two enzymes: phenylalaninammoniumlyase (PAL-lyase) which is the key enzyme in phenolic metabolism, and cinnamic acid hydroxylase (CAH). The biosynthetic pathway leading through trans-cinnamic acid terminates with the formation of such secondary products as flavonoids and lignin. Their synthesis in plant cells is illustrated below. Smith (1972) points out about the existence of other photo dependent reactions in polyphenol biosynthesis. Scheme and Table 2.10 below lists enzymes which regulate the biosynthesis of flavonoids in different plants under various light conditions (photo dependent, light-induced and those independent to the effect of light). Harper et al., 1970 supported the idea that phytochrome may act as a photoactive trigger which promotes the incorporation of the benzene ring of the phenylalanine molecule into the flavonol structure of quercetine. At the same time, the biosynthesis of p-coumaric acid (acyl residue of quercetin-glycosyl - coumarate, QGC) does not appear to be light-controlled.

Photodependent flavonoids and lignin formation from phenolic compounds in separate cell compartments. From Mohr (1972).

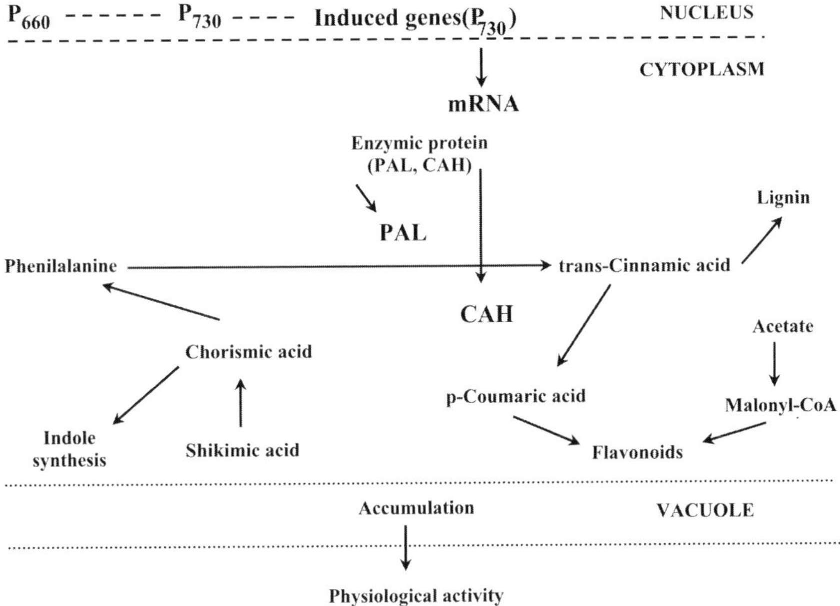

*PAL – phenilalanine-ammonium-lyase;
CAH – cinnamic acid hydroxylase.

Table 2.10. Polyphenols biosynthesis and specific enzymes involved. From Smith (1972).

Metabolic pathway section	*Enzyme*	*Plant*	*Light effect*
Shikimic pathway	5-Dehydroquinqse	*Phaseolus aurens* Pea	I; PL
Shikimic pathway	Shikimate: NADP oxydoreductase	*Phaseolus aurens* Pea * Pea *	I PL I
Shikimic pathway	PAL-lyase	Potato, pumpkin, buckwheat, mustard, pea	PL
B ring synthesis	PAL-lyase	Sunflower, coclebur, (*Xanthium*), strawberry, *Phaseolus aureus*, Citrus cabbage, radish	PL
B ring synthesis	Cinnamic acid hydroxylase (CAH)	Pea, buckwheat, soybean	PL
B ring synthesis	ρ-Coumarate coenzyme A ligase	Soybean, parsley	PL
A ring synthesis	Acetate: coenzyme A liase	Soybean, parsley	PL
Flavonoid modification	Chalcone-flavone isomerase	Parsley	PL**
Glycosiding	UDP-apiososynthetase	Parsley	PL
Glycosiding	Apiosyl-transferase	Parsley	PL**
Glycosiding	Glycosyl-transferase	Parsley	PL**

Legend:

* Contradictory data from different authors.

** Introduction of metabolite formation, though the effect of light on the synthesis of enzymes is not proved by direct experiments.

For this reason, the synthesis of the flavonol skeleton and the acyl residue may occur in different cell compartments.

The compartmentalization of the biosynthesis of individual types of polyphenols seems to be associated with the localization of specific photoreceptors. This issue had not been adequately studied. The hypothesis is that chloroplasts contain specific phenols different from the phenols of cytoplasm sap, so that the phytochrome and phenol synthesis may conjugate here. Kolonkova et al (1968) showed that if leaves are placed in the atmosphere of $^{14}CO_2$ for tea and willow chloroplasts, the biosynthesis of phenolic compounds starts as early as 30 minutes from the start of the exposure. It is not yet clear whether the biosynthesis occurs independently or with the participation of the cytoplasm enzymes. The interpretation of these data is further complicated by the fact that no PAL-lyase has so far been detected in chloroplasts. Strictly speaking, PAL-lyase and CAH cannot be called light-controlled enzymes. At the same time, these two enzymes are exceptionally responsive to changes in the intensity and quality of light. Most research in PAL-lyase, CAH and phenol biosynthesis had been made using whole seedlings. Lately the use of separate parts of organs or tissue cultures took place. Both in experiments with intact plants and tissue culture it was demonstrated that light is a powerful agent inducing the biosynthesis of phenolic compounds some of which may act as allosteric effectors that inhibited synthesis of enzymes. The problem of light-controlled phenol

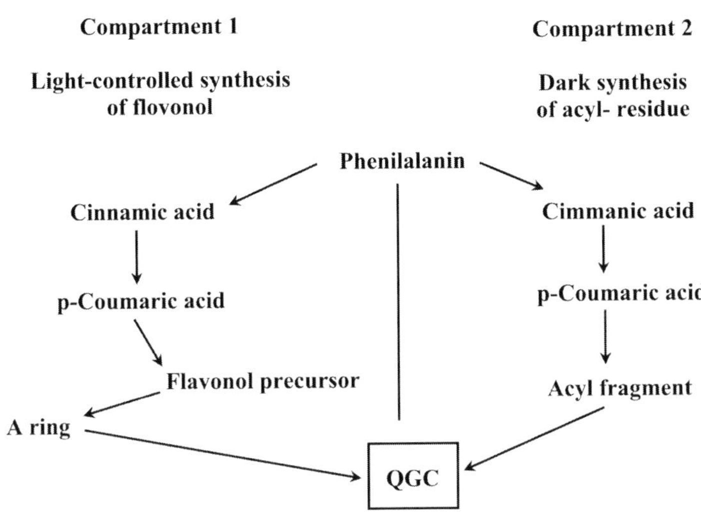

biosynthesis had a number of aspects that are not totally clear. In particular, we do not know the role of chlorophyll in regulating phenol biosynthesis, nor are we fully aware of the details of compartmentalization of photoreceptors and light-induced enzymes. This makes it impossible to provide an unambiguous answer to many problems of compartmentalization of final products in the biosynthesis of polyphenolic compounds; one of such issues was the contribution of phenols to photomorphogenesis.

Earlier we considered the function of phytohormones as mediators in photomorphogenesis and came to the conclusion that the hypothesis of a direct mediator role of phytohormones in photomorphogenesis is scarcely probable. Changes in phytohormone biosynthesis observed under the influence of light are too slow to take place, and it is not always possible to verify that exogenous phytohormones can act as mediators of the phytochrome effect. Nevertheless, phenols and such enzymes as PAL-lyase and CAH exhibit a very fast response to the dose of irradiation, even though this response is sometimes short. Often phenol synthesis continues to intensify while the activity of PAL-lyase and CAH is weakening. A reduction in the phenol content accompanies such morphogenetic processes as the opening of the hypocotyl hook, while an increased phenol content is the consequence of stem growth inhibition. Some authors assert that correlation between an increase in the phenol content and stem growth inhibition is not yet a reason for relating flavonoids to morphogenesis. However, Galston et al, 1965 presented direct experimental data which showed that the amount of some flavonoids, for example QGC, droped abruptly when the hypocotyl hook is straightened. The addition of QGC inhibits straightening of the hook.

It was demonstrated that QGC introduced into an excised pea tip in a 10 μM concentration reduced the curvature of the hypocotyl hook by almost 1.5 times. It is remarkable that that biosynthesis of abscisic acid in pea plants did not display light dependence.

2.5. Light and Hormones.

The change of the QGC (quercetin-glucosil-coumarate) content in pea plants which growth was inhibited under the effect of high intensity light from xenon lamps has been already discussed. The more inhibited the growth of 20- days old plants, the higher the QGC content. It may be asked how sure we can be that the stem growth inhibition is really associated with changes in the QGC content. One of the techniques, to prove the existence of this relation, which is actually is an indirect proof, is to exclude light as an exogenous factor from the light - phenolic inhibitor - growth triad system. Experimentally, this was done in the already-described pea mutants, in which stem growth is inhibited through chemical mutation of one or two genes (mono- and di-gene mutants). It was found in these experiments that the QGC content in mutant pea seedlings was reduced with an increase in the stem length of these seedlings. These facts allow us to conclude that morphogenetic effects are closely connected with the function of quercetin – glycosyl - coumarate and other polyphenolic inhibitors. Although the mechanism of influence on their growth is still little known, the idea of these inhibitors acting as mediators in growth inhibition by light is becoming more popular among researchers. The results of the experiments with pea mutants and with high-intensity light is summarized in Figure 2.2.

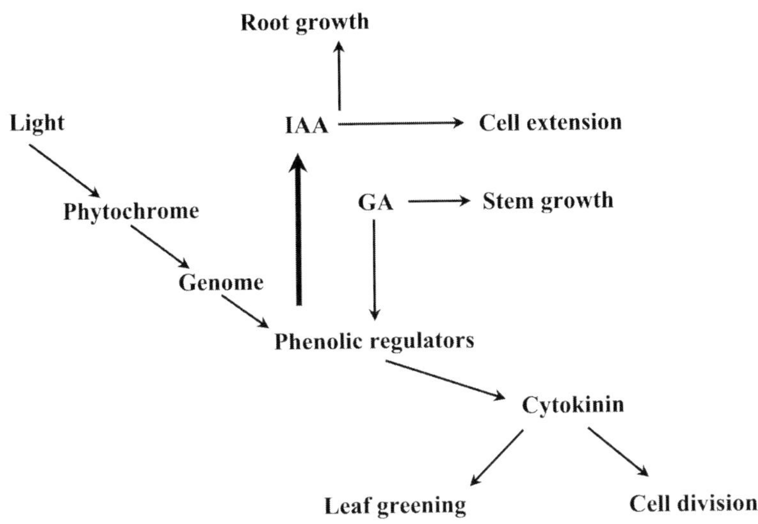

Figure2.2. Effect of light on pea mutants growth (hypothetic shema).

According to this model, both the exogenous and endogenous factors (light and mutated genes, respectively) could induce the synthesis of p-coumaric acid which is detected in pea in the form of glycoside, or as a QGC acyl component. p-Coumaric acid itself is able to increase the activity of IAA-oxidase and to accelerate its decomposition. At the same time, the synthesis of IAA derivatives is strongly inhibited in the presence of p-coumaric acid and thus the total IAA level in the tissue is reduced. Also p-coumaric acid can suppress gibberellin activity. As a result, stem growth is retarded. The accumulation of p-coumaric acid in the form of QGC could also be accounted for by the fact that it is a C6-C3 fragment of lignin, but it is not used for lignification under the effect of an exogenous factor (light), of an endogenous factor (mutated genes), nor it remains in a relatively labile equilibrium form (QGC = p-coumaric acid). Thus, photomorphogenesis is such a complicated process that involves a number of endogenous systems: a photoreceptor phytochrome, genome, light-induced phenols, like acylated quercetin derivatives and phenylcarboxylic acids and, finally, phytohormones; the latter being located rather far away from the photoreceptor and producing a pleiotropic effect on various growth processes (Figure 2.3).

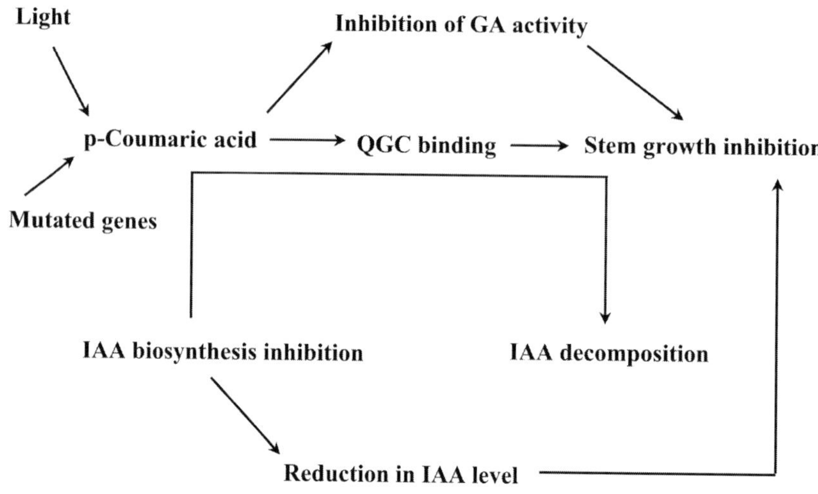

Figure2.3 Endogenous Systems of Photomorphogenesis (possible steps).

The two light-regulated processes - morphogenesis and lignification of tissues seemed to occur in parallel and to be regulated by one group of phenolic metabolites that included also pheholic growth inhibitors. The direct effect of light on the biosynthesis of the quercetin fragment existed, but the synthesis of the acyl residue was not light-induced. It is known that exposure of an etiolated plants to light does not induce the biosynthesis of the phenolic compounds. The biosynthesis of phenols may directly involve the conversion of shikimic acid into C6- compounds, or to a deamination (removal of $-NH_2$ group) of phenylalanine with formation of C3- compounds. Another pathway is an acetate-malonate conversion.

The determination of phenolic compounds in light and in the dark of the grown willow shoots had shown that, under light, the content of phenolcarbonic acids and flavonol glycosides increased, whereas etiolated shoots contained almost no flavononoids in albino mutants (Table 2.11).

Table 2.11 Photohormones in pea mutant

Forms	*Auxin**	*Inhibitors***
Green	124	85
Etiolated	127	94
Albino (growth in light)	122	80
Albino (growth in dark)	130	108

* Biological activity of IAA Rf region (% of control);

** Biological activity of ABA Rf region (% of control);

As was discussed previously the growth of pea albino mutants, 7-days old, was similar to the green ones. Levels of phytohormones and inhibitors were also similar in the plants. Chlorophylls in etiolated and in light-grown albino plants were practically absent. Yellow (albino) pea forms are not able to photosynthesize. The activity of photosynthesis was determined by $C0_2$-evaluation. Albino plants didn't possess remarkable fluorescence, which could denote changes in plastid structure. The albino plants cultivated in light didn't not form any noticeable amount of chlorophyll and they had a highly reduced level of carotenoids. It is necessary to mention that these values do not correlate with the level of phenolics.

The content of phenolics calculated on fresh and dry mass showed similar levels; the content of QGC and p-coumarate was equal in albino (xantha) and green plants. No distinct differences were observed in the level of auxins and inhibitors. Etiolated plants didn't contain any significant amount of natural inhibitors whereas only auxins were present in a significant concentration. These facts were in coincidence with results obtained with cotton albino mutants.

The xantha mutants and original green plants had similar characteristics in their reactions to light. Xantha mutants preserved all morphogenetic reactions concerning growth, although they lost all their ability to synthesize chlorophyll.

In albino mutants the level of auxin and growth-inhibitors was not altered remarkably. It is necessary to mention that the level of ABA in albino cotton plants increased 2.5 times in respect to normal green plants. In pea plants there was no suppression of synthesis of phenolics Only anthocyanins in cotton albino plants decreased twice. Thus, there is no correlation between the level of chloroplast pigments in albino plants of pea and cotton and the activity of phytohormones (auxin and ABA) and level of phenolics.

Photomorphogenetic studies had demonstrated that the formation of flavonoids in plants is regulated mainly by low- and high-energy systems. However, in greening shoots with functioning chloroplasts photosynthetic products began to participate in the biosynthesis of phenolic compounds (Table 2. 12)

Table 2.12 Effect of light on synthesis of flavonoid compounds (in mg/g of dry matter)

Plant	*Substance*	*Darkness*	*Light*
Willow	Isosalipurposide	0.198	1.10
Apple-tree	Phloridzin	13.5	61.20
Lettuce	Flavonoid	0.09	1.09

Zolotareva, 1989 and co-authors assumed that the relationships between the photosynthesis and biosynthesis of phenolic compounds is marked by the presence of secondary products of photosynthesis, that exert a direct as well as an indirect effect on the process of solar energy accumulation. So, flavonoids as multifunctional compounds in green plastids had three major functions:

- they serve as substrates (use of polyphenols and products of their catabolism for other biosynthetic pathways);
- they are energy supply sources (electron and proton transport, effect on ion exchange and membrane potential, formation of radicals);
- they also served as regulators (involvement in enzyme reactions as inhibitors or activators)

At the light stage of photosynthesis flavonoids change the rate of electron transport and photophosphorylation bringing about the change of the ATP/NADPH ratio. In the reactions of carbon metabolism they can shift the dynamic equilibrium of the pento-sephosphate reduction cycle to enhance the synthesis of certain metabolites, both due to the change of energy substrate intake and to the interaction with the enzymes of the pathway, flavonoids exercise a feed-back control over their own biosynthesis. If we regard flavonoids as metabolites inherent not only in the cell but in the chloroplast as well, then two basic points remain questionable. One, is that until now the biosynthesis of the entire flavonoid structure in plastids had not been yet proved, since not all the enzymes of their biosynthesis have been yet discovered. The other point is concerned with the lack of direct evidence for flavonoid transport both inside the cell and in the whole plant. Nevertheless, a large variety of phenolic compounds present in organelles and the cell at the same time, that are able to influence the rate and direction of biochemical reactions. Any change in the flavonoid structure or qualitative composition of the phenol complex resulted in the change of the mechanism of cell energy exchange.

Considering the regulatory function of flavonoids as being much wider we can note the main stages of regulation in the whole plant. They are as follows:

- enzymatic (individual enzymes and multienzymatic complexes);
- membrane, including energy exchange processes;

- subcellular (biosynthesis of cell organelles and cell metabolism in general).

Further, control of photosynthesis and respiration by endogenous phenols is important in maintaining the donor-acceptor relationships as the hormone inhibitory balance controlled growth and photosynthetic functions of the plant.

Thus, flavonoids together with phytohormones constitute a specialized system of regulation, which on the cell level provides an effective control over photosynthesis and other biosynthetic processes. Of fundamental importance is the fact that these compounds also have an adequate effect on epigenesis, proximal and distant transport of assimilates. The accumulation and storage of assimilate acceptors is known to be controlled by hormones and inhibitors. Therefore, plant growth and storige of substances determine metabolite transport direction and rate, and they in their turn determine the yield of photosynthesis from chloroplasts. Consequently, the effect of flavonoids on photosynthesis can be mediated by their influence on plant growth and development. The correlations observed between light intensity, content of phenolic compounds in leaves and chloroplasts, rate of photosynthesis and growth parameters suggested that the quantitative and qualitative content of the phenolic complex in the plant may determine dynamic parameters of the plant development.

2.6 Natural Growth Regulators and Functions of Chloroplasts.

The long chain of growth inhibitors functions start from photobiological processes that occur in the plant cell components - chloroplasts and cytoplasm (Figure 2.4).

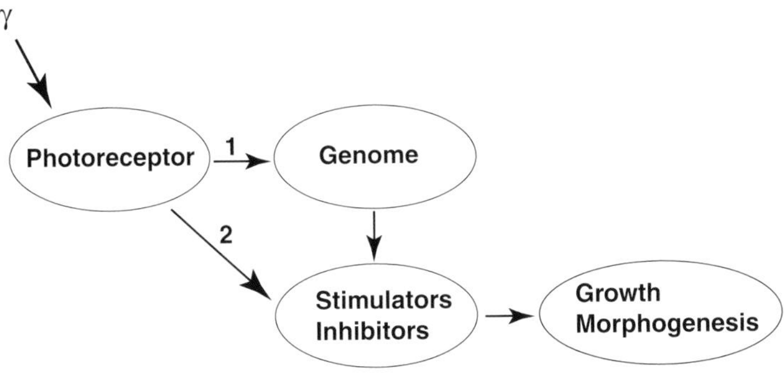

Figure 2.4 Common principle of transformation of light effect on the plants

Non-photosynthetic light effects are of key importance for the regulation of phytohormonal balance, growth, and morphogenesis in plants. Chlorophyll-deficient mutants as it was shown above are an appropriate tool for studying this phenomenon.

Two groups of chlorophyll-deficient mutants are recognized : (a) pale-green mutants they because of their low chlorophyll content, are viable and can grow further and develop; and (b) albino and xantha mutants that lack chlorophyll synthesis and photosynthetic activity. Mutant seedlings of the second group usually die while they are still juvenile; for this reason, such lethal mutations are usually maintained as heterozygous plants.

Chlorophyll-deficient mutants are of considerable interest for both geneticists and plant physiologists, especially where the hormonal regulation of plant growth is concerned because some growth hormones are synthesized and exhibit their activity in the chloroplasts. This is why the relationships between photomorphogenesis and photosynthesis can be investigated in terms of a relationship between the level of phytohormones and chloroplast photosynthetic activity. The effect of light on the content and location of phytohormones controlling growth and morphogenesis can be suitably examined using chlorophyll-deficient mutants, in which the light effects are not mediated via plastid biogenesis, greening, or photosynthesis.

The effects of light and chlorophyll deficiency on growth, morphogenesis, and phytohormone content in pea mutants were studied. The chlorophyll-deficient, non-photosynthesizing, yellow pea mutant (*Pisum sativum* 1, line 14, M-14) of the xantha type was used. The mutant was produced from the pea seeds (cv. Svoboda), by treating them with N-nitrose-N-methylurea, a chemical mutagen . The mutant phenotype was determined by a recessive mutation in a nuclear gene. The mutant genotype was maintained just like the hetorozygous plants. This mutation was lethal, and the plants died after their seed reserves were exhausted. For this reason, we sampled 8-day-old pea seedlings before the plants of chlorophyll-deficient line M-14 retarded their growth.

The plants were grown in glass pots filled daily with tap water (10 - 15 plants per pot). The water medium was aerated for 5 minutes every 2 hours throughout the period of plants exposure to light.

Growth was measured; pigment and phytohormone contents were determined using three to five analytical assays. The experiments were conducted in triplicates; the means are given with their standard deviations.

Light-grown, 8-day-old mutant pea seedlings were similar to wild type (WT) pea seedlings in the number of nodes and fully developed leaves (five or six leaves) as well as in the lengths of their stems and roots. Dark-grown seedlings of both forms showed a hook curvature, and their leaves were reduced; their fresh weight was greater and their stems and roots were much longer than in the light-grown seedlings. No difference was found in the etiolated seedlings between the WT and mutant forms. The fresh weight in the dark-grown seedlings increased because of their higher water content. The dry matter was 14 - 15% of the fresh weight in the light-grown seedlings and 7% in etiolated plants. No significant difference was observed between WT and mutant lines in dry matter content in any group of seedlings tested (Table 2.13).

Table 2.13 Growth characteristics in seedlings

Pea form	*Growth conditions*	*Plant height, cm*	*Shoot weight, g*		*Length of the longest root, cm*	*Root weight, g*	
			fresh	*dry*		*fresh*	*dry*
Wild type	Illumination	3.8±0.2	4.1±0.1	0.57±0.02	11.1±0.5	2.8±0.1	0.19±0.01
	Darkness	15.4±0.6	7.1±0.3	054±0.02	10.5±0.4	2.5±0.1	0.17±0.01
M-14	Illumination	3.0±0.1	3.3±0.1	0.52±0.02	10.3±0.4	2.4±0.1	0.16±0.01
	Darkness	15.7±0.6	7.4±0.3	0.55±0.02	10.9±0.4	2.0±0.1	0.14±0.01

In the light, unlike the wild form, the leaves and stems of the mutant seedlings were yellow, contained minute amounts of chlorophyll and were rich in carotenoids (Table 2.14). The amount was about 12% higher of as compared to their content in the green, WT seedlings. All of the dark-grown seedlings, however, were almost free of both chlorophylls and carotenoids.

The ABA content in the shoots of the light-grown xantha and dark-grown WT seedlings was lower than in the shoots of the light-grown WT peas. In all of the seedlings, the ABA content in the roots was three to four times lower than in the shoots. At the same time, the ABA level in the roots of the green plants was 2.5 to 3.5 times higher in the light-grown seedlings than in the light-grown mutants or dark-grown WT seedlings. IAA content in both the roots and shoots of WT and M-14 was shown to be similar under illumination. The IAA level in shoots was usually somewhat lower than in the roots. A considerably higher IAA content was found in the etiolated seedlings: it was 1.5 times higher in shoots and four times higher in roots than in the corresponding organs of the light-grown seedlings.

Table 2.14 Pigment content in seedlings

Pea forms	*Growth conditions*	*Chlorophill a+b*		*Carotenoids*	
		μg/g dry wt	*μg/g dry wt*	*μg/g dry wt*	*μg/g dry wt*
Wild type	Illumination	1318 ± 40	9400 ± 280	356 ± 10	2540 ± 70
	Darkness	2 ± 0	26 ± 1	2 ± 0	26 ± 0
M-14	Illumination	19± 0	122± 3	44 ± 1	290 ± 6
	Darkness	2± 0	26 ± 1	5 ± 0	66 ± 1

While illuminated, both the WT and mutant seedlings were fairly close in the level and shoot-to-root ratio of cytokinins. Unlike IAA, the cytokinin level in the shoots of the light-grown seedlings was higher than in their roots. In the dark-grown seedlings, however, the cytokinin content was higher than in the roots.

The obtained results indicate that, despite the blocked photosynthesis, the early growth and morphogenesis in the chlorophyll-deficient mutant and WT seedlings were similar under both light and dark conditions.

Photosynthetic membranes in the chloroplasts from the light-grown green plants consisted of grana lamellae that formed stacks of 5 - 15 distinct discs of thylakoids named grana. Small ribosomes, starch grains, and scarce osmiophilic globules were observed in their stroma. The chloroplasts were lenslike in shape, and varied in size (Figure2.5).

Much smaller and rounder plastids were found in the light-grown mutant seedlings. Their photosynthetic membranes represent either non-stacked, rather long lamellae or vesicles (vesicles were absent in previously discussed plant varieties). Ribosomes, numerous osmiophilic globules, and vacuoles of various size were observed in the stroma of the mutant chloroplasts.

A similar, but slightly varying structure was typical of chloroplasts in other chlorophyll-deficient mutants. Apparently, blocked chloroplast development results in preventing chlorophyll synthesis. Under illumination, total chlorophyll content in the mutant seedlings did not exceed 1.5% of their level in the WT seedlings, whereas the etiolated plants almost lacked chlorophylls. For this reason, the light-grown mutants, like the etiolated seedlings, were unable to exhibit photosynthetic activity. In other words, the interference of the mutation with the chlorophyll synthesis and photosynthesis is similar to the effect of darkness. Nevertheless, in contrast to the etiolated plants, the light-grown mutant proceeded to produce carotenoids, but more slowly, thus accounting for a yellow pigmentation of their leaves and stems. That is why the mutant studied is referred to as a xantha mutant.

A colorless precursor of yellow and orange carotenoids, phytoene, is known to accumulate in the colorless tissues of mosaic mutants. This phenomenon appears to result from the suppression of either a phytoene dehydrogenase or another enzyme of carotenoid biosynthesis. Protection

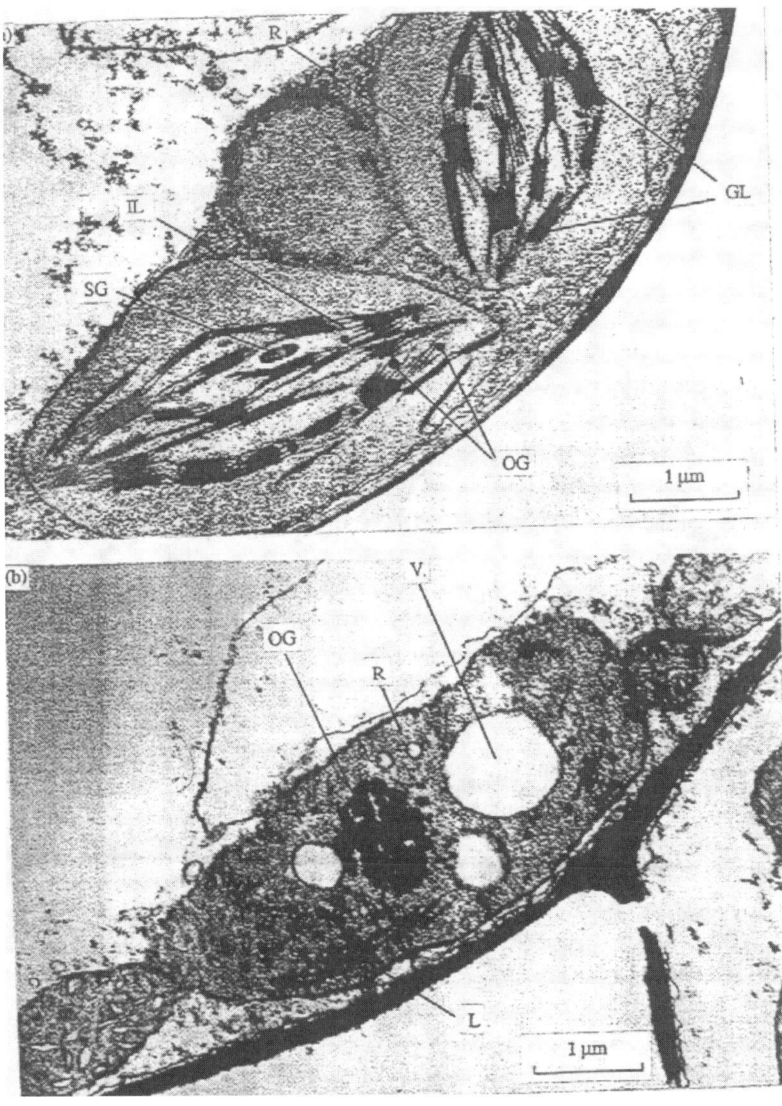

Figure 2.5 Chloroplasts from the leaves of (a) wild-type and (b) mutant M-14 pea seedlings. V – vacuole, Gl – grana lamellae, IL –integranal lamellae, SG – starch grain, OG – osmiophilic globules, L – lamellae, R – ribosomes.(Kof, Gostimskii, Kefeli, 1994)

against photooxidation is known to be a major function of carotenoids in cells, especially under high light conditions. The site of carotenoid biosynthesis in the xantha mutant appears to be unaffected (or only slightly changed), whereas the chlorophyll biosynthesis was primarily blocked. Hence, in the xantha mutant plants, the direct light effect (that was not mediated by photosynthesis) on the morphogenesis and level of growth substances was observed. Needless to say, in our case, the light-dependent response can occur only if the reserve substances and the photoreceptors phytochrome and cytochrome are available.

Light - dependent processes involved in the biosynthesis and action of phytohormones abd inhibitors (ABA, IAA, and cytokinins) operate in chloroplasts in parallel to the typical light dependent photosynthetic reactions. It was demonstrated that neither levels nor shoot-to-root ratios of both IAA and cytokinins were determined by greening and chloroplast photosynthetic activity. Under illumination, the levels of IAA and cytokinins in the seedlings of chlorophyll-deficient line M-14 were similar to those in WT seedlings.

The IAA content in the dark-grown WT seedlings however, was much higher. Apparently, in the dark, either IAA catabolism or its conjugation were diminished. Special consideration must be given to the fact that the auxin content in the roots was always higher than in the above ground parts. The enzyme systems of auxin formation in roots, which grow usually in darkness, can exhibit the properties typical of etiolated seedlings, resulting in attenuation of IAA catabolism and conjugation.

Cytokinins play a key role in the regulation of greening and chloroplast biogenesis, influencing the activity of chlorophyll synthetase as well as the protein synthesis and membrane development in chloroplasts. The mutation, inducing a chlorophyll deficiency did not influence the level and distribution of cytokinins in the plant. In both pea forms, the level of cytokinins was higher in the shoots under illumination, while in the roots it was much higher in the dark-grown plants, although the total level of cytokinins in the dark-and light-grown seedlings was similar. 1 - 4 hours after the beginning of the illumination, the cytokinin level decreased in the roots and rose in the above-ground parts of etiolated pea seedlings. The effect was more evident when the light period was prolonged. This fact indicated that, in darkness, smaller amounts of cytokinins synthesized in roots, then moved to the shoots, where they exhibit their functional activity. Nevertheless, we cannot eliminate the possibility that cytokinins might also be produced in the shoots. The illumination seems to be insignificant for cytokinin biosynthesis, if the seed reserves are still not exhausted.

The light, however, is important for the export of cytokinins from the roots to the shoots. However, cytokinins can induce greening and plastid biogenesis only in green plants of WT peas. They are unable to do it in a rantha mutant, where these processes are blocked at the gene level.

As opposed to IAA and cytokinins, the ABA level in peas depended on the light conditions, photosynthetic activity, and chlorophyll presence in chloroplasts. ABA content in non-photosynthesizing, chlorophyll-deficient, light-grown seedlings and in the etiolated pea seedlings was considerably lower than in the green, WT seedlings. Low ABA level had previously been shown in a chlorophyll-deficient cotton mutant. Moreover, exogenous ABA was reported to decrease the rate of photosynthesis and suppress chloroplast biogenesis, chlorophyll synthesis, the activity of chlorophyll synthetase and function of other enzymes in chloroplasts.

Simultaneous reduction in both ABA and carotenoids in a xantha mutant can result from a slow production of their common biosynthetic precursor: mevalonic acid.

In all of the seedlings tested, the ABA level in roots was several times lower than in the shoots. In the roots of the dark-grown seedlings, the IAA content was higher, and the ABA content was lower than under illumination. A similar pattern of hormone distribution and their levels was observed when the etiolated pea epicotyls and roots were grown in axenic culture. Hence, the contribution of epiphytic root microorganisms to the hormonal distribution seems to be negligible. In our experiments, the illumination changed the hormonal pattern, markedly lowering IAA level and elevating the ABA content. Nevertheless, under illumination, a similar decrease in IAA was observed in the seedlings of both forms, whereas the ABA level rose only in the WT plants. This indicates that photosynthetically active chloroplasts, in addition to illumination, are required for significant ABA synthesis. Nevertheless, the decrease of ABA synthesis induced by the mutation or darkness was probably incomplete, and the synthesis proceeded, only more slowly.

The results obtained suggest that the mutation blocked plastid biogenesis and, thereby, greening and photosynthesis in chlorophyll-deficient mutant M-14 didn't effect the growth and morphogenesis of pea seedlings and did not interfere with the content and distribution of IAA and cytokinins between their organs. Nevertheless, this mutation induced a decrease in the ABA level that could result from a pleiotropic effect of the mutant gene.

CHAPTER 3

NATURAL INHIBITORS AND PHYTOHORMONES DURING LEAVES GROWTH AND DEVELOPMENT.

Plant leaves become the center of accumulation of abscisic acid and phenolic inhibitors during ontogenesis. Our review (Kefeli et al., 1989) and direct experiments showed that ABA in free and bound forms is accumulated in leaves 4-7 times higher than in the root. These data were obtained by our research student using sterile culture of grape seedlings. Thus, leaves might have a center for the biosynthesis of ABA, in the chloroplast of green leaves.

However, before addressing the issue of ABA biosynthesis in chloroplasts, let us concentrate on the biogenesis of chlorophyll in it, using genetic models – like albino mutants.

3.1 Chlorophyll Biosynthesis in the Green Chloroplasts.

N.G. Averina et al., (1992) investigated chlorophyll biosynthesis in the green leaves of parent cotton plants and in the mutants with etiolated leaves . General observations showed that an exceptionally complex physiological process entails the biosynthesis and assembly of a whole series of membrane components into supramolecular groups. Thus, the biogenesis of the photosynthetic apparatus is under the control of both the nuclear genome and the genome of the chloroplast itself. A wide range of chlorophyll mutations have been used to decipher the mechanisms taking part in the formation of pigment-protein complexes of the photosynthetic apparatus. At the present time, fairly thorough research was done on given of mutants carrying injuries (defects) in structural genes. Genes that code separate enzymes in the chain of chlorophyll biosynthesis, and the apoproteins binding chlorophylls, as well as in genes responsible for the regulation of the biosynthesis of individual protein and pigment components of these complexes. It is certain that the given approach in the final analysis will lead to deciphering the mechanisms involved in the genetic control of photosynthetic apparatus, so the creation of mutant systems of varying complexity seems to be exceptionally important..

In etiolated 4-days old cotyledons leaves of a Chlorophyll (Chl)-deficient cotton mutant (*Gossypium hirsutum L.*), the content of chlorophyllide (Pd) and the ability to accumulate 5-aminolevulinic acid (ALA) turned out to be the same as in leaves of the parent forms. On the other hand, the amount of ALA and Chl (a + b) detected in mutant leaves greening for 17-20 h under light of medium intensity (12 W/ m^{-2}) were only 26 and 5%, respectively of their content in leaves of the parent type. Use of low intensity light(0.2W/m^{-2}) {which does not promote development of photodestructive processes under conditions of carotenoid deficiency in the mutant}, removal of limitation of Chl synthesis at the stage of ALA formation achieved with the aid of exogenous ALA, and analysis of the activity of ALA dehydratase. This enzyme catalyses conversion of ALA to profobilinogen (aka porphobilinogen) enables us to conclude that low Chl content in the mutant is not a result of pigment photodestruction or decline in activity of enzyme systems converting ALA to Pd. Mutation induces breakdown of light-mediated regulatory processes controlling synthesis of ALA molecules.

In keeping with the difference observed in synthesis of ALA under light, the content of Chl in cotton leaves of both types was markedly different (Table 3.1). Thus, the content of Chl (a+b) in leaves of mutant plants greening for 20 h under continuous light of medium intensity, constituted only 5% of its value in leaves of the parent form, on average.

Table 3.1 Content of Chl a and Chl b and Ration between them in 4-Day Cotyledon Leaves of cotton of the parent type and mutant greening after etiolation under continuous light (the data cites for exposure of 24 and 48 were obtained in different series of experiments)(mg·g^{-1} of raw mass).

Experiment #	*Parent type*		*Mutant*	
	Chl (a+b)·10-3	*Chl a/Chl b*	*Chl (a+b)·10^{-3}*	*Chl a/Chl b*
	20 h of greening. 12 W·m^{-2}*			
1	845.4	2.7	43.2	1.0
2	652.7	1.9	38.6	1.5
3	921.2	2.1	51.4	1.4
4	1210.0	1.6	33.8	1.6
X	907.3±115.6	2.3±0.3	44.3±6.0	1.4±0.1
	24 h of greening. 0.2 W·m^{-2}*			
1	268.6	2.8	17.9	1.8
2	261.4	2.9	25.9	1.0
3	385.7	2.7	20.0	1.5
X	305.2±40.3	2.8±0.1	21.3±2.4	1.4±0.2
	48 h of greening. 0.2 W·m^{-2}*			
1	340.0	2.5	18.3	1.5
2	309.5	2.7	21.7	1.8
3	292.1	2.6	23.4	1.7
X	313.8±19.8	2.6±0.1	21.1±2.1	1.6±0.1

* *Extracts of leaves in 100% acetone*

We studied also the possibility of photoinactivation of enzyme systems that perform conversion of ALA to Pd in mutant plants greening after etiolation for 20 h., under light of medium intensity (12 W/m^{-2}). For this purpose, greening plants were kept for 17 h in the dark on a 5-day etiolated plants received exogenous ALA, and end products of their metabolism that took place in the dark were analyzed. It was shown that both the amounts of resynthesized endogenous Pd in leaves that did not receive ALA and the amounts of Pd formed from exogenous ALA turned out to be the same in the two types of plants. This means that neither the mutation itself nor illumination of plants carrying the given mutation affected the activity of enzyme systems performing conversion of ALA to Pd.

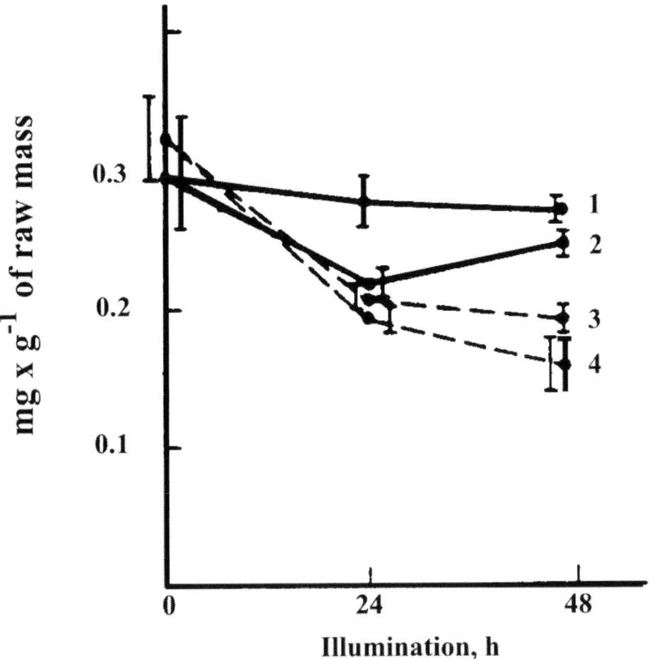

Figure 3.1 Changes in content of carotenoids in 4-day etiolated cotyledon leaves of cotton of the parent type (1,2) and mutant (3,4) greening after etiolation under continuous light for 20 h (12 W·m^{-2}) or 24 h and 48 h (0.2 W·m^{-2}). The figure gives arithmetic means and their standard deviations in four repeated experiments.

These results, were corroborated by experiments involving direct measuring of the activity of one of such enzymes, 5-aminolevulinate dehydratase ,which determined the condensation of two ALA molecules with formation of a monopyrrole porphobilinogen . Activity of the enzyme estimated from the amount of porphobilinogen formed from exogenous ALA per unit of protein of homogenates obtained from cotyledon leaves greening for 20 h differed little for the two types of plants: 1.1 to 0.6 μmoles of porphobilinogen in leaves of the parent form and 0.9 to 0.3 μmoles in leaves of the mutant. Normalization conducted in each of five individual experiments in relation to activity of the enzyme in the control plants revealed no differences in the relative value characterizing activity in the mutant as compared with activity obtained for the control plants, which was adopted as the unit. This means that the enzyme in converting ALA to porphobilinogen under conditions of carotenoid deficiency (Figure 3.1) did not undergo photoinactivation at the given level of illumination in mutant plants.

The lack of any influence of light or mutation on the activity of enzymes synthesizing Pd from ALA and clear suppression of the formation of ALA molecules under these conditions suggest a direct connection between the observed genetic injury and the breakdown of ALA synthesis.

Thus, the study clearly showed that the given mutation affect certain photoregulatory mechanisms of Chl synthesis in its initial link associated with formation of ALA molecules without exerting any influence in the process on enzyme systems that convert ALA to Chl. Data indicated no influence of the mutation on synthesis of ALA and Chl. precursors in the dark and clear suppression of ALA synthesis only under light in different regimes of illumination of mutant plants didn't support the idea that the mutation is located in the structural region of a gene coding information for an enzyme of ALA synthesis. The results indicate also a breakdown in light-mediated regulatory mechanisms controlling synthesis of ALA molecules. It is possible that the given mutation is associated with a breakdown in function of the phytohormone controlling the formation of ALA molecules under light. A second hypothesis supports the idea of a breakdown in synthesis of one of the apoproteins of pigment-protein complexes of the photosynthetic apparatus. However, it should be noted that absence of a pigment-protein complex in the latter case could not have such a dramatic effect on the Chl content in leaves of the mutant plants. This is also indicated by the less sharp decline of the ratio between Chl a and Chl b during the greening of the mutant. It should be noted that the last observation may indicate greater influence of the genetic disturbance on synthesis of Chl molecules and consequently, on photosystem development, rather than on formation of light-harvesting complexes. It is certain that the level of phytochrome in leaves of the mutant and parent form must be studied further to establish more precisely the function of the disturbed gene, and that pigment-protein composition of the photosynthetic apparatus.

These data showed that pigments biosynthesis is not tightly connected with the level of phytohormones and inhibitors in leaves, however, chloroplasts, according to review of Lichtenthaler (1998) are centers of isoprenoids biosynthesis (Figure 3.2), the probable precursor of ABA.

According to this model the shikimate-isoprenoids pathways are considered as precursors of secondary substances, including natural phenolic inhibitors and ABA (Figure 3.3).

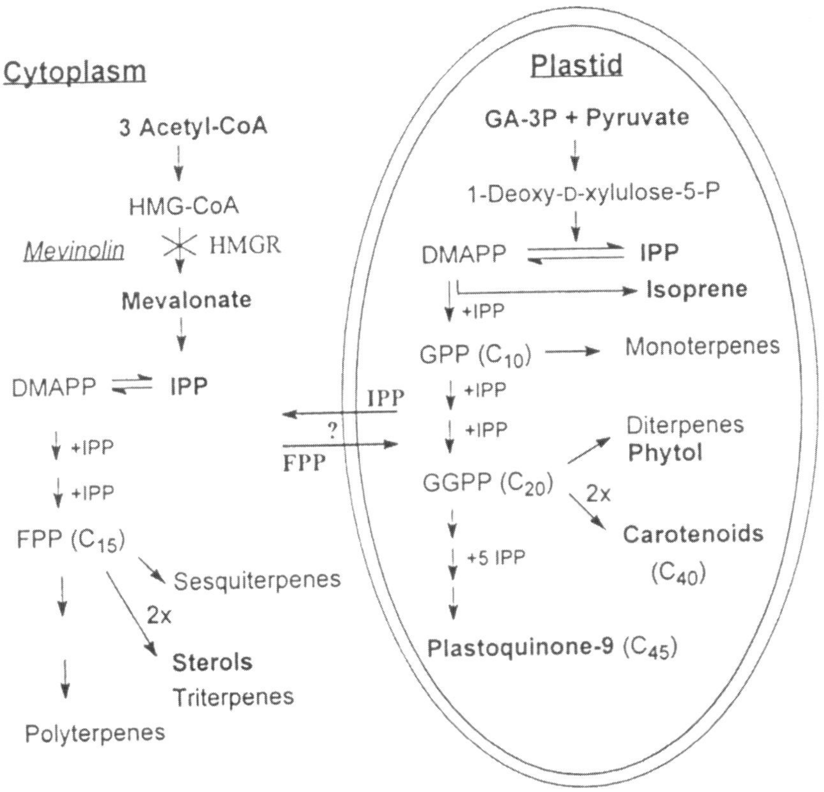

Figure 3.2 Supposed compartmentations of the IPP and isoprenoid biosynthesis in higher plants between cytosol (acetate/MVA pathway) and plastid (DOX-P route). The specific block of the enzyme HMG-CoA reductase (HMGR) by the antibiotic mevonolin is indicated. Abbreviations used: DMAPP = dimethylallyl-diphosphate, GPP = geranyl-diphosphate, FPP = farnesyl-diphosphate, GGPP = geranylgeranyl-diphosphate.(Lichtenthaler, 1998)

K. Hermann and L.M.Weaver (1999) also confirmed, that the shikimate pathway links metabolism of carbohydrates to biosynthesis of aromatic compounds. A sequence of seven metabolic steps, phosphoenolpyruvate and erytlirose 4-phosphate are convened to chorismate, the precursor of the aromatic amino acids and many aromatic secondary metabolites. This pathway intermediates can also be considered branch point compounds that may serve as substrates for other metabolic pathways.

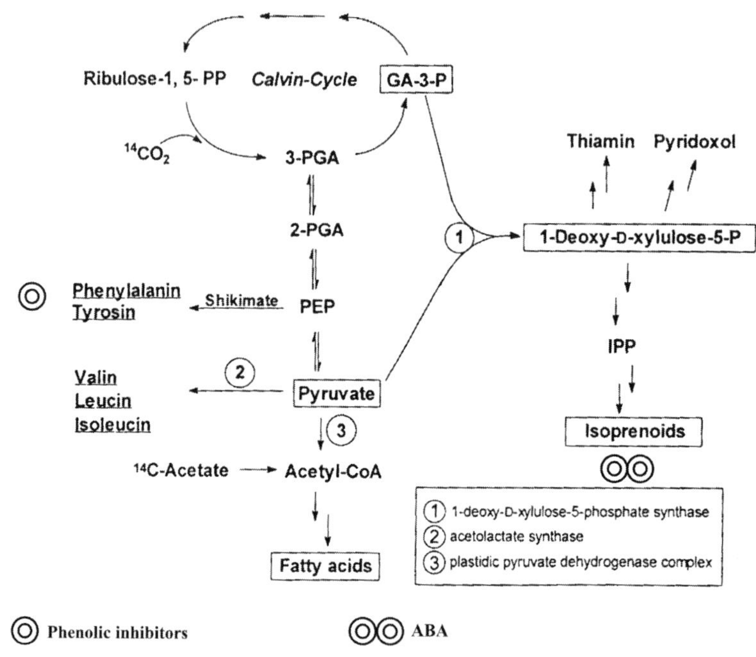

Figure 3.3 Flow of metabolites from the photosynthetic reductive pentosephosphate cycle (Calvin cycle) into different end products, such as isoprenoids, fatty acids, amino acids as well as thiamin and pyridoxol. All processes appear to proceed in plastids. The central role of GA-3-P and pyruvate in the formation of 1-deoxy-D-xylulose-5-phosphate, IPP and isoprenoids is emphasized. (one circle with dot represent phenolic inhibitors, two – ABA)

The shikimate pathway is found only in microorganisms and plants, never in animals. All enzymes of this pathway have been obtained in pure form from prokaryotic and eukaryotic sources and their respective DNAs have been characterized from several organisms. The cDNAs of higher plants encode proteins with amino terminal signal sequences for plastid import, suggesting that plastids are the exclusive place for chorismate biosynthesis. In microorganisms, the shikimate pathway is regulated by feedback inhibition and by suppression of the first enzyme. In higher plants, no physiological feedback inhibitor has been identified, suggesting that pathway regulation may occur exclusively at the genetic level. This difference between microorganisms and plants is reflected in the unusually large variation in the primary structures of the respective first enzymes. Several of the enzymatic pathways occur in isoenzymic forms which expression varies with changing environmental conditions and, within the plant, from organ to organ. The ultimate enzyme of the pathway is the sole target for the herbicide glyphosate. Glyphosate-tolerant transgenic plants are at the core of novel weed control systems for several crop plants.

3.2. Phytohormones and Inhibitors in Mutants.

The role of the photosynthetically active leaves and light in morphogenesis and phytohormone synthesis is of fundamental importance. The chloroplast is one of the main sites of the photosynthesis or the action of the inhibitors and phytohormones precursors of abscisic acid (ABA), indole-acetic acid (IAA), and cytokinins, which take part in the regulation of growth, chloroplast biogenesis, greening, and stomata movement. The modern approach to the resolution of this research issue is to use chlorophyll-deficient mutants to decipher the role of the light effects on hormonal synthesis and growth.

For this purpose the pea mutant of "xantha"-type was used in our experiments. It was obtained after treating seeds of cv. Svoboda by NMU (urine derivative). The mutant yellow phenotype of line N14 is caused by a nuclear recessive mutation. The plastids of this mutant are smaller than chloroplasts of the wild-type green form. Their lamellar structure consists of infrequent vesicles and they are unable to carry out photosynthesis. Mutant plants die before producing seeds and the allele is maintained through the heterozygotes. Homozygous seeds for the mutant allele can be distinguished by their light-yellow color compared with the light-green color of seeds, bearing the wild-type allele. Cv Svoboda was used as the wild-type in all experiments.

ABA level was determined by a GC (gas chromatograph) equipped with an electron capture detector. IAA content was estimated by HPLC PYE (Unicom PU 4002 liquid chromatograph). Extraction and determination of cytokinins was performed using the bioassay method. The content of chloroplast pigments was calculated. Dry weight was determined from duplicate samples not used for extraction. The fresh-weight to dry-weight ratio was 7.1:1 and 13.3:1 for light-grown and dark-grown plants, respectively. Plant samples for phytohormone analysis were placed in liquid nitrogen prior to the extraction procedures. The upper part of the plant included the shoot only. The cotyledons were discarded. The data are based on the combined results from three experiments. Eight to ten plants were harvested for each treatment and five analyses were used to get the data. The plants were raised in a growth chamber at a temperature of 24-26^{0}C. Light-grown plants received a 12 h photoperiod under LB-65 lamps providing an intensity of 10,000 lux (light intensity). The plants were grown in 0.5 L glass containers filled with water which was changed daily. The 8-day-old seedlings were harvested at the stage when they had 6-7 nodes (4-5 normally developed leaves) and before the onset of retarded growth in the mutant plants

Dark-grown plants of wild-type and mutant lines had reduced leaves and a closed apical hook. However, they did not differ in their stem and root length and weight. Likewise stem growth and leaf morphogenesis of light-grown wild-type and mutant seedlings were similar. Their leaves developed normally and elongation of the stem was depressed compared with dark-grown plants. However, light-grown mutant plants, like dark-grown wild-type and mutant seedlings had no chlorophylls. The carotenoid content of light grown mutant plants was about 12% of the level in the wild-type green line. Carotenoids were practically absent in the dark-grown seedlings of both lines .

The content of ABA in the upper part of light-grown wild-type plants was much higher than in the upper part of light-grown mutant plants and of dark-grown seedlings of the wild-type (Figure 3.4). The ABA content was 3-4 times lower in the roots of all forms in comparison with their shoots. The ABA content was 2-3 times as high in the roots of light-grown wild-type plants as in the roots of the chlorophyll-deficient light-grown mutant plants and dark-grown seedlings. It is possible that the decreased ABA content in mutant plants is related to decreased synthesis of mevalonic acid, a precursor common to ABA and carotenoids, or with a change of mevalonic acid metabolism to the carotenoid C40 pathway. In the latter case our results could be taken as indirect evidence of the carotenoid pathway of ABA synthesis.

The light-grown wild-type and mutant plants (both upper part and roots) did not differ in the content of IAA. Moreover, the IAA content was 3-4 times higher in the roots than in the upper part of the seedlings. The level of IAA was higher in dark-grown seedlings than in light-grown plants, especially in their roots where the difference exceeded 3-fold. Thus the blockage of chloroplastogenesis did not change IAA level but IAA level was influenced by whether the plants were grown in light or darkness.

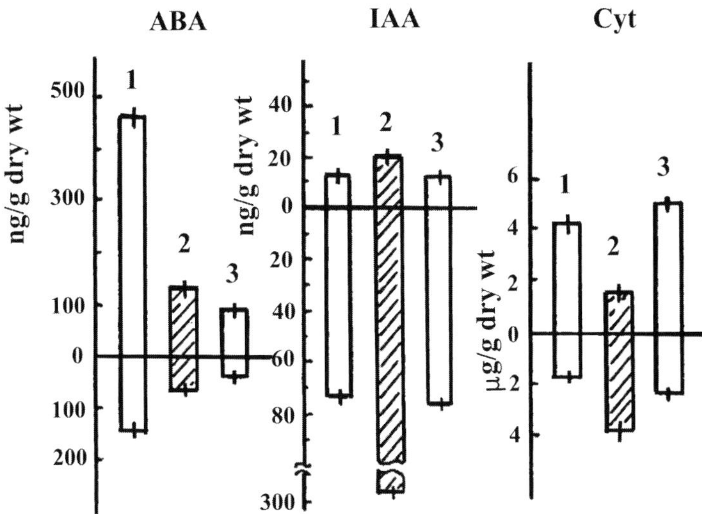

Figure 3.4 The content of ABA, IAA and cytokinins in light-grown wild-type (1) and mutant (3) plants, and in dark-grown wild-type plant (2), A – upper part, b – roots. The bar lines indicate one standard deviation

The level and distribution of cytokinins was approximately the same in the light-grown wild-type and mutant plants. In contrast to IAA levels in light-grown plants, the level of cytokinins was higher in the upper part of the plant than in the roots. However, in dark-grown seedlings the level of cytokinins was higher in the roots than in he shoots. The content of cytokinins decreased in the roots and increased in the upper part of the seedlings after 4 hours of exposure to light. It is possible that cytokinin transport from the roots to the upper part of the plant was suppressed in the darkness. After exposure to the light, cytokinin transport appeared equally intense in both the wild-type green and the mutant yellow lines. However, in the mutant plants the stimulating effects of cytokinins on greening could not be estimated as of the genetic block in chloroplast biogenesis.

These results show that growth and morphogenesis, and the content and distribution of plant growth regulators depend on light. However, the level of IAA and cytokinins is not related to chloroplastogenesis, greening and photosynthesis. The level of ABA depends both on light and on the presence of green photosynthetically active chloroplasts. The blockage of greening and photosynthesis in the chlorophyll-deficient mutant leads to a substantial decrease in ABA content. Naturally, these data are correct only when storage substrates are available from the cotyledons.

In conclusion albino mutants, which are deficient in the chlorophyll and carotenoids contents, have also certain fluctuations in the level of natural hormones. This assumption is also based organ compartmentalization of these substances. Further experiments showed that roots could be a separate center of ABA biosynthesis from mevalonate.

Similar research was done with birch tree. Vlasov and .Lemann (1985) investigated the transport and transformation of growth regulators in bleeding sap, bark and xylem of the birch tree. Cytokinins, gibberellins auxins and ABA, but not phenolic inhibitors were examined. The main growth inhibitor found in plant juices and bark is ABA. The highest amount of this compound was observed during the period of dormancy in the fall, and the least amount after buds opening in the spring. During period of dormancy and buds flowering, the bark had a high content of "neutral" inhibitors, one of which was derivative of ABA-glycosylic ether of ABA. It's content was 0.3 μg/kg of green mass.

During the bud formation the content of gibberellins increased in the bark of the plants compared with the dormancy period. However, during the bud blossoming the content of these compounds decreased significantly, leading to almost minute amounts of gibberellins present in the bark at that time.

The tree juices collected during the imbibition and blossoming of the buds contained almost the equal amount of auxins during each stage – 4.43 μg IAA. Cytokinins were also found at that time, especially zeatin.

Thus, the tree juices while leaving the dormancy period had such biologically active substances as: auxins, gibberellins, cytokinins and abscisic acid. Major fluctuations in the amounts of phytohormones were observed for ABA and gibberellins. Thus emphasizing their role during spring active growth and development. To determine the physiological effects of these substances, the certain experiments were put together. The seedlings of birch (*Betula pendula*) with nondeveloped yet buds were treated with phytohormones. In this case only gibberellin stimulated bud blossoming when ABA demonstrated opposite effect. The simultaneous application of gibberellin and ABA showed that ABA slows down the effectiveness of gibberellins.

In order to demonstrate the transport of ABA in the birch seedling the labeled ABA compound was used as a marker.

The year-old seedlings of birch from spring and fall periods were pre-treated first with IAA and gibberellin, and then they were washed from even residue of these phytohormones, and were submerged into radioactive solution of ABA for 44 hours. After that each seedling was cut into six parts, bark with buds was removed. The radioactivity marker was identified and measured with the help of liquid-scintillated counter "Tricarb", Packard, USA. (see Figure 3.6). Different methods of chromatography were used for further purification of the sample.

In the fall the ABA marker didn't move into the upper part of the seedling where the buds were located, almost 95% of radioactivity localized in its lower part, and mostly in cambium and bark. Phytohormones such as IAA and gibberellin didn't affect at all the movement of the marker.

The opposite effect took place in the spring seedlings: only 50% of the marker was in the lower part of the seedling, other 50% were found in the cambium, bark and buds of the upper part of the seedling.

Further experiments on biochemical properties of ABA designated that both types of seedlings contained the conjugated form of ABA: abscisyl-b-D-glucopyranozoid.

These data show that the amounts of endogenic conjugated form of ABA in cambium and bark of the birch increased during the spring period. During the fall bark is a depot or storage of conjugated ABA. Conjugated ABA can be easily transformed in plant to its free form when it is metabolically necessary, as well as during stress periods such as drought or warming up period in the middle of the winter.

Since, ABA mostly found in cambium and bark, goes through the process of conjugation there very intensively, and significantly affects the periods of dormancy and active growth, including the photosynthetic function of the leaf. A.Mokronosov (1985) in his review summarized our data in the general concept of the formation of hormonal-inhibitory system of growing leaves (Figure 3.5).

Thus, some regulating substances are moved on the short distance, like phenolic inhibitors and other transport on the long distance (ABA, auxins, gibberellins, cytokinins). Some phenolic substances mostly connected with different types of growth retardations including cell wall lignification. Let us consider their biological properties in details.

3.3. Phenolics in Chloroplasts.

An approach that makes it possible to elucidate the role of photosynthetic processes in the formation of phenols is to use antibiotics and growth inhibitors that block the functions of the chloroplasts either totally, or partially. The introduction of the inhibitor simazine into etiolated willow shoots and their subsequent exposure to light inhibited the accumulation of chlorophyll by 40 per cent. The formation of the chalcone isosalipurposide was also inhibited (Kefeli, 1978).

The experiments performed by us on application of other inhibitors (chloramphenicol, diurone) in low concentrations into etiolated apple-tree and willow shoots demonstrated that under light, greening of shoots and accumulation of phenols in them are strongly retarded.

A logical question is how closely phenol biosynthesis is connected with individual cell structures. Differentiated centrifugation of cell organelles and obtaining of pure cytoplasm sap from green and etiolated shoots of willow

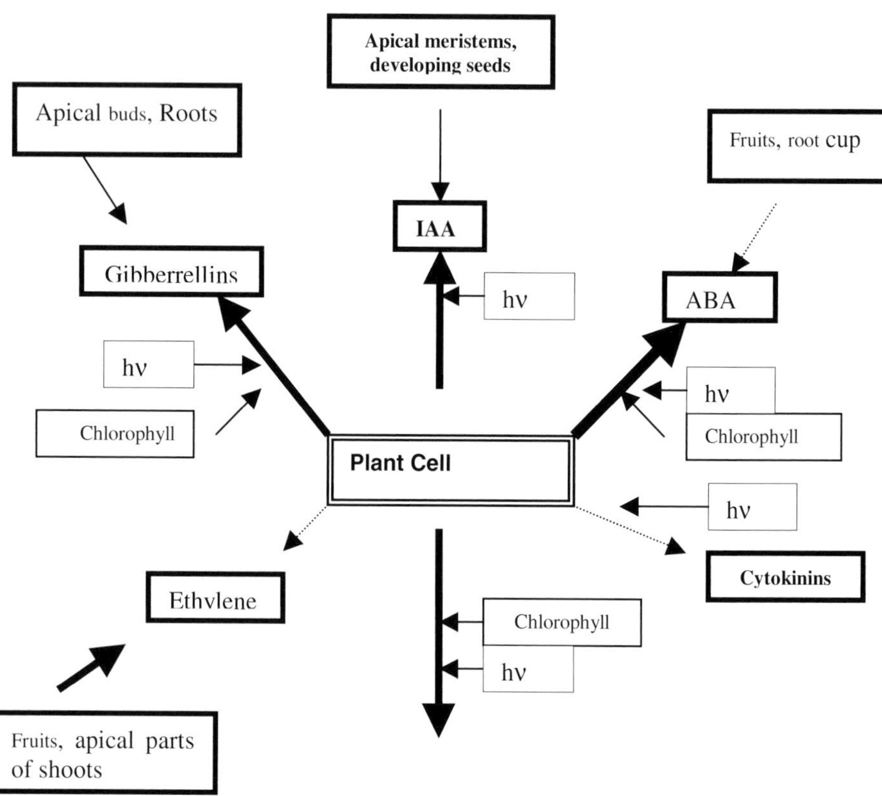

Figure 3.5 The concept of formation of hormonal-inhibitor system in leaves (Mokronosov, 1985)

(*Salix rubra L.*) made it possible to come to the conclusion that the sap from etiolated shoots contains the chalcone isosalipurposide and phenolcarbonic acids, whereas the sap of green leaves contains flavonol-glycoside and traces of flavonol and does not contain some phenolcarbonic acids.

Chalcone and phenolcarbonic acids present in etiolated willow shoots can be viewed as the potential precursors of light-synthesized flavonoids.

Flavonoids are known to be localized mainly in vacuoles, and phenolcarbonic acids and other lignin precursors - in plant cell walls (Lewis, Yamamoto, 1990). As to chloroplasts, there were practically no data pointing to chloroplasts as the site of biosynthesis of phenolcarbonic acids and flavonoids.

At the same time experiments on the retarding action of growth inhibitors on the phenol biosynthesis in greening willow and apple tree shoots had shown that chloroplasts could be directly involved in the formation of phenols in light, the more so as they had been found to be able to oxidize p-coumaric acid to caffeic acid and to oxidize phloridzin.

In order to study the role of light in the formation of phenolic compounds in cell structures, lignified shoots of willow (*Salix rubra L.*) of the last year's growth were cut in spring, the cuttings placed in a dark, and after one month young shoots forced from willow buds were used in experiments as etiolated material. At the same time similar cuttings were kept under light and green shoots that appeared from willow buds were analyzed. Besides, young shoots were cut in spring directly from willow bushes, and used in the experiment.

The chloroplasts of the spring green leaves isolated by centrifugation revealed the presence of flavonoids and also traces of some unknown phenolic compound. It is interesting to note that the chloroplasts contained also flavonol which could not be detected neither in the wash water from the chloroplasts nor in the supernatant above the chloroplasts destroyed by an osmotic shock (Figure 3.6).

Flavonol was also localized in the supernatant, though in a smaller amount. The chloroplasts from green spring leaves contained also phenols inherent in the cytoplasm, for example, isosalipurposide. The proplastids of the etiolated willow shoots precipitated by centrifugation contained no phenols whatsoever. The chloroplasts of autumn willow leaves contained much less phenols and their composition was qualitatively different; there were no isosalipurposide and phenol, but some types of flavonols were present as previously.

Since phenolic compounds are mostly water-soluble, it can be assumed that the presence of phenols in willow chloroplasts is the result of their readsorption from the cytoplasm which occurs during homogenization and centrifugation. In order to verify this assumption, a mixture of quercetin and quercitrin solutions on a sucrosephosphate buffer was prepared; this mixture of solutions was intended for adding to a homogenate containing pea chloroplasts in which the above compounds could not be detected. When the solution was cooled down, quercetin partially precipitated. The precipitate was filtered out, and thus the homogenate could be considered as containing mainly quercitrin with a small amounts of quercetin present in mixture. The fact that quercetin precipitates on cooling makes its ability to adsorb on chloroplasts questionable. Chromatographic analysis after centrifugation of the pea chloroplasts with a quercitrin-quercetin mixture also evidenced that the chloroplasts were free from these compounds, whereas both quercitin and traces of quercetin were detected in the cytoplasm.

These data on phenols readsorption from the cytoplasm into the chloroplasts seemed to be doubdtful. The chloroplasts of spring and autumn willow leaves contain specific phenols, the chloroplasts of spring leaves contain chalcone isosalipurposide present mainly in the cytoplasm sap.

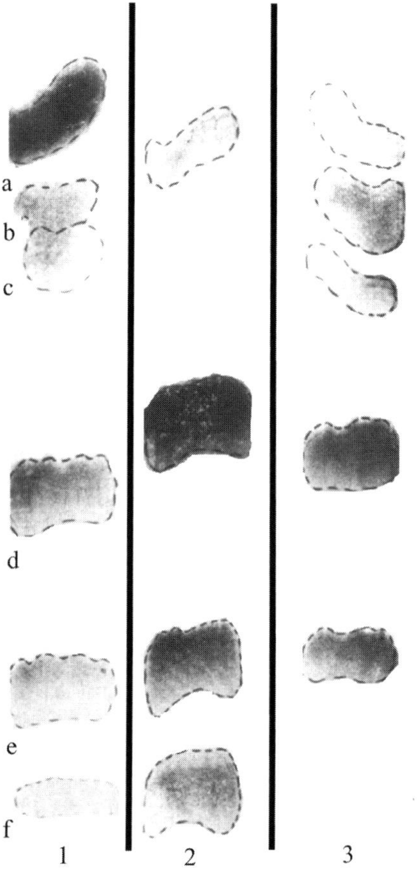

Figure 3.6. Occurrence of phenolic compounds in chloroplasts of spring willow leaves. 1 – cytoplasm (1:200 dilution); 2 – degraded chloroplasts; 3 – supernatant over degraded chloroplasts: a, b, c – phenolcarbonic acids; f – flavonol-glycosides; e – isosalipurposide.

A reduced amount of phenolic compounds in the chloroplasts of autumn leaves can be attributed to the attenuation of their photosynthetic functions, while the absence of isosalipurposide in these leaves can be accounted for by its transport into the cytoplasm. These data on localization of phenols in chloroplasts were confirmed by several researches. Nevertheless, the mechanism of transport of phenols from chloroplasts to cytoplasm is not yet quite clear.

The fact that phenolic growth inhibitors of the isosalipurposide type have been discovered only in the cytoplasm of autumn leaves can be explained by changes in the chloroplast membranes caused by attenuation of their functions which results in a gradual release of metabolites through 'weakened' membranes into the cytoplasm. In general, the accumulation of synthesized phenols in cytoplasm sap seems to be able to alter the permeability of cell organelles thereby causing the evacuation of not only phenols, but also of a number of important metabolic products.

Several aspects of phenol localization in a cell remains vague even until now. Though we already know that phenols are present in cell walls, vacuoles and chloroplasts, their role in the cytoplasm is completely unknown. Can phenols have some effectson mitochondria? How easily do they move in a cell when leaving one organelle and entering another? What is the degree of interaction between phenols and the enzyme systems of a cell?
All these questions need experimental investigation and experiments with radioactively marked (labeled) metabolites may be tremendously important, and help to solve these problems.

To summarize the results of studies on the biosynthesis of phenolic compounds in green plant leaves, the following can be stated. The synthesis of phenol-carboxylic acids and flavonoids is strongly stimulated by light. Metabolic inhibitors that suppress the photosynthetic activity of chloroplasts (simazine, diurone, chloramphenicol) inhibit the biosynthesis of flavonoids as well. The chloroplasts of green leaves have phenol compounds present, some of which are specific to these organelles only. The chloroplasts of spring willow leaves contain more phenols and in greater amounts than the chloroplasts of autumn leaves. The synthesis of phenols in chloroplasts depends on the presence of the light. The proplastids of etiolated willow shoots contained no phenols. Phenolcarboxylic acids and chalcones occur in cell sap. Light causes the appearance of flavonols in chloroplasts and cytoplasm. Chalcone (isosalipurposide) and phenolcarbonic acid present in etiolated willow shoots can be considered as metabolic precursors of light-synthesized flavonols.

There are parts of the cell where phenols are contained in significant amounts (cell walls, vacuoles); however, there is still a number of major questions to be answered, dealing with the ability of phenols to move inside the cell and affect, the functions of mitochondria, ribosomes and other cell organelles found within the cytoplasm.

Since the natural growth inhibitors - hydroxy derivatives of cinnamic acid, coumarin and naringenin - are phenolic substances, and they are formed through the same major pathways as other phenolics , latest have no inhibitory effect on the growth of plants. The synthesis of the growth inhibitors derivatives of hydroxycinnamic acids follows the scheme: shikimic acid - chorismic acid - prephenic acid - cinnamic acid - p-coumaric acid.

In this way, indole auxins (IAA, indoleacetonitrile, etc.) as well as henolic inhibitors (p-coumaric acid, coumarin, naringenin and others) are formed rom the common precursors shikimic and chorismic acids.

CHAPTER 4

NATURAL GROWTH INHIBITORS AND PHYTOHORMONES IN THE PROCESS OF CELLS AND ORGANS ELONGATION

The formation and transport of natural growth inhibitors and phytohormones is the beginning of a long process, responsible for their mode of action within the plant and plants community (Figure 4.1).

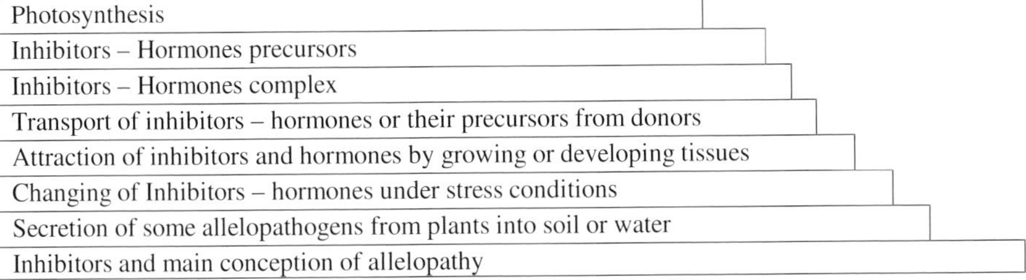

Photosynthesis
Inhibitors – Hormones precursors
Inhibitors – Hormones complex
Transport of inhibitors – hormones or their precursors from donors
Attraction of inhibitors and hormones by growing or developing tissues
Changing of Inhibitors – hormones under stress conditions
Secretion of some allelopathogens from plants into soil or water
Inhibitors and main conception of allelopathy

Figure 4.1 Natural growth regulators in one plant and plant community

Phenolics usually have the short transport pathways in the cell while other regulators have long distance transportation.

The formation of phytohormones is closely connected with the formation of natural growth inhibitors in plant tissues. Such an interrelation becomes possible through their common precursors, which are a starting point for the metabolic pathways of the synthesis of these growth-regulating substances. The previous chapter was devoted to the consideration of a number of contacts between the intermediate products of the biosynthesis of phytohormones and natural growth inhibitors. Such contacts may be conventionally called interaction of phytohormones and natural inhibitors at the level of their biosynthesis (Kefeli et al., 1988; Kefeli, Sidorenko, 1991).

In this chapter we shall discuss peculiar features in the behavior of natural growth regulators during the next stage, i.e. when they affect growth processes in plants direct Corn is sensitive to gibberellins A_3. The same tests are sensitive to natural growth inhibitors. Irradiation of these test block the growth of hypocotyl and msocotyl growth and activize the ABA and phenolic biosythesis in the green part of seedlings. In each organ these are a certain centers (target zones) for stimulatory and inhibitory effects. Thus cytokinin inhibitor interactions you can observe on the growth of isolated squash cotyledons (Figure 4.2). Hypocotyl of cucucmber is sensitive to gibberellin A_4. (Figure 4.3).

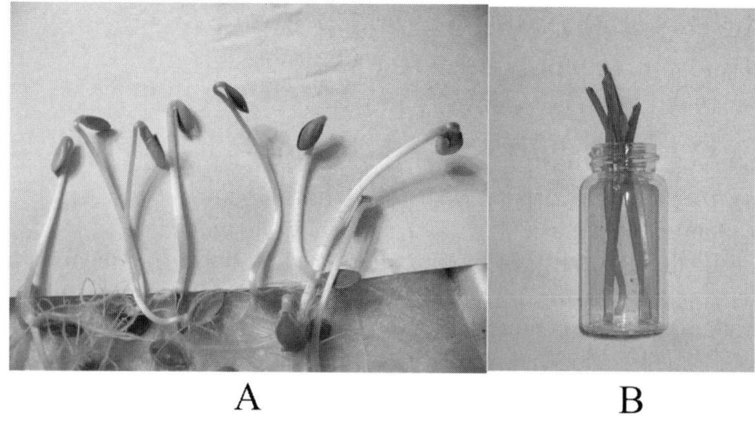

Figure 4.2 Test plants for gibberellin investigation.
A – Cucumber hypocotyl test for GA_4; B – Upper part of corn seedlings, mesocotyl test for GA_3

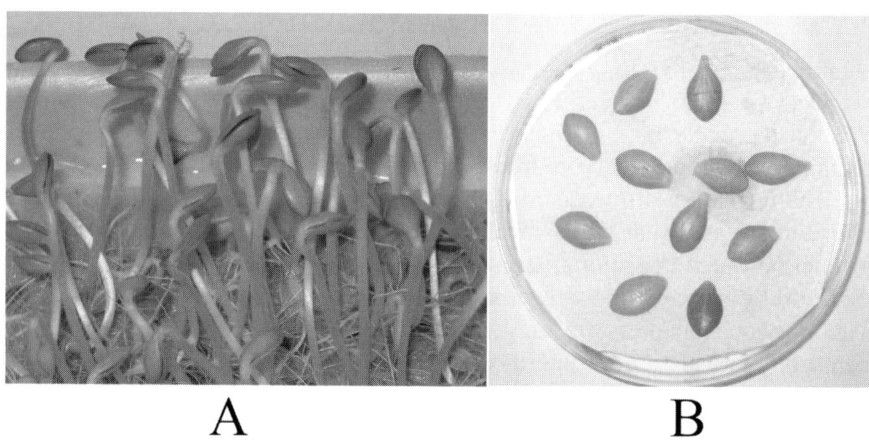

Figure 4.3 Squash test for cytokinins investigation. A – squash seedlings – 10 day old; B – squash cotyledons-test on cytokinins.

4.1 The role of phytohormones and natural inhibitors in growth of aboveground organs of plants.

Besides apex-produced auxins, the auxins of cambium cells also play an important role in the growth of a stem (these auxins are responsible for radial growth). IAA introduced exogenously can imitate cambial activity promoting tile development of broadbundle xylem in perennial plants.

Usually, the appearance of natural auxin in the cambium coincides with the termination of dormancy in woody plants. This is the period of vigorous radial growth of the shoots. The cambium itself is not always responsive to exogenous auxins and synthetic stimulators. Such a negative response of the cambium observed for coleus, oxalis, crab, goosefoot and other plants may be attributed to an excessive auxin content in cambium cells. Whether auxins are synthesized in the cambium or apical meristem, they are transported within the plant to active growth sites.

Gibberellins promote stem growth mainly through accelerating the development of internodes rather than through increasing their number. Similar to auxins gibberellin can also stimulate cell division in a plant, but it affects only cell division in meristemic areas, and not in differentiated tissue. Gibberellin-induced mitoses are oriented along the main axis which eventually causes elongation of the stem (Figure 4.2). Data are also available now on the participation of gibberellins in the geotropic response of plants in the signaling processes (Richards et al., 2001).

The stems of dwarf plants are known to be most sensitive to gibberellins. Early investigators explained such a strong response to occur due to the deficiency of endogenous gibberellins in the plant itself. The comparison of the action of gibberellin and IAA has shown that the former is more active with respect to an intact plant, and the latter with respect to segments of stems, leaves and other organs.

Cytokinins constitute another group of factors regulating the growth of a stem and axial organs. If they applied to isolated leaves, they attract nutrients. This effect of cytokinins seems to be closely associated with their ability to retard yellowing of leaves, and to intensify the synthesis of protein, nucleic acids, lipids, starch and other compounds. Usually, young leaves do not respond to kinetin treatment, whereas in senescent and particularly isolated leaves this response to kinetin is extremely strong. The main site of Icinin synthesis is in the roots. Experiments on leaf rooting have convincingly demonstrated that if a leaf has rooted, it resists such a rapid exhaustion and destruction as a leaf without roots (Mothes, 1964). According to Mothes, however, kinins are synthesized not only in roots, but may also originate in young leaves and in buds. Thus, each phytohormone displays its own specific properties in plant growth regulation. Auxins, for instance, promote differentiation of a growing shoot stem and inhibits the growth of lateral buds. Gibberellins cause extension of stem internodes, induces bolting of rosette plants and change the shape of leaves. Kinetin, which retards senescence of leaves, produces a mobilizing effect on the transport of nutrients and hormone substances and, if applied to a plant, induces the growth of its lateral buds. Cytokinins are able to increase the size of isolated pumpkin and squash cotyledons (Figure 4.3).

In a growing plant, phytohormones may act as factors causing new morphogenetic or organogenetic processes, but at the same time they may simply maintain and direct the process in a plant organ already formed. Treatment of intact plants with phytohormones has shown that only gibberellin can strongly promote plant growth, whereas neither auxin nor kinetin exhibited this ability. At the same time, the growth of isolated segments of stems and tissues is induced by auxin and kinetin, while gibberellin produces no regulatory effect in these growth models.

Phytohormones are able to intensify the growth of stems and leaves and also the tissues of these organs and the effect of natural growth inhibitors is quite the reverse. The functions of natural growth inhibitors in plants, as we know, are performed by some phenolic compounds and abscisic acid. The common property of phytohormonres as well as cytokinins is their active oxidation (Bilyeu et al., 2001). Some phenols (coumarins, phenolcarboxylic acids, benzoquinones and chalcones) are more stable. They accumulate in plant tissues and may then inhibit the growth of the plants. Plants have no systems for taking toxic products out of the organism, therefore, in the course of evolution they have developed capabilities for localizing toxic substances in separate, relatively isolated sites: vacuoles, secretory vesicles. Phenolic compounds may accumulate in great amounts in greening tissues of young shoots and seedlings, for example, in a tea plant. Large quantities of phloridzin are produced in apple seedlings or in the tips of young shoots. The content of phenols undergoes a radical change during the growth process.

The content of flavonoids in a *Bupleum* plant increased appreciably during the transition from the growing period (rosette phase) to flowering. It is significant that qualitative differences in the composition of flavonoids were observed for old and young leaves of mountain cranberry Vaccinium; the arbutin analog pyroside was present in old leaves, whereas arbutin itself occurred in young leaves and buds.

In spring shoot growth vigorous condensation of phenolic compounds takes place in plants. Experiments on the content of phenols in rooting spring shoots of willow demonstrated that the catechin content in such shoots reduces abruptly as compared with the dormant period whereas the amount of condensed products (insoluble and soluble tannins) rises appreciably.

When growth processes terminate phenolcarbonic acids and phloridzin are accumulated in the tissues (Sarapuu, 1 970). It is known that p-coumaric acid and coumarin strongly inhibit the growth of a stem and its segments. A great number of data has been accumulated on the seasonal fluctuation of flavonoids and phenol carbonic acids in the tissues of woody plants. Usually, the investigators observe the accumulation of flavonoids and coumarins, one in spring, after buds have opened and shoots start growing, and the other in autumn before the inception of dormancy. It is proposed that the phenolic acid form, is the best known group of inhibitors among the secondary plant chemicals. Most of them are synthesized via the shikimic acid pathway, through the domination of phenylalanine or tyrosine, although a few are synthesized from acetic acid units. Members of the cinnamic acid series can be converted into the benzoic acid counterparts by oxidation. Elaboration of more complicated hydroxylation patterns can follow the formation of the simple phenolic acids. The lactones of the coumarin series are formed by the ready lactonization of the phenylpropene equivalents, especially after the formation of the o-glucoside and the conversion of the transcinnamic acid to the cis configuration by the action of light . The phenolic lactones ordinarily are formed via hydrolysis of the glycosides.

The phenolics ordinarily occur in plants as glycosides, usually involving glucose but sometimes several sugars in series rupture of plant tissues ordinarily, leads to the hydrolysis of phenolic glucosides to the aglycones, which are markedly more reactive.

The inhibitory effects of phenolics on growth are commonly attributed to the enhancement of indoleacetic oxidase, but it is very probable that other actions such as an interference with oxidative phosphorylation are involved. The effectiveness of phenolics as co-factors for IAA oxidase is fairly specific for monophenols such as p-coumaric acid, diphenols such as caffeic acid can serve as inhibitors of IAA oxidase, hence acting as stimulants of growth under some conditions.

Regulation of growth and development by phenolics and other secondary plant chemicals is markedly less clear than with ABA. Certainly, the best documented regulatory effect relates to the responses of seedlings to light. The kaempferol increase continued to expand over a 15-h period following the light treatment. It was noted a similar increase in p-coumaric acid and other phenolics in seedlings after illumination with red or blue light and until the early 1960's the growth inhibition process was most commonly associated only with the accumulation of phenolic growth inhibitors. At the same time, as far back as 1955, Osborne reported that senescent leaves contain a substance, which diffuses into agar and induces defoliation of plants.

Further studies on substances contained in fruits and leaves, which induce formation of an abscission layer, were carried out in the USA. There were isolated two compounds from cotton: abscisin I and abscisin II, which were called so because the experimenters thought them responsible for regulation of leaf abscission. After the Ottawa Conference on Plant Growth Regulators (1967) the name "abscisins" or "dormins" was substituted by "abscisic acid'. It was shown that abscisic acid retards cell division in the abscission zone, accelerates lysogenous destruction of stem parenchyma and destruction of cell walls, inhibits callose synthesis in abscission zone vessels, all this resulting in the abscission of a leaf stalk 24 hours after treating it with abscisic acid. Unlike abscisin, IAA as a typical inhibitor of leaf and fruit abscission accelerates leaf stalk growth and cell division, promotes callose bi6synthesis and prolongs the life of a stalk on the plant up to 120 hours. It can move with phloem exudates in a willow stem. Besides this ABA (abscisic acid) has a broad range of activity. I is able not only to accelerate leaf abscission (antiauxin) but also to retard flowering, inhibit bud breaking (antigibberellin). delay cell division in tissue cultures (antikinin), and depress transpiration.

Abscisic acid is contained in pea seedlings, on which basis it was assumed that it is involved in correlative inhibition of bud growing. Exogenous abscisic acid inhibited the growth of lettuce seedlings. Gibberellic acid promoted the growth of pea seedlings and reduced the content of abscisic acid in the top part of the seedling, whereas the amount of abscisic acid in the bottom part of the seedling went up.

First leaf of seedlings growing in the light retard their grow in the darkness, where as coleoptiles elongate easily. These coleoptiles are the source of sections for cell elongation investigations (Figure 4.4).

The data available permit to conclude that abscisic acid inhibits the growth of a number of bio-assays, including tests associated with seed germination and bud opening, in concentrations 2 or 3 orders lower than those of phenolic inhibitors. This, however, in no way means that abscisic acid possesses some unusual properties. Some investigations showed that apart from abscisic acid, the in inhibitor complex includes also several strong growth-inhibiting phenolic substances.

Abscisic acid can be detected with the help of bio-assays and by rechro-matography of the eluates in solvent systems for which its position (Rf) is known. Below certain data are given on the mobility of abscisic acid in different mixtures of solvents. The data available on the colour reactions of natural auxins and growth inhibitors point rather to the subordinate nature of this analysis procedure which, in the first stage, can give only a general description of the position of substances on chromatograms (indole auxins and abscisic acid are not indentifiable on chromatograms with the aid of colour reactions). Therefore, biological methods of investigation (bio-assays) become still more important for detecting physiologically active substances. Usually after detecting natural growth regulators by means of paper chromatography, colour reactions and bio-assays, it becomes necessary to isolate some most active growth regulators in larger quantities of the order of tens of milligrams (for phenolic inhibitors) or several fractions of a milligram (for IAA and abscisic acid), to purify them from foreign admixtures and describe them in detail from the physico-chemical and biological viewpoints. Chromatographic separation on columns with a caprone, silica gel or cellulose adsorbent or on Whatman 3MM drawing paper is commonly used for this purpose.

Figure 4.4 Wheat Genova seedlings. 10 days greening in day light 9background), yellow in darkness (front)

Chemical analysis of a natural compound involves a number of general procedures such as extraction, isolation, purification and determination of the chemical structure of the compound. Unknown compounds are usually extracted in steps by solvents of an increasing polarity: petroleum ether, benzene, ethyl ether, ethylacetate, alcohols and water. The advantage of alcohol extractions over water extractions is a low boiling point of these solvents; however, such extraction may involve the break-down of ester bonds, for example, those of phenol esters with sugars. Limitations of water extraction are extraction of large amounts of foreign substances, and the impossibility of preventing the action of all enzymes whose activity cannot be stopped even by boiling. A differentiated approach should also be applied for selecting a method of substance separation. Separation on Whatman 3 MM drawing paper is convenient where less than 0.5 g of a substance is to he obtained, whereas separation on columns is more appropriate for extracting larger amounts of substances. Below the scheme of preparative isolation is given which was used for isolating and identifying the natural growth inhibitor of cabbage (phenolic substance, 12 mg from 500 g of leaves). willow (chalcone glucoside, 75 mg from 1000 g of leaves). pea (quercetin-glycosyl-coumarate, 500 mg, p-coumaric acid, 10 mg) and maize (p-coumaric acid, 4 mg from 250 g of leaves). Caffeic acid was also isolated from pea leaves.

For extracting natural phenolic growth inhibitors. green leaves (or other plant organs) were pulverized in liquid nitrogen, placed in 96% ethanol heated up to 70^0 and digested during two or three days. Then, the extract was decanted, and the extraction repeated until the entire chlorophyll was extracted. After that, ethanol was evaporated, the water phase decanted, (i.e. released from pigments) and extracted three times with ether, ethyacetate and n-butanol. The extracted fractions were evaporated until dry and dissolved in 70% ethanol. The first chromatographic separation of the fraction was performed on Whatman 3 MM drawing paper in a selective solvent mixture and the fraction which contained the largest amount of the inhibitor was determined by colour reactions and a bio-assay. This can be conveniently accomplished by associated chromatographic analysis of small amounts of the fraction extracts on Whatman 2 paper. Active zones are cut out of the primary separation chrornatograms and rechromatographed on Whatman 3 MM paper using specially selected solvents out of those listed above. Rechromatography on Whatman 3 MM paper together with associated substances on Whatman 2 paper is repeated until a separate spot of the required substance is obtained. Then, the spot is eluated with 70% ethanol, the ethanol is evaporated, the dry residue dried over $CaCl_2$, weighed and used for taking the UV and IR spectra, hydrolysis, chromatography with standard substances and for assessing the activity be means of bio-assays. The nuclear magnetic resonance technique has become extensively used for identifying the structure of flavonoid glycosides (Figure 4.5).

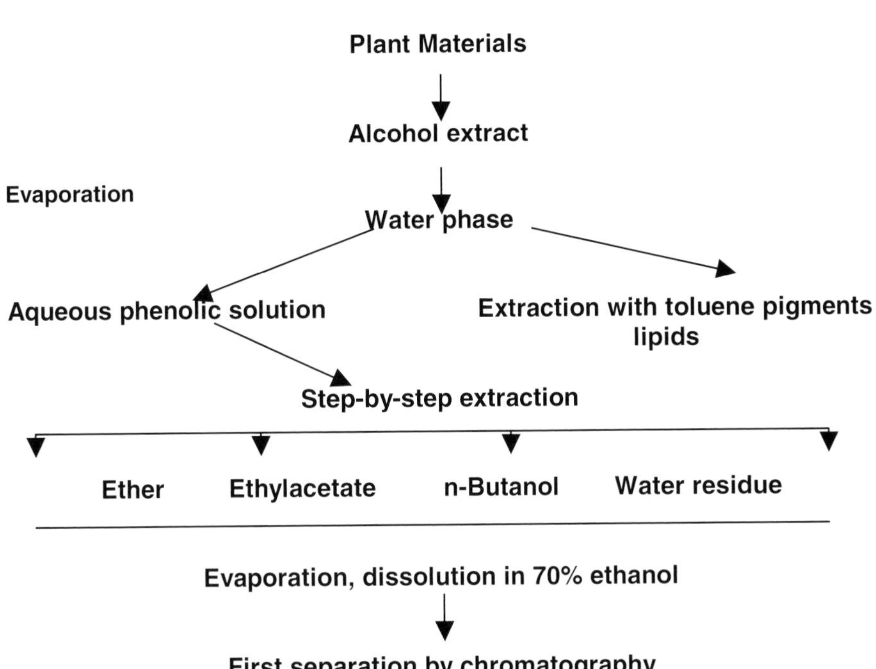

First separation by chromatography
On Whatman 3 MM paper in 15% CH_3COOH in n-butanol-CH_3COOH – H_2O, 40:12:28 or in other solvent mixure

Control over separation on Whatman 2 paper, UV light color reaction

Secondary separation on Whatman 3 MM paper in specially selected optimum solvent mixture

Elution, purification of individual substances, drying over $CaCl_2$, weighing

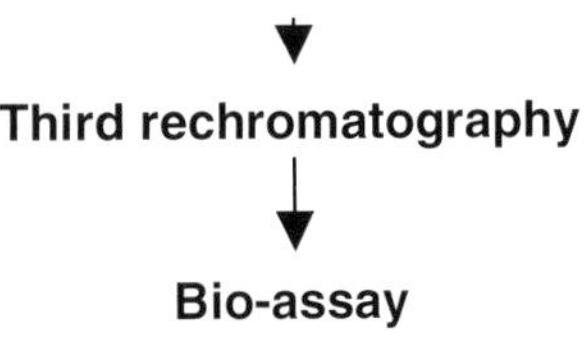

Third rechromatography

Bio-assay

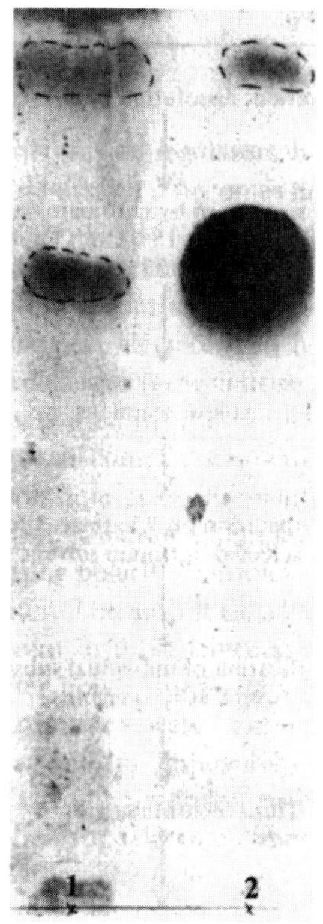

Figure 4.5. Chromatograph detection of p-coumarine in quercetine-glycosyl-coumarate (QGC) molecule. Products contained in pea flavonoid hydrolyzate (1) and standard (2) p-coumaric acid. (Chumakovskiy & Kefeli, 1968)

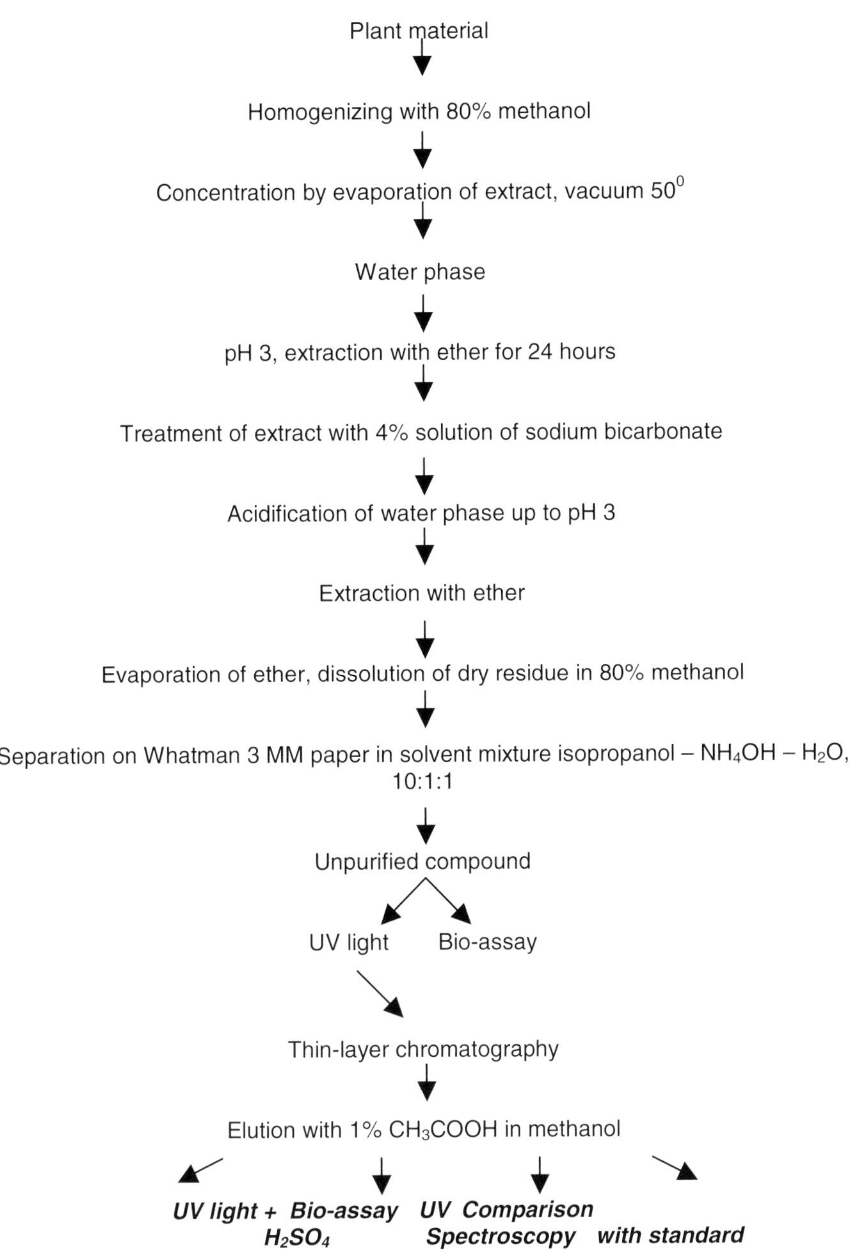

Let us consider in greater detail the preparative isolation of phenolic compounds from pea. 500 mg of flavonoid and two phenolcarbonic acids were extracted from 1500 g of green pea leaves. Following the general identification scheme, flavonoid was identified as flavonol. It did not show fluorescence in UV light and, therefore, was identified as flavonol-glycoside. This was confirmed by the fact that, after being treated with an $AlCl_3$ solution, flavonoid fluoresced in UV light (which is the basic property of flavonol-glycosides); flavonol-glycoside was hardly soluble in ether and ethylacetate and dissolved readily in butanol. This circumstance evidences that the extracted substance was polyglycoside. Rf data and the above properties suggest that the substance was quercetin-3-glycosyl-p-coumarate (Furaya & Galston, 1965).

The UV spectrum of the extracted compound was identical to the UV spectrum of quercetin-glycosyl-coumarate (QGC). The acid hydrolysis of this product confirmed the presence of p-coumaric acid. We did not estimate the amount of sugar residues in a QGC molecule, there are only published data to this effect (Furuya & Galston, 1965).

QGC can be considered a bound form of p-coumaric acid. Its inhibitory effect was 20 times as weak as that of p-coumaric acid. Apart from QGC, free p-coumaric and caffeic acids were also detected in pea seedlings.

Chalcone isosalipurposide was preparatively isolated from willow shoots (700 mg from 1.5 kg of the shoots) and identified; the leaves of an apple tree contained dihydrochalcone phloridzin which was isolated and identified by the Sarapuu method (1970).

The leaves of maize contained derivatives of p-coumaric acid and p-coumaric acid itself (or its glycosi4e) which was isolated in the amount of 4 mg from 100 g of the leaves. The properties of preparatively isolated phenolcarbonic acid, chalcone, dihydrochalcone and QGC are shown in Table 4.

Preparative extraction of phenolic growth inhibitors was paralleled by the study of their action on the growth of wheat coleoptile segments and the growth of etiolated stem segments in specific bio-assays.

Preparative extraction of abscisic acid is much more complicated experimentally as compared with extraction of phenolic growth inhibitors, since the amounts of abscisic acid do not exceed 1-2 mg per kilogram of plant material.

1. Primary extraction is carried out with the mixture methanol - water, 4:1, the extract is concentrated by evaporation until a water residue is obtained and is acidified to pH 3, after which it is extracted with ether, and the ether extract containing organic acids. is repeatedly extracted with a sodium bicarbonate solution at pH 8.

2. To remove neutral material from the fraction of organic acids dissolved in sodium bicarbonate, the solution is acidified up to pH 3 and extracted again

with ether. As a result of this procedure, abscisic acid passes into the final ether extract and is then separated from the admixtures by chromatography.
3. The chromatographic mobility (Rf) of abscisic acid contained in the extract and Rf of the labeling substance are compared by their joint chromatography in two or three mixtures of solvents. Let us consider the activity of some natural inhibitors and phytohormones in the cell elongation models.

The functions of natural growth inhibitors and phytohormones are closely interrelated. Abundant data are available on the antiauxin action of phenolic inhibitors and abscisic acid directed at inhibiting cell extension processes and leaf stalk abscission and manifested in retarding bud growth in tissue cultures. However, it has recently been found that natural inhibitors can also reduce the activity of gibberellins through inhibiting the formation and activity of hydrolytic enzymes and depressing stem growth. The antagonistic action of natural growth inhibitors was also observed with respect to cytokinins, in particular, relating to their effect on leaf greening and seed germination. The rhythmic process of plant growth in cells, tissues and organs was investigated by E. Bunning and D.Sabinin. However, the role of phytohormones and inhibitors in these processes is still unclear. For this purpose Le Thi Muoi (1985) investigated the effect of phytohormones and inhibitors on the rhythmic growth of the intact wheat coleoptile and wheat coleoptile sections (Figure 4.2).

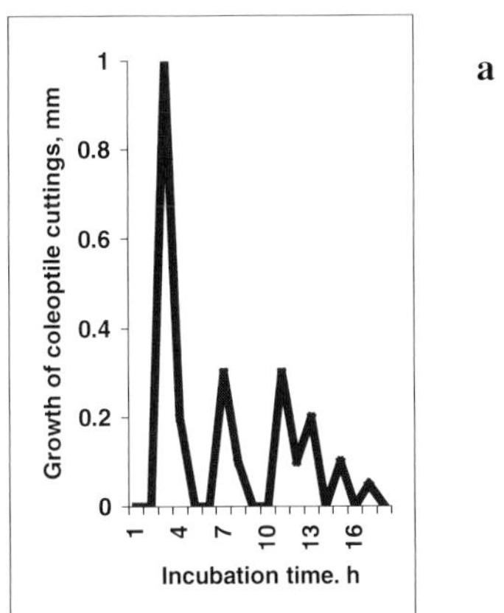

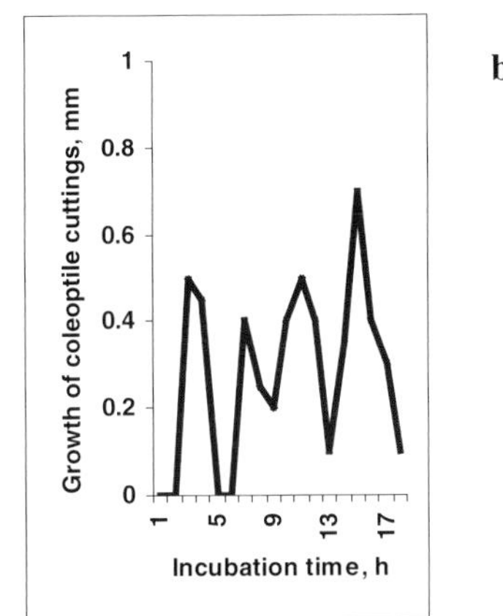

Figure 4.6 Growth rhythms of subapical zone of wheat coleoptile (a) and it's isolated cutting (b). Growth media – 20% sucrose.

The intact wheat coleoptile 18-20 mm length was grown and had four main maximum of growth intensity-3,9,13 and 17 hours. If subapical region of these coleoptile was isolated and incubated in 2% sucrose-the same type of rhythms was observed (Figure 4.6)

If in the sucrose solution was added auxin IAA or natural inhibitors - coumarin and ABA the type of rhythms was preserved, but IAA enhanced amplitude of the pikes during the beginning of the elongation process. Later the amplitude was reduced. Inhibitors decrease the amplitude of pikes in the comparison of auxin without further reduction of the height of the amplitudes (Figure 4.7)

Taking into consideration such a type of effect of hormone-IAA and inhibitors on the growth of wheat coleoptile sections it was decided to apply these regulators just during the periods of the activation of the growth processes.

It was observed that if inhibitors were applied for 3 hours and than sections were incubated with auxin-the effect of inhibitors was brought close to nothing by auxin. If the sections were incubated in auxin during 3 and more during 9 hours and than incubated in the solutions of inhibitors-the auxin effect was brought close to nothing (Figure 4.8).

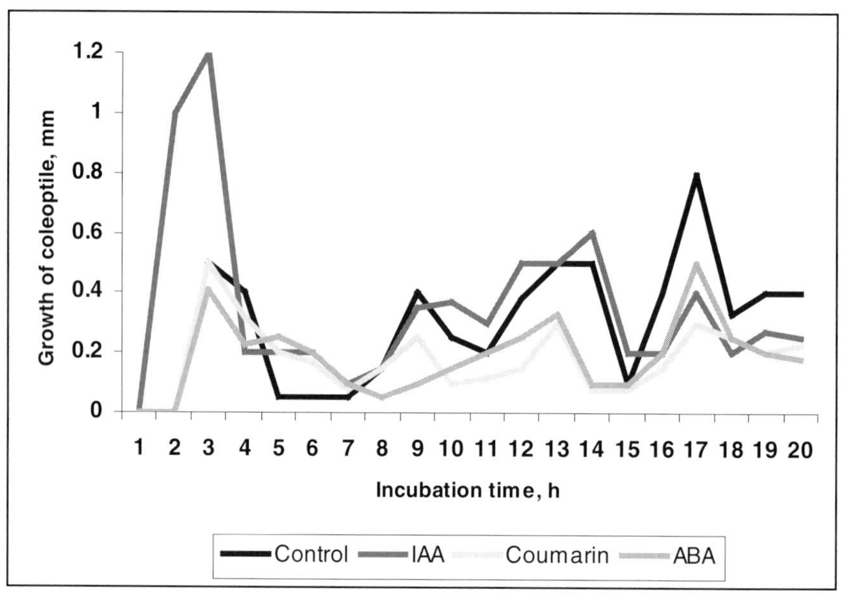

Figure 4.7 Growth rhythms of wheat coleoptiles in presence of IAA, ABA and coumarin. 1 – control; 2 – IAA (28.5×10^{-6}M); 3 – coumarin (13.7×10^{-5}M); 4 – ABA (18.9×10^{-7}M)

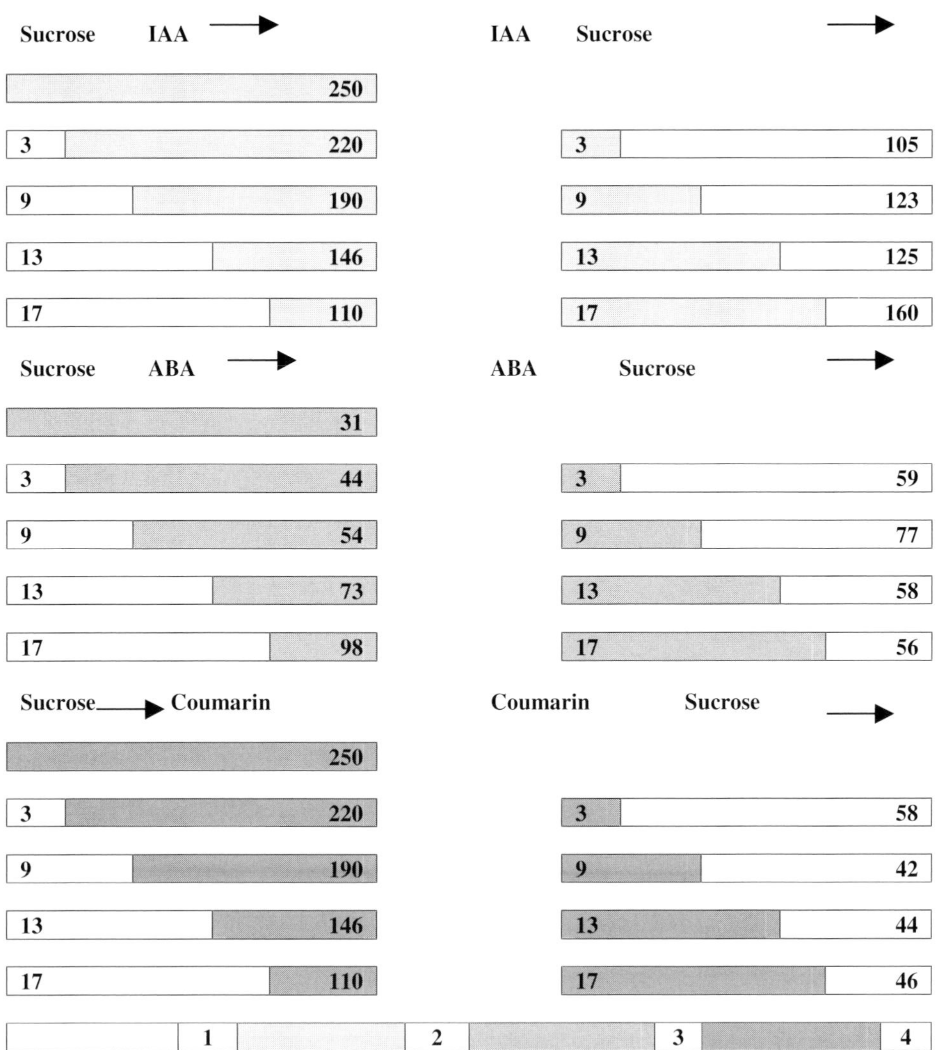

Figure 4.8 Effect of auxins and inhibitors on the growth of wheat coleoptiles. Time of introduction of biologically active regulators into the media – 3,9,13 and 17 hours. Numbers on the right side indicate the growth of coleoptiles (compare to control, %); 1 – control, 2% sucrose; 2 – IAA (29.5×10^{-6}M); 3 – coumarin (13.7×10^{-4}M); 4 – ABA ($18.9 \mathrm{X} 10^{-7}$M).. White blocks – incubation in sucrose, gray – in IAA or ABA.

If auxins or inhibitors preincubated with coleoptile sections longer than 9 hours the y retained their effect and the other factor could not minimize it (Figure 4.9). Contemporary conception of cell wall loosing was presented by Cosgrove (1998).

Such a combination of hormonal and inhibitor factors could observe in nature during seasonal changes in plant growth.

→	Sucrose	ABA	
	31		
3	44	3	59
9	54	9	77
13	73	13	58
17	98	17	56

→	Sucrose	Coumarin	
	250		
3	220	3	58
9	190	9	42
13	146	13	44
17	110	17	46

1 2 3 4

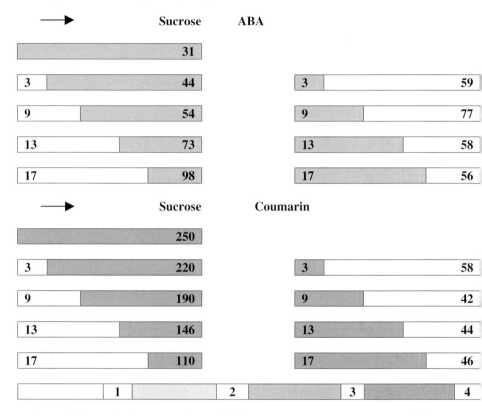

Figure 4.9 Interaction between auxins and inhibitors. (See details on experiment conditions on Figure 4.8)

The process of cell growth proceeds in a certain rhitmic order and is under steadily hormonal-inhibitory control.

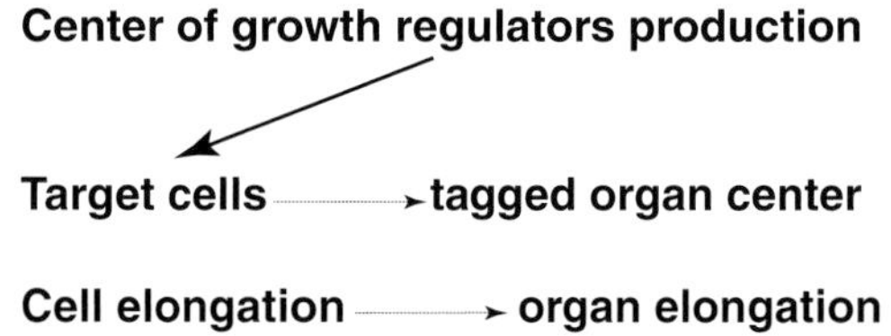

However the biological clocks are not directly regulated by growth inhibiting systems and climax of rhythms do not change under natural inhibitors effects.

CHAPTER 5

PHENOLIC INHIBITORS AND ABSCISIC ACID DURING DORMANCY.

Growth processes appear to occur rhythmically, as it was established by short-interval measurements (every hour, day, week), and even by measurements made with rather long time intervals (a month, a year). These observations compelled botanists to come to the conclusion that there must be long periods when the growth of shoots in perennial plants is inhibited and there is a delay in seed germination in annual plants whose other organs die off with the coming of winter. This inhibition of plant growth in the temperate zones of the northern hemisphere, coincides with the end of summer, and the beginning of the first winter months. In literature this absence of visible growth in plants is called dormancy. It is due to dormancy that plants, or their parts, fail to grow despite morphogenesis. The definition of the dormancy condition was suggested as far back as the 1940's by Sabinin who wrote that the cessation of apparent growth is the period of particularly high activity of the meristem in the shoot growing point.

Some authors regard bud dormancy not only as seasonal inhibition of growth, but also as a process associated, for example, with the interaction of organs and with correlated growth inhibition patterns.

The retardation of growth processes starts in the summer and is accompanied by the inhibition of bud opening in a plant. Leaves and buds play an important role in growth inhibition. Doorenbos (1953) defined such a correlative inhibition of bud growth, which occurs on account of vigorous activity of other plant organs, as a period of summer dormancy. It is also important to compare the high level of natural inhibitors with the phenolics as components of the system of plant resistance (Ribnicky et al., 1998).

Nevertheless, there are many authors who do not regard summer dormancy as an independent phase. Numerous investigators are of the opinion that the growth deceleration period, i.e. dormancy of plants, starts from two main stages: spontaneous and imposed dormancy. By spontaneous dormancy is meant the state of the growing point in a period when visible growth has ceased despite favorable external conditions (autumn-winter growth inhibition period). If, on the other hand, visible growth cessation is imposed by environment conditions and a shoot resumes visible growth when transferred to favorable conditions, this state is called imposed dormancy. Besides the two main dormancy periods, some researchers distinguish also an organic dormancy phase which is called predormancy.

This phase either precedes spontaneous dormancy, or is concomitant with it. The organic dormancy, or predormancy period, similar to spontaneous dormancy, is a phase in which inhibiting activity is exerted by endogenous factors present in the buds and seeds of woody and herbaceous plants. Among internal factors determining the condition of dormancy the investigators usually single out a group of endogenous regulators which, if changes qualitatively or quantitatively, cause the inception or termination of the dormancy period. Other internal factors involved in dormancy include a deficiency in easily available nutrients. However, this latter group of factors is of rather subordinate nature and depends on the mobilizing functions of natural regulators. The description of dormant organs and tissues is given in a number of special reviews and monographs. Let us consider the processes, which occur in the autumn growth inhibition period. Fukuda, 1996 describes the process of xylogenesis in some woody plants as the result of cell death and formation of tracheary elements. Lignin formation as a second most abundant natural product after cellulose is deposited mainly in the cell walls of fibers and tracheary elements (Zong et. al. 1998).

In higher plants the dormancy period usually separates two growth forms: embryonic growth -organ formation in embryos and enlargement of organs in size. The complex of external and internal inhibiting factors serves here as a block interfering with the continuity of the growth process.

Among external factors which affect the inception of the dormancy period are: temperature, light of various quality and intensity, supply of oxygen, water and mineral nutrients.

5.1 Natural Inhibitors, Auxins and Plant's Dormancy.

One typical symptom indicating the preparation of plants for dormancy is leaf abscission. It cannot be regarded exclusively as a form of plant transition to dormancy as a part of the organs always drop in the course of growth.

Leaf abscission is a sign of senescence and is preceded by important metabolic changes such as: changes in the composition of leaf pigments and a reduction in their amounts. Senescence of a leaf changes its turgor, the changes depending on the plant species. Certain mineral and organic elements are transported from leaves to stems and buds. Senescent leaf stalks contain smaller quantities of protein, nitrogen, asparagine, tannides, while the content of most aminoacids, glutamine, glucose and amylase rises appreciably.

Abscission involves radical changes in the regulatory factors of plant ontogenesis. At first, the amount of auxins decreases in the aging tissues. The absolute content of auxins in a leaf is not fully correlated with the abscission process (Addicott, 1965). A much more pronounced correlation is observed between leaf abscission and an auxin gradient along the leaf blade and the stalk. Treatment of a leaf with exogenous auxins helps retain it on the plant for a longer time. The content of such substances as lignin and its components increases in the abscission zone, and all these substances accumulated at the base of the leaf stalk before abscission. The activity of cellulase regulated by abscisic acid also rises in this zone. Auxin antagonists which accelerate abscission of leaf stalks in explants include ethylene (gaseous factor) and the natural growth inhibitors (abscisic and gibberellic acids). Osborne (1968) reported that the ethylene-auxin ratio controls defoliation. It would appear that excessive amounts of auxin inactivate high ethylene concentrations and a young leaf remains firmly attached to the plant. On the contrary, in a senescent leaf the ratio is shifted towards ethylene, though its amount is appreciably reduced. Jackson & Osborne (1970) analyzed leaf abscission as related to the ethylene concentration on *Phaseolus* explants. They found that the concentration of ethylene rose abruptly before formation of the abscission layer at the stalk. At the same time, plant tissues, which are not directly involved in leaf abscission, did not exhibit great changes in ethylene concentration.

Addicott and co-workers (Addicott et al., 1964; Addicott, 1965) attributed an important role in leaf abscission to abscisic acid. Abscisic acid is known to be widespread in plant tissues. It readily causes abscission of leaf stalks in explants of a number of species, but in contrast to ethylene it does not affect leaf abscission in intact cotton plants, i.e. exogenous treatment of plants with abscisic acid didn't result in defoliation of the plants. Abscisic acid is not accumulated in the autumn leaf-fall period and is not correlated with this process.

Abscisic acid, gibberellic acid and IAA produce distinctly different effects on abscission in cotton explants. Abscisic acid promotes leaf stack abscission two times faster as compared with gibberellic acid. It should be noted that abscisic acid strongly depresses cell division, which results in a weak development of the abscission layer, however, leaf stalks drop after 24 hours.

Gibberellic acid forms an abscission layer by stimulating cell division in this zone. The activity of the procambium in the abscission zone is attenuated, leaf stalks drop after 48 hours. IAA causes active cell division in the entire stalk, and an abscission zone is formed without a distinct abscission layer. In this case leaf abscission occurs not earlier than after 120 hours. The action of abscisic acid on excised leaves causes senescence of the leaf tissue and its degradation.

Defoliation in autumn occurs before or concurrently with the inception of the dormancy period. It is characterized by an abrupt reduction in the content of natural growth regulators and accumulation of growth inhibitors in the seeds of annual and perennial plants and in the branches and buds of perennial trees and shrubs. The phenol concentration rises significantly, particularly in buds and tubers large amounts of natural growth inhibitors are also present in storage organs: potato tubers or bulbs in the spontaneous dormancy condition. Growth inhibitors isolated from these organs by means of water or ether extraction suppressed the growth of carrot tissue (Hemberg, 1961). Steward and Caplin (1952) isolated growth inhibitors mainly from parenchyma tissues which could no longer proliferate, although it contained all the metabolites required for normal growth. The authors put forth a hypothesis that storage loses the ability of further growth either on account of the lack of auxins, or due to the presence of growth inhibiting substances. Auxins were not detected in dormant organs, but growth inhibitors occurred in excessive amounts. The cessation of growth leads usually to a high concentration of lignin in cell walls, which diminishes when growth is resumed. Xin Xu et al, 1998 assumed that ABA was able to stimulate the formation of tubers by counteracting with gibberellic acid (GA).

Considerable amounts of acid growth inhibitors are contained also in mature seeds and fruits, i.e. organs that had terminated their growth. One of such inhibitors is abscisic acid, but, besides this acid, there are many other strong growth inhibiting substances. In the course of fruit ripening the content of growth stimulating substances drops rapidly, while the concentrations of natural growth inhibitors rise abruptly. A mature fruit contains no auxins whatsoever, and growth inhibitors become most active.

External factors play a leading role in transition of plants to dormancy and act as specific 'regulators' of endogenous regulators by changing their qualitative, and sometimes also quantitative composition. Thus, for example, shortening of the day length caused an increase of the content of the growth inhibitors and formation of dormant buds in several plant species. Artificial reduction of the oxygen concentration in the atmosphere inhibited strongly the rate of seed germination.

This blockage was controlled by environmental factors, which actually determined dormancy, and was affected by natural growth inhibitors, which acted as switches cutting off a required biosynthetic element.

Since natural growth inhibitors actually accumulate vigorously before the inception of dormancy in tubers, seeds and buds, the decomposition of these growth inhibitors is associated with strong stimulation of growth processes. Studying the seasonal dynamics of growth regulators in *Acer pseudoplatanus L.*, Dorffling (1963) came to the conclusion that natural growth inhibitors start to function in autumn only when natural auxins disappeared or became inactive. Autumn attenuation of growth processes and accumulation of storage substances, lignification of cell walls intensified, for example, in apple trees. They found that the content of growth promoting substance decreased and the activity of the inhibitor phloridzin was promoted in this penod.

The mechanism of the retarding effect of growth inhibitors at the inception of the dormancy period appears to be a rather common phenomenon. Indeed, a large number of natural growth inhibitors were discovered not only in ash but also in peach buds in spontaneous dormancy.

One of the major arguments in favor of the participation of natural auxins and growth inhibiting substances in the onset of dormancy was their dynamics, which correlated with changes in the rate of growth processes. Leaves and buds were the sites of localization of auxins and natural growth inhibitors, which were in continuous interaction. Observation of the growth of leaves and buds in a green willow shoot convinced us that leaf growth preceded the growth of an axillary bud and occurs predominantly in May and June (Kefeli &Turetskaya, 1965). By July the growth of the leaves and elongation of the shoot ceased and buds started to grow more vigorously. From the data below it can be seen that during one month, between July and August, the bud grew three times (311 per cent), and during the next month, between 20 August and 20 September, its weight increased another 1.9 times. In autumn, between September and October or between October and November, the growth of the bud decelerated considerably.

The table shows that the buds grew most actively in July-August.

Month	*Dry weight of one bud, mg*	*Weight increase, % of previous month growth*
July	0.7	---
August	2.2	311
September	4.1	187
October	6.2	151
November	7.0	131

Chromatographic analysis in combination with a biotest for the growth of coleoptile segments evidenced that auxin Rf on the chromatograms of leaf and bud extracts was 0.5 (mixture n-butanol - CH_3COOH - H_20, 40:12:28). Auxin exhibited light yellow fluorescence in UV light.

The increase of the coleoptile segment length on the eluate from the auxin zone expressed in percent of the control was taken to evaluate the biological activity of auxin. The maximum activity was observed in August leaves, i.e. in the period of accelerated bud growth (Kefeli, 1965). At that time free auxin was also detected in the buds themselves and in the bark.

Month	*Leaves*	*Buds*	*Bark*
July	120	100	97
August	162	120	166
September	137	90	81
October	125	95	85

The formation of natural growth inhibitors in leaves was studied by the same technique, but without a detailed analysis of their chemical nature. The growth inhibitor with Rf 0.7 appeared in leaves in July and was stored in increasing quantities in leaves and buds in September-October. The percentage increase of coleoptile segments relative to the control was as follows:

Month	*Leaves*	*Buds*
July	85	112
August	75	70
September	70	60
October	55	70

So, auxin and growth inhibitors acted alternatively in leaves and buds. The activity of auxin was stimulated in the period of intensive bud growth; the retarding effect of the growth inhibitor had become strongest in the period when bud growth was decelerated.
It is noteworthy to observe that summer defoliation caused breaking of buds on cuttings which gave rise to underdeveloped dwarf shoots. In the autumn, period when growth inhibitors have accumulated in buds, defoliation didn't affect their dormancy and buds didn't open. One could assume that leaves were the sites of the primary synthesis of growth inhibitors, which are then translocated to buds.

The following facts might be taken to be indirect evidence as to the mobility of auxins and growth inhibitors.

1. The appearance of a high concentration of auxins in a leaf was accompanied by
their occurrence in the buds and the bark.
2. Active synthesis of phenolic inhibitors and abscisic acid started in young photosynthesizing leaves.
3. Decreased activity of auxin and accelerated formation of a growth inhibitor in
autumn leaves caused its simultaneous accumulation in the buds.
4. Finally, the last example confirming our assumption of the mobility of auxins and inhibitors was a defoliation experiment (Table 5.1):

Table 5.1 Role of leaves in rooting of different age willow cuttings

Month	*Number of roots on one cutting*		*Length of one root, cm*	
	with leaves	*without leaves*	*with leaves*	*without leaves*
July	5.0	2.4	4.0	1.8
August	1.9	0.2	1.0	0.3
September	1.6	2.6	1.1	2.7
October	8.3	1.2	1.9	2.7

It was obvious that removal of leaves on summer cuttings strongly depressed rooting: the length and number of roots on the cutting became much smaller. This inhibition of root formation in summer in the absence of leaves might be accounted for by the fact that summer leaves contained a large amount of metabolites and inducing substances required for the initiation and growth of a root. The amount of nutrients in autumn leaves was also large, but autumn leaves contained also natural growth inhibitors known as agents retarding the activity of hydrolytic enzymes and preventing the formation of easily available substrates (hydrolyzable sugars, amino acids) in leaves. By removing autumn leaves from their stalks we thereby eliminated a large quantity of growth inhibitors which canceled out their inhibiting effect on root formation.

Consequently, the inception of dormancy in willow plants was paralleled by a decrease of the auxin content and accumulation in leaves of natural growth inhibitors which were then stored in large quantities in buds which was well illustrated by the growth inhibitor isosalipurposide (Figure 5.1). This did not affect meristemic processes in buds as only organ extension processes were inhibited.

5.2 Role of Natural Growth Inhibitors During Plant's Dormancy.

For a long time one of the serious drawbacks of the inhibition dormancy theory was a lack of data on the chemical nature of compounds involved in growth inhibition.

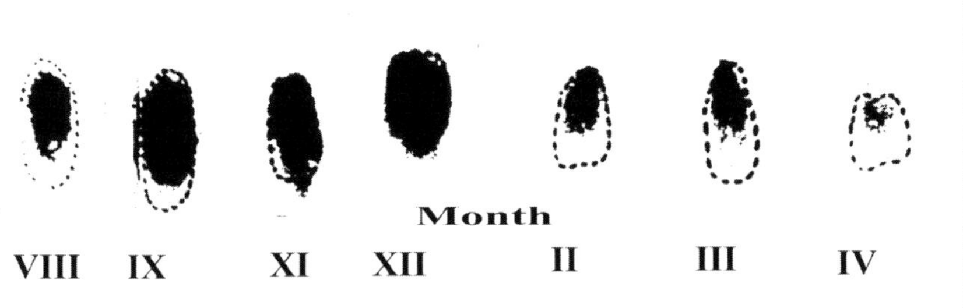

Figure 5.1 Seasonal fluctuation in content of natural growth inhibitor isosalipurposide in willow buds (fragment of chromotogram, separation mixture 15% CH_3COOH)

Since 1965, however, several inhibitors have been isolated from plants. Below is the list of endogenous substances from fruits and seeds which inhibited seed germination.

Substance	*Plant from which substance was isolated*
Cyanides	Prunus, Crataegus
Mustard oils	Sinapis, Brassica
Trans-Cinnamic acid	Parthenium
Ferulic acid	Beta
p-Sorbic acid	Sorbus
Coumarin	Various plants
Essential oils	Citrus, Forniculum
Alkaloids	Nicotiana
n-Butylidene hexahydrophthalide	Umberliferae

The strongest inhibitors were considered to be salicylic acid, derivatives of phenolcarboxylic acids and coumarin among lactones.

Mayer and Poljakoff-Mayber, 1963, presented the comparison between the inhibitory effect of aromatics and phenolics on the germination of lettuce seeds.

Substance	*Concentration, M*
Pyrogallol	10^{-2}
Pyrocatechol	10^{-2}
Resorcin	$5*10^{-3}$
Salicyclic acid	$1.5*10^{-3}$
Gallic acid	$5*10^{-3}$
Ferulic acid	$5*10^{-3}$
Caffeic acid	10^{-2}
p-Coumaric acid	$5*10^{-3}$
Coumarin	$5\text{-}10^{-4}$

The same authors observed that the seeds of different plants responded to coumarin. Regarding their sensitivity to coumarin, these plants may be arranged in the following sequence; flax, cabbage, carrot, dandelion, radish, garden peppergrass, beet, onion and mustard, lettuce , wheat, and the last- rice.

It was shown that neither the effect of 2,4-DNP nor the effect of 2,4-D inhibiting lettuce seed germination was neutralized by light. This property of coumarin might be explained by its chemical nature and its ability of inactivation in the presence of the enzyme systems from the seeds which undergo modifications under light. However, despite significant progress achieved in the investigation of the dormancy, a number of problems remain completely vague till now. One of the most mysterious aspects is the termination of dormancy in plants. There is no visible reason, for example, for such a phenomenon when growth inhibitors abruptly disappeared and phytohormones appeared in buds in an imposed dormancy condition in plants that were exposed to below zero temperatures in winter. External factors, for instance, warm baths could accelerate this process and even induced the formation of a hormonal factor which action extended along the stem to buds which had not been treated thermally. Let us briefly summarize possible factors that might disturb dormancy in plants.

External factors: 1) the effect of low temperatures (stratification); 2) the effect of light on seed germination; 3) promotion of oxygen supply to a seed embryo.

Internal factors: 1) decomposition of naturally occurring inhibitors; 2) production of phytohormones; 3) activation of enzymes with a low temperature maximum; 4) transport of hormones and hydrolytic enzymes from the roots into the above-ground part of plants.

None of the above factors, however, could offer a complete interpretation of the dormancy regulation process. The dominating endogenous factor, which reduced the rate of growth and was accumulated in a plant during dormancy, were natural growth inhibitors substances. Xin Xu et al (1998) observed correlation between balance of GA, ABA, and tuber formation.While functioning in a plant together with growth stimulators, the inhibitors could interfere with the normal growth process and limit the activity of phytohormones.

5.3 Functions of Phytohormones and Growth Inhibitors During Dormancy Termination.

The termination of the dormancy period in plants was accompanied by changes in a number of endogenous factors, including the ratio between the phytohormones and inhibitors. The cuttings of growing and dormant willow shoots exhibited opposite responses to exogenous phytohormones. Neither indolylacetic acid (150 mg/I) nor gibberellin (150 mg/L) could induce growth and formation processes in autumn willow cuttings (Table 5.2). Ether extracts obtained from such cuttings contained natural inhibitors, which retarded the growth of wheat coleoptile segments. It might be considered that the presence of such inhibitors retarded the manifestation of the stimulating effects of the phytohormones.

There might be three ways to verify this assumption: I) to introduce phytohormones in spring cuttings containing no natural growth inhibitors; 2) to isolate natural growth inhibitors from autumn cuttings and introduce them into spring cuttings, i.e. to use a short-time imitation of autumn dormancy conditions; 3) to introduce natural growth inhibitors together with phytohormones in order to reproduce the effect of neutralizing stimulation caused by IAA and gibberellin.

Table. 5.2 Effect of natural willow growth inhibitors (INH) on retardation of growth of shoots and roots in willow (in per cent of control)

Version	*Number of open buds*	*Number of roots*	*Number of open buds*	*Number of roots*
	Autumn		*Spring*	
Water (control)	100	100	100	100
IAA	80	138	60	250
GA	50	10	144	50
INH	--	--	70	50
INH+GA	--	--	114	10
INH+IAA	--	--	56	50

When starting studies on the growth retarding effect of the inhibitors we used a complex of inhibitors designated INH rather than individual isolated compounds. These complexes were isolated from one autumn cutting and introduced into one spring cutting. The results of the experiments are summarized in Table 5.3.

Table 5.3 Effect of complex of natural willow growth inhibitors (INH) on growth of bioassay

Version	*Growth of coleoptile segments, % of control*	*Growth of pea seedlings, mm*	*Number of roots per Phaseolus cutting*	*Length of Phaseolus epicotyls, mm*
Water (control)	100	37	6	17
IAA	175	41	40	18
GA	130	108	1	29
INH	19	41	4	16
INH+GA	57	58	1	18
INH+IAA	27	49	0.6	20

It was found that spring cuttings with buds that had emerged from dormancy could respond to phytohormones. Root formation in these cuttings was promoted by the action of IAA; gibberellic acid (GA) induced bud opening and shoot growth. The complex of natural growth inhibitors retarded native growth processes in the cuttings, and when the inhibitors were introduced into the spring cutting together with IAA or GA, the inhibitors depressed both the auxin-induced rooting effect and the gibberellin-induced bud breaking process.

In order to break up complex organogenesis processes occurred in a woody plant cutting into simpler components, we have used such additional tests as wheat coleoptile segments (cell expansion), pea seedlings (stem elongation), and, finally, Phaseolus cuttings as a complex root and shoot formation test. In Table 5.3 it can be seen that the complex of growth inhibitors depressed IAA-induced growth of coleoptile segments, GA-induced growth of a pea stem, and also root and stem formation processes which were promoted by IAA and GA.

The above data permit a general conclusion that natural growth inhibiting substances accumulated in autumn shoots were able to retard growth processes occurring in these shoots. Autumn willow cuttings in our experiments and birch and currant plants in the experiments carried out by Tumanov and associates, 1970, showed no response to phytohormones and became sensitive only with the termination of the dormancy period.

However, no matter how correct this conclusion might be, it should be emphasized that although autumn buds had lost the ability to open due to the accumulation of natural inhibitors, they continue to grow in size. The same applied to seeds in which organ-forming processes were not retarded by natural growth inhibitors appearing there in maturity. Natural growth inhibitors were localized mainly in the external coats of buds and seeds. It might be that this localization actually ensured the normal course of organogenesis. Besides, it should be mentioned that formation processes were less sensitive to natural growth inhibitors than cell and organ extension processes. This seemed to be the most likely explanation of the fact why buds, though increasing in size, had lost the stem elongation ability in the presence of growth inhibitors. At the same time the inhibitor concentrations, which depressed stem extension, couldn't retard formation processes. It would be wrong, however, to ascribe stem growth inhibition only to the inhibitors present in the buds. The absence of phytohormones in autumn buds was one of the factors, which determined the loss of their ability to open.

Comparison of the biological activity of isolated abscisic acid with the inhibiting action of the synthetic compound showed that 1 kg of green leaves could contain from 8 to 16 mg of abscisic acid, as abscisic acid in a 1 mg/I concentration caused 30% inhibition, and a 2.5 mg/l solution of this acid - 50% inhibition of growth. Natural abscisic acid from 125 g of green leaves caused a similar 30-50% inhibition of coleoptile segment growth. The above calculation is highly approximate, but it clearly demonstrated that the phenol content in willow leaves exceeded the content of abscisic acid by at least 2-2.5 times.

Treatment of spring willow cuttings with abscisic acid solutions showed that in a concentration of 20 mg/l abscisic acid could inhibit 70% of bud breaking, but only 24% of root growth and that this inhibitory effect of abscisic acid was short-time only. Thus, bud breaking and other processes involved in the termination of dormancy in plants are controlled by the phytohormones-inhibitors system.

Above we considered the physiological activity of growth inhibitors assuming that this complex consists of several substances with the ability to inhibit coleoptile segment growth.

One of these inhibitors of willow identified is isosalipurposide. This substance was purified from admixtures and its identity with the radioactive labeling substance was established.

Characteristic	*Chalcone-glucoside from Salix purpurea*	*Chalconaringenin-2'-glucoside (iso-salipurposide)*
Rf		
2% - CH_3COOH	0.08	0.08
2% - CH_3COOH	0.10	0.10
2% - CH_3COOH	0.16	0.16
n-butanol-C_2H_5OH-H_2O 4:1:2.2	0.56	0.56
Colour reaction		
DSA	Orange	Orange
$AlCl_3$	Green	Green
CHI	Red	Red
Spot fluorescence		
UV	Dark brown	Dark brown
UV + NH_3	Red-orange	Red-orange
Daylight + NH_3	Orange	Orange
λmax of UV spectrum, nm	240,372	240,369

Testing of isosalipurposide for wheat coleoptile segment growth showed that it could totally suppress the growth of elongating cells in coleoptile segments up to a concentration of I mg/ml, and the retarding effect of this inhibitor in the dose of 0.5 mg/ml was 60 per cent. In a concentration of 0.25 mg/mL, isosalipurposide lost its inhibiting properties.

Another natural inhibitor, which depressed coleoptile segment growth, was phenolcarbonic acid. It was also isolated preparatively and subjected to four subsequent rechromatography analyses. The elemental composition of phenolcarbonic acid was C:H:O, 53.5:6.42:40.03, its melting point 79-80^{0}C. The mobility (Rf) of the acid in 15% CIA_3COOH was 0.90. It reacted positively with DSA and $FeCl_3$. The analysis of its IR absorption spectrum proved that it contained medium-intensity glycoside bond (1254 cm^{-1}), an aromatic ring (1500-1600 cm^{-1}), sugar residue (1080, 1040 cm^{-1}) and carboxyl group (1720 cm^{-1}).

Isosalipurposide and phenolcarbonic acid exhibited an equal ability of accumu1ate in autumn buds in the maximum amount. Isolated and purified isosalipurposide and phenolcarbonic acid were able to depress organogenesis in spring willow cuttings. Although the concentration of the phenolic inhibitors was only (2 mg/l), it still performed a physiological function, as one leaf and bud of a cutting contained about I to 3 mg of these substances depending on the month when cuttings were made.

Below is given the effect of isosalipurposide and phenolcarbonic acid on bud breaking and root formation in *Salix purpurea* willow cuttings (in per cent per cutting):

Inhibitors isolated from willow	*Buds*	*Roots*
Water (control)	42	100
Isosalipurposide, 2 mg/ml	10	0.6
Phenolcarbonic acid, 2 mg/ml	17	38

Version	*Bud breaking*	*Number of roots per cutting*
Water (control)	100	100
IAA	35	220
GA	150	20
Isosalipurposide, 2 mg/ml (I)	70	80
I + IAA	35	56
I + GA	30	35
Phenolcarbonic acid, 2 mg/ml (P)	40	38
P + IAA	52	35
P + GA	80	50

The accumulation of such concentrations of the phenolic inhibitors in the willow autumn shoot could retard the growth processes in the shoot. The same phenolic compounds produced an inhibitors' effect on phytohormone-induced growth processes (in per cent of the control).

By using natural phenolic inhibitors, we have managed to obtain a short-time imitation of dormancy, i.e. depressed growth processes and eliminated the activity of phytohormones.

Apart from phenolic inhibitors, willow leaves contained also a growth inhibitor with the properties of abscisic acid.

Our student N.Chernobrovkina also showed the phenomena of buds growth of birch tree since August up to October. In the same time they lost their ability to bloom (Figure 5.2, 5.3); this period corresponded with the decrease of auxin level and the increase of the abscisic acid (Figure 5.4)

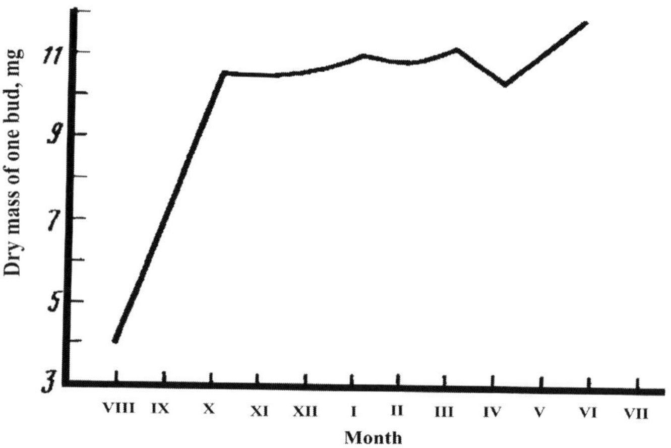

Figure 5.2. Annual growth of birch buds.

Special research of ABA content in birch tree during the rest period and transition to the spring blooming was carried on by W.Dathe and Vlasov. They observed the decrease of free ABA level in buds, bark and xylem (Figure 5.5, 5.6). These data showed the definite organ compartmentalization of free and bound ABA in different organs of birch-tree and in its bleeding sap. The end of bud dormancy brings to the decrease of ABA in the bleeding sap.

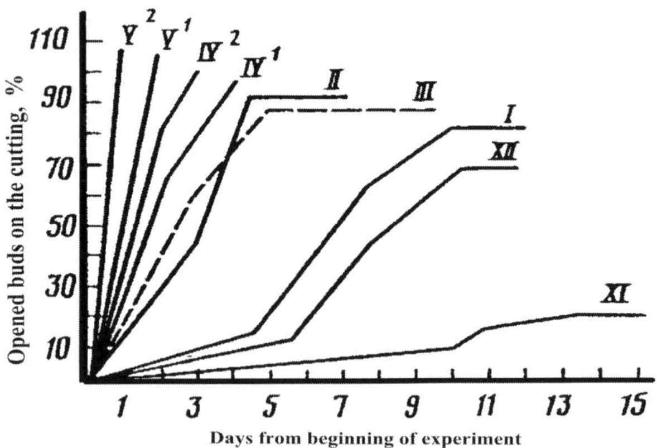

Figure 5.3 Seasonal fluctuations in birch buds opening.
I, II, III and IX, XII-months, IV^1- April, 16, IV^2- April, 26; V^1- May 5, V^2- May 16.

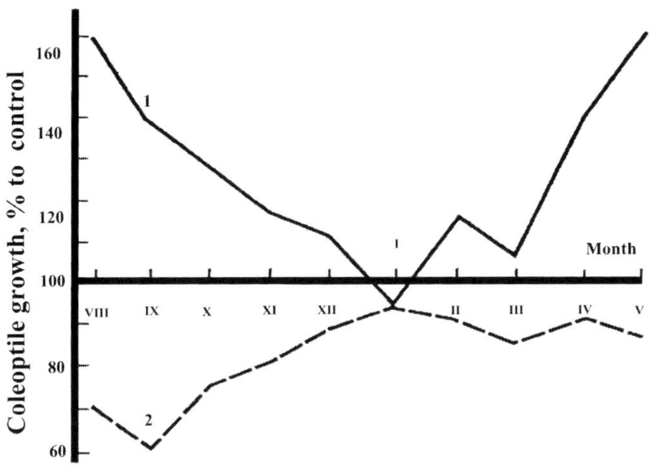

Figure 5.4 Seasonal fluctuations in biological activity of auxins and inhibitors (ABA) in birch buds. 1 – substance with Rf (IAA); 2 – substance with Rf (ABA)

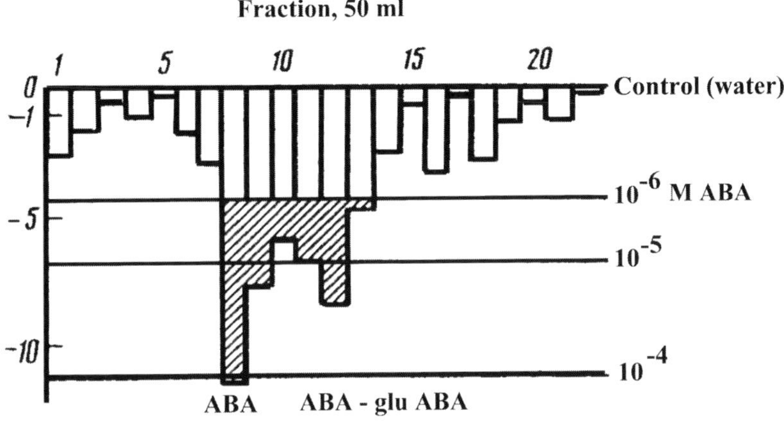

Figure 5.5 Content of free and bounded (ABA-glu) ABA in bark of dormant birch plant.

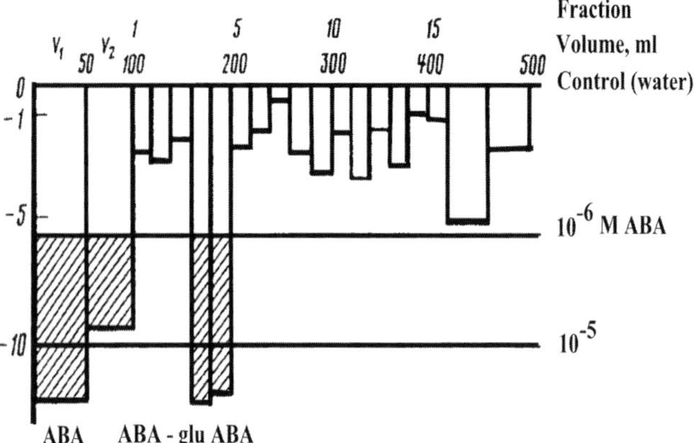

Figure 5.6 Content of free and bounded ABA in wood of dormant birch plant.

5.4. Natural growth inhibitors and seeds germination

Seeds germination also depended on the concentration of natural growth inhibitors in their parts -endosperm, embryo, seeds coat. There were different type of such inhibitors-phenolics, cyanide- derivatives and abscisic acid. Some inhibitors were easily leaching from the seeds and could induce the effect of self- inhibition like in the experiments with the tobacco (*Nicotiana tabacum* L.) (Figure 5.7).

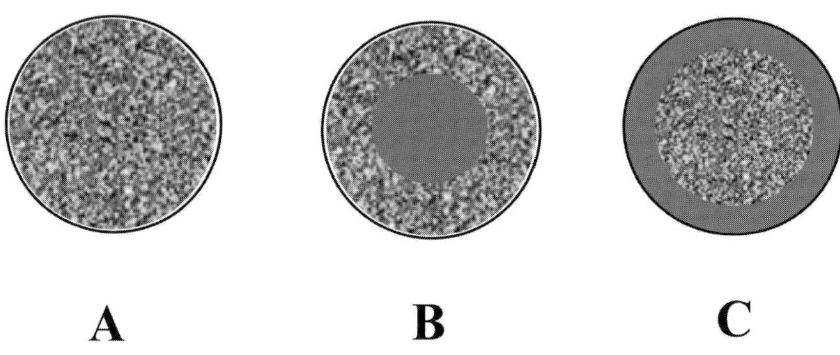

Figure 5.7 Germination of tobacco seeds on moist filter paper. A – uniform soaking; B – soaking from center to edges ; C – soaking from edges to center.

These data showed that tobacco seeds produce ABA, which was leaking. Further experiments with the synthetic ABA confirmed these data on the seeds of wheat, apricot and apple (Figs. 5.8, 5.9). Different organs of the germinated seeds possessed different sensitivity to ABA and other inhibitors.

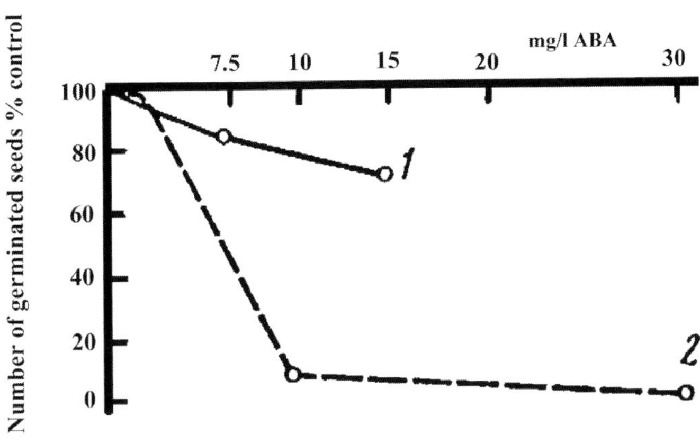

Figure 5.8. Germination of wheat seeds (1) and apricot (2) in ABA solutions.

Considering the processes: seed germination-seedling growth-seedling (coleoptile) segment growth, we found ourselves moving from a complicated growth process to a simpler one and than to the simplest, which is a coleoptile segment with extending cells. 50% inhibition of a complicated process (seed germination) requires 5 times as much abscisic acid as the amount needed for inhibiting the seedling growth process, and 150 times as much as that for inhibiting coleoptile segment growth. It would appear that the more complicated is a process, the higher the dose of inhibitor required for its inhibition. Indeed, 1000 mg/l of p-coumaric acid will be required for 50% inhibition: of wheat seed germination; the dose needed for inhibiting seedling growth is two times lower, and that for coleoptile segment growth inhibition three times lower. However, this statement could not be regarded as universal. The comparative estimation of the effect of abscisic and p-coumaric acid on stem and root growth made it obvious that the two growth processes of the same complexity are suppressed by the terpenoid inhibitor in various degrees.

Evidently, in the latter case we have dealt with the metabolic specificity of a stem as an object more sensitive to abscisic acid.

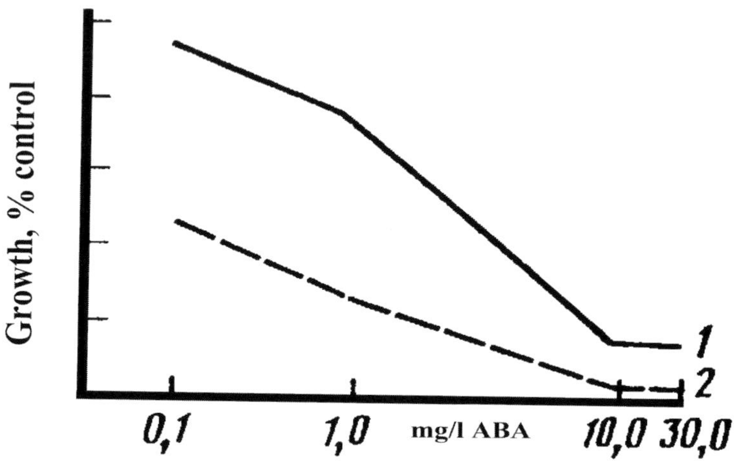

Figure 5.9 Effect of ABA on root growth (1) and stem growth (2) of apple seeds after 7 days of germination.

Differences in the sensitivity of some forms of growth processes to abscisic acid are illustrated in Figure 5.10. Abscisic acid could inhibit seed germination in a number of woody and herbaceous plants, however, an inhibitory effect was also exhibited by a number of other compounds. The certain fact should be considered that natural growth inhibitors, abscisic acid and phenolics, are readily washed out by water and may pass into secretions affecting seed germination and seedling growth in plant communities.

Thus, growth inhibitors of various classes have the following common properties: accumulation of these substances in the growth inhibition period; the ability of preparatively isolated and purified inhibitors to depress the growth of coleoptile segments or etiolated stems of donor plants which contained minimum amounts of these substances; the ability to accumulate in conditions adverse to normal growth, the ability of decomposition at the termination of the dormancy period.

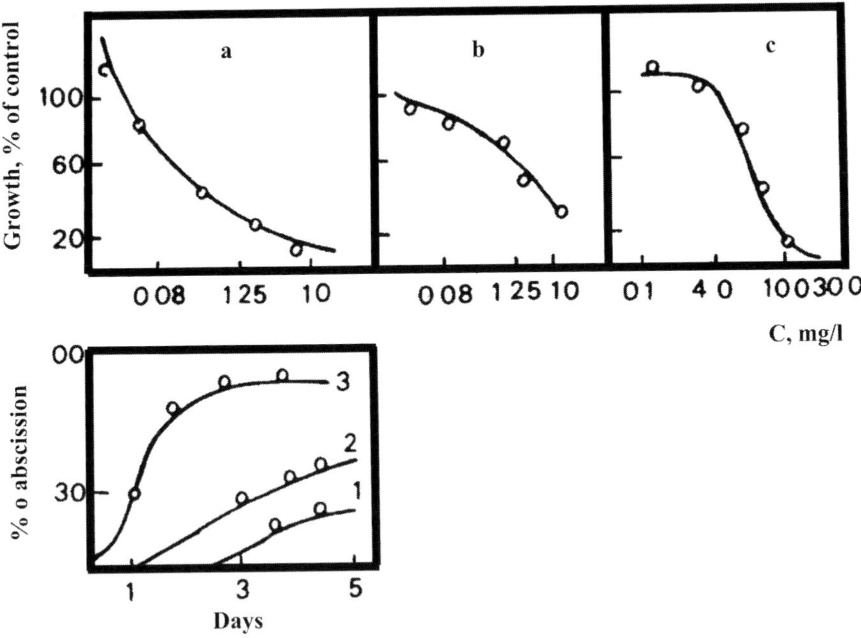

Figure 5.10 Properties of abscisic acid. a – growth of wheat coleoptile segments; b – pea seedling growth; c – peach seed germination; 1 – abscission of cotton leaves under effect of ABA concentrations (μg): 1 – control, 2 – 0.001 and 3 – 1.0

So far mainly indirect evidence is available on the participation of phytohormones and growth retardants in the growth and dormancy of a vegetating plant. However, the above formulated common properties suggested the following conclusion: various growth forms are controlled by a balance of inducing and inhibiting factors. The inhibitor dominance in plant tissues caused retardation of visible growth processes and the inception of dormancy in plants but the main effect of natural growth inhibitors consisted in retarding the extension rather than meristemic processes in plant organs.

Thus natural growth inhibitors were accumulated in the buds and leaves of plants in the spontaneous dormancy period. The removal of inhibitor-saturated leaves and buds in a willow cutting resulted in the stimulation of the root formation process.

The inception of dormancy in plants was usually accompanied by an abrupt decrease in the level of phytohormones or even their complete disappearance. In this period plant cuttings (willow) did not respond to exogenous IAA and gibberellic acid, while the action of these phytohormones on spring cuttings induced, respectively, the root formation and bud opening processes.

A complex of growth inhibitors isolated from autumn willow buds and introduced into spring shoot cuttings imitated temporarily autumn dormancy.

Some phenolic compounds (isosalipurposide and phenolcarbonic acid contained in willow exhibited the ability to depress bud opening and root formation in spring willow cuttings, and also to neutralize the stimulating effect of IAA and GA on these processes.

Abscisic acid present in willow tissues inhibited (in a concentration of 20 mg/l) bud opening in spring shoot cuttings, but, in contrast to natural phenolic inhibitors, exerted a weak retarding action on root initiation.

The comparison of the retarding properties of phenolic growth inhibitors and abscisic acid with respect to the germination of wheat, apple and peach seeds had shown that abscisic acid inhibited the growth process in a concentration two or three times lower than the phenolic inhibitors.

Larger doses of natural inhibitors would be required for retarding more complex forms of growth processes (seed germination) than for inhibiting relatively simple growth forms (growth of coleoptile segments, wheat seedling or root growth).

One of the factors responsible for termination of dormancy in plants was the reduction in the natural growth inhibitor content: isosalipurposide changes into salipurposide, naringenin and phenolcarbonic acid, and phioridzin is decomposed to produce phioretin, yellow colored chalconenype substances, and phioretic acid. Such decomposition occured vigorously when the above growth inhibitors came in contact with the buds of willow (isosalipurposide) and apple-tree (phioridzin) upon the termination of dormancy. Dormant buds decomposed natural growth inhibitors in a much weaker degree.

Dormancy in plants was accompanied with a decrease in the phytohormone content in buds and falling leaves and the accumulation of natural growth inhibitors. Exogenous introduction of natural growth inhibitors into a growing plant or germinating seeds caused a temporary retardation in plant growth. In tropical and subtropical plants the level of ABA increased during the autumn (Chkaidze et al., 1993).

The bleeding sap is one of the natural transporting systems which brought ABA and some other inhibitors from roots to the upper part of plants and bark.

Natural growth inhibitors of both phenolic and terpenoid nature were also very important for seed dormancy. It should be pointed out that although each of these classes of the inhibitors was studied by a large number of researchers, there were almost no comparative studies carried out. In an attempt to bridge this gap, we decided to compare the action of abscisic acid (terpenoid

inhibitor) and that of p-coumaric acid (phenolic inhibitor) on growth processes occurring in germinating seeds of wheat, apple and peach
The seeds of *Triticum vulgare L.* wheat, Moskovka were germinated in solutions of abscisic and p-coumaric acids.

The seeds of crab apple-tree *Malus domestica L.* with coats removed after stratification were placed for germination on filter paper into Petri dishes (d = 3 cm). 2 ml of the solution was placed into each dish. Non-stratified seeds of peach were soaked for three days in hard coats, then the hard coats were removed and the seeds soaked again overnight in tap water. After taking off soft coats, the seeds were germinated on experimental solutions in Petri dishes. All experiments on seed germination were made in the dark at 26C.

Seeds of woody plants without coats were used, since preliminary experiments with germinating coated seeds yielded similar results. The only difference was that the coated seeds displayed a somewhat weaker development of roots at the first stages and were strongly infected with microorganisms. Such an inhibiting effect of the seed coats resulted in a considerable scatter in the results and increased in the root mean square error, therefore, in pure subsequent experiments the use of coated seeds was given up.

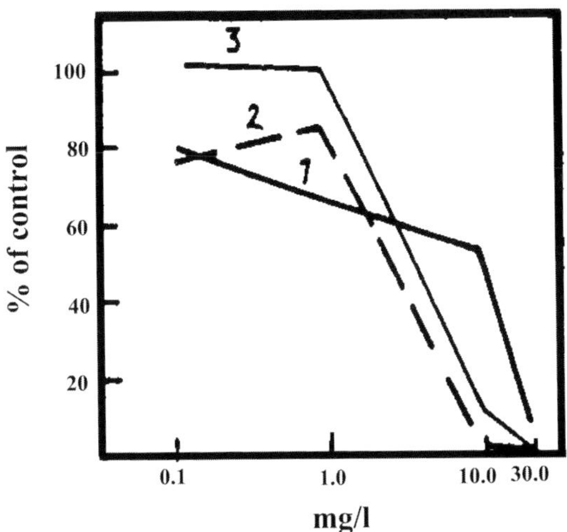

Figure 5.11 Effect of abscisic acid (ABA) on growth of root (1), stem (2), seed germination (3) of peach.

The inhibitory effect of abscisic and p-coumaric acids on the germination of wheat and peach seeds was observed as early as one day after the start of the incubation and persisted till the end of the experiment (Figure 5.9,5.10). Since the curves for different days did not intersect, growth data for one experiment day will be presented in further text for the sake of convenience. Germination of apple seeds was not taken into account, because only stratified seeds ready to germinate were taken for the experiment.

Comparing the action of abscisic and p-coumaric acids, it should be pointed out that the former inhibits seed germination in doses 80 to 100 times lower than the latter. This, however, is not the only difference between the representatives of the two classes of growth inhibitors. In germinating wheat and apple seeds, abscisic acid inhibited the growth of a seedling stronger than root growth (Figure 5.11).

Below are given the concentrations of abscisic and p-coumaric acids, which caused 50% growth suppression in germinating wheat and apple seeds.

	Growth of above-ground part of seedlings	*Growth of root*
Wheat		
ABA	3.8	11.0
p-Coumaric acid	500	125
Apple tree		
ABA	0.1	5
p-Coumaric acid	1000	750

For inhibiting the growth of an apple seedling by 50%, 0.1 mg/I of abscisic acid was required; 50% inhibition of root growth required 5 mg/I of abscisic acid. The growth of a peach seedling on the 9th day after germination was 50% depressed by almost the same concentrations of abscisic acid as those for root growth, however, concentrations needed for 100% inhibition of the growth of seedling and root were, respectively, less than 10 and more than 30 mg/I. Thus, a peach stem was also more responsive to abscisic acid than the root. p-Coumaric acid caused 50% inhibition of root growth in almost the same concentrations as those required for seedling growth inhibition, but the doses required were 100 times and more higher than the doses of abscisic acid (Figure 5.12). Differences in the doses of abscisic acid required for 50% growth inhibition had been observed not only for such independent organs as a seedling and a root, but also for one organ and its part. For example, 3.8 mg/I and 0.1 mg/l of abscisic acid were required, respectively, for 50% inhibition of wheat seedling growth and coleoptile segments growth, the same figures for p-coumaric acid being -500 and 300 mg/I.

Inactivation of abscisic acid is associated with the formation of new products.

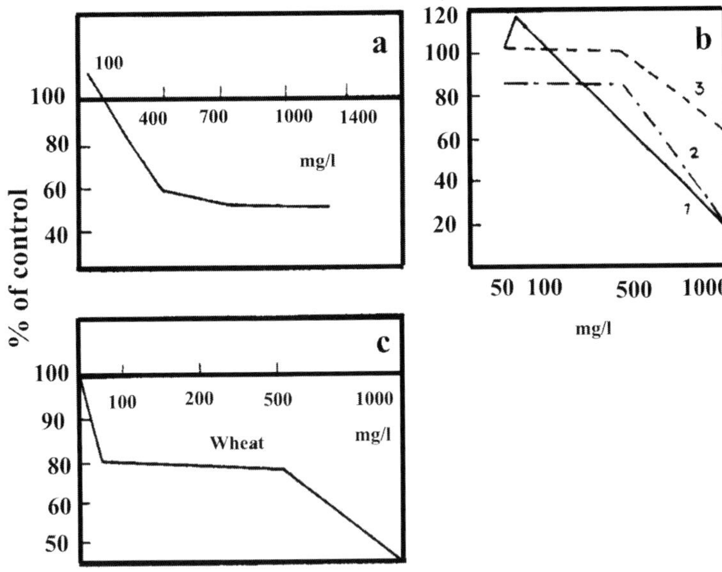

Figure 5.12 Effect of p-coumaric acid on growth processes (in % of control); a – coleoptile segment growth; b – growth of root (1), stem (2) and seed germination (3) of peach.

In our experiments with coleoptile segments, natural abscisic acid was not stable either. It inhibited the growth of coleoptile segments by 17%, but its inhibitory activity weakened after 24-hour contact with the tissues of these segments.

Thus, at present natural growth inhibitors are recognized as a class of growth regulating substances with clearly defined physiological functions. In investigations major emphasis is made on ABA as the strongest growth inhibitor. But the more profoundly the properties of this inhibitor are studied, the more difficult it becomes to establish the direct relationships in its content in tissues and plant growth. Even such a seemingly indisputable phenomenon as accumulation of ABA at the inception of dormancy in a birch plant has been revised and given a different interpretation (Hillman et al., 1974). The authors showed that ABA applied to birch leaves did not affect the dormancy of the buds. Studying the action of exogenous ABA on the growth of plant tissues in detail made it possible to divide the entire inhibition process into two stages:

reduction in water absorption, and in the entire growth process. The same idea is developed by Raschke (1974) who stated that ABA was formed in leaves in case of water deficiency which led to closing of stomata. This mechanism, however, did not took place in young leaves. Such leaves contained much ABA but their stomata were wide opened.

Accumulation of ABA in autumn was associated with growth inhibition. However, the cause of ABA accumulation is not based on photoperiodic control , but changes in water deficiency.

Studying the behavior of ABA in dormant and germinating seeds of *Coryltis avelana L.*, Williams et al. (1973) pointed out that the inception of dormancy in seeds was connected with embryo growth inhibition and accompanied by loss of ABA. Stratified and non-stratified seeds lost the same amount of ABA, but the former emerged from dormancy, while the latter remained in this condition.

Dehydration and stress effected by temperature rise increased the ABA content in bleeding sap and reduced the rate of growth. In a plant, ABA inhibited uptake of potassium ions and weakly stimulated their release from the tissues. No correlation was established between the inhibitory action of ABA on the growth and ion supply. Introduction of ABA (4-20 mg/I) through the roots into pea seedlings inhibited the growth of lateral buds. Application of ABA to the bud inhibited bud opening and induced the growth of a bud in the axil of the leaf which was not removed. A conclusion was made about direct participation of ABA in correlative interaction between organs.

It is clear, that bud dormancy and seed dormancy had some common and specific properties. Among common and specific features are morphogenetic structures development, block of elongation of shoots, leaves and roots, increase of the level of free natural inhibitors and decrease of the level of auxins and gibberellins.

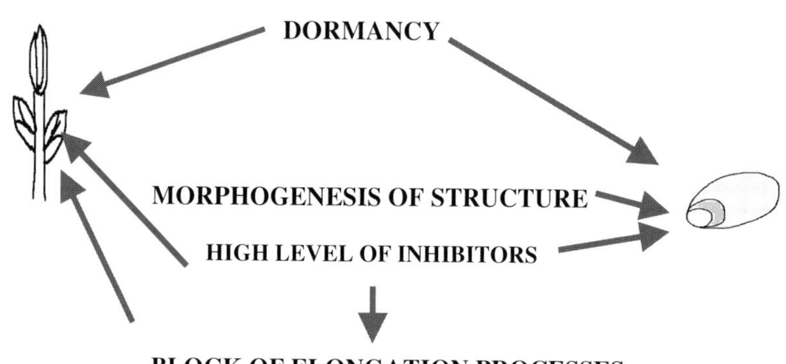

CHAPTER 6

CATABOLISM OF NATURAL GROWTH INHIBITORS IN THE PLANT AND IN THE ABSCISED LEAVES

Natural growth inhibitors can be actively metabolized in plants tissues and in the abscised organs. In the last case the great role is given to microorganisms as the decomposition agents.

6.1 Transformations of natural growth inhibitors.

The basic property of natural inhibitors is their ability to inhibit temporarily plant growth and then decompose them.

As it was mentioned, the analysis of a *Salix purpurea* willow inhibitors showed that all the properties (Rf, colour reactions, IR spectrum, water and ether solubility) are identical with isosalipurposide (chalco-naringenin-2-glycoside). Similar to the dihydrochalcone phloridzin (apple tree inhibitor), isosalipurposide inhibited the growth of wheat coleoptile segments. However, after 24 hours of incubation, this willow inhibitor lost all its inhibiting properties. A new portion of coleoptiles immersed in the solution of the "vanished" willow inhibitor grew in the same way as control coleoptiles. Such a reversible and easily vanishing inhibiting effect of the natural growth inhibitor needed a more detailed study. With this goal in mind, wheat coleoptile segments were grown in willow inhibitor solution, then extracted with alcohol. The extract was mixed with the used inhibitor solution and the mixture was evaporated. The dry residue was dissolved in alcohol and separated chromatographically. The results of chromatographic separation showed that incubation of the growth inhibitor with coleoptiles causes its decomposition. The spots were developed in diazotized sulfanilic acid and ferric chloride and were identified as phenolic compounds. It is interesting that the products of isosalipurposide decomposition possessed properties were similar to those of substances occurred in the bark and buds of the willow .

The seasonal fluctuations of the isosalipurposide content in willow buds showedthat its maximum quantity occurred in September through December, while in February-April its content in buds diminished .

A question might be asked why phenolic inhibitors disappear in a period preceding the growth process? The convincing evidence was provided that plant tissue homogenates and enzyme substances isolated from them were able to decompose phloridzin (Figure 6.1). Decomposition of isosalipurposide

(a) by willow buds (1) and stems (2), and decomposition of phloridzin (b) by apple tree buds (3) and stems (4) also shown on this figure.

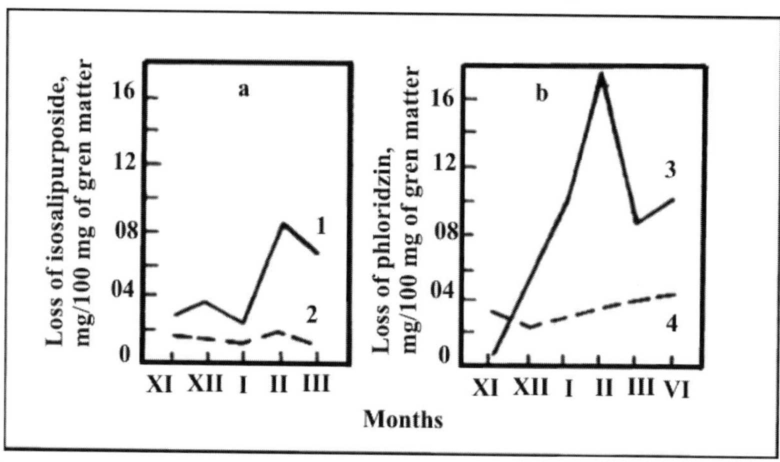

Figure 6.1 Decomposition of isosalipurposide (a) by willow buds (1) and stems (2), and decomposition of phloridzin (b) by apple tree buds (3) and stems (4)

6.2 Dynamics of decomposition of isosalipurposide arid phloridzin by plant tissues

In November-December, willow buds decomposed and released two or three times less isosalipurposide than in February-March (Figure 6.1a). The same is true for phloridzin decomposition by apple tree buds (Figure 6.1b). It is interesting that in March the phenol decomposing ability of willow and apple tree buds became somewhat weaker as compared with February, although it still remained at a rather high level. Lignified stems of a willow and an apple tree decomposed the same amounts of isosalipurposide and phloridzin in November and in February, though an apple tree developed a slightly stronger ability to decompose phloridzin by April.

Certain seasonal variations could also be observed in the release of phenolics from incubated segments into the medium. Thus, the most vigorous release of phenolic compounds from willow stems and buds started in January, while a release of phenols from the segments of apple tree stems and buds occured only in March. This is probably due to the longer dormancy period in an apple tree as compared with a willow plant. Chromatographic analysis on products isolated from the slices of willow stems and buds revealed the absence of isosalipurposide, and the presence of phenols with Rf = 0.5-0.9 (the solvent used was 15 % solution of CH_3 COOH). The phenols identified on chromatograms with the aid of diazotized sulfonilic acid might be the products of isosalipurposide conversion. The analysis of exudates from sliced buds and stems of an apple tree indicated that they included phloridzin which content was taken into account during decomposition experiments.

The qualitative and quantitative composition of inhibitor decomposition products depended on the type of an incubated organ (roots, buds, stems, and tips of stem leaves).

The decomposition of phenolic inhibitors by segments of various willow and apple tree organs (in mg weight of a decomposed compound per 100 g of wet tissue) is shown in Table 6.1, 6.2

Table 6.1. Decomposition of Secondary Metabolites in Different Plant Organs.

Plant organ	*Decomposition of isosalipurposide by willow*	*Decomposition of pliloridzin by apple-tree*
Leaves	0.31	1.09
Stems	0.32	0.46
Buds	1.25	1.08
Tips	1.42	4.53

Note: The experiments were made in July.

The inactivation of inhibitors by stems was at a much slower rate than by tips or buds. The age of the organ was also important for the rate of decomposition. April buds had broken down 2.09 mg of isosalipurposide, while in July they decomposed only 1.22 mg. Green stems decomposed 1.32 mg of isosalipurposide in April and 0.32 mg in July. The qualitative composition of products originating from the decomposition of the tested phenolic inhibitors was different for different organs.

Table 6.2. Changes in Phloridzin and Phioretin Content (Apple Leaf Homogenates).

Incubation, min	*Phloridzin*		*Phloretin*
	Content	*Formation of phloretin*	
0	1.35	0	0.90
30	0.45	0.50	0.80
60	0.40	0.35	0.75
120	0.30	0.25	0.50

Isosalipurposide and phloridzin, respectively, always gave rise to a much larger number of by-products in the presence of willow and apple tree leaves than in other organs; for instance, stems (for an apple tree) or adventitious roots grown on a willow cutting.

It should be noted that if isosalipurposide was decomposed faster in the leaves of an apple tree rather than willow leaves, and phloridzin, vice versa-in willow leaves more than in the leaves of an apple tree. The originated compounds resembled the compounds produced in the conventional decomposition pattern.

6.3 Decomposition products of isosalipurposide and phloridzin

Incubation of isosalipurposide with willow leaf tissues produced four substances. According to the colour reaction data, none of the products was a yellow pigment-flavonol or chalcone. The colourless substances were similar in their properties to such flavonols as naringenin and its derivatives. For identifying the isosalipurposide decomposition product with Rf 0.44 on the chromatogram, naringin (naringenin-glycoside) and isosalipurposide were subjected to acid hydrolysis and the products of the hydrolysis were compared. Both substances were found to cleave off a sugar residue to give away the aglycone naringenin with Rf 0.43 in 15%CH_3COOH solution.

Isosalipurposide was easily transformable into its colourless isomer salipurposide which was present in water solution in two stereo-forms. This transformation was a simple process and was rather of chemical, than enzymatic nature. The isomers of salipurposide may supposedly be products with Rf 0.32 and 0.55. Apart from these substances, the chromatogram revealed a compound with Rf 0.83 which properties make it similar to phenolcarboxylic acid (Rf = 0.83). The chemical nature of the above isosalipurposide transformation products needed further investigations.

Incubation of the solutions of phloridzin and its aglycone phloretin with the homogenates of apple leaves taken in May shows that splitting off of glucose from phloridzin and formation of phloretin are very rapid processes, whereas the sub-sequent transformation of phloretin occurs much slower.

The phloretin and phloridzin content in control samples in which the homogenate was substituted for a buffer solution did not change in the course of incubation (Table 6.3).

Then the amount of phloretin reduced due to its conversion into other products. Chromatography of phloridzin and phloretin transformation products reveals the above yellow pigments of chalcone or aurone type formed from both compounds.

Final identification of the yellow pigments is of extreme importance for studying the phloridzin transformation pathway.

Table 6.3. Identification of isosalipurposide and phloridzin decomposition products.

Rf in 15% CH_3COOH	*Spot colour*				*Substance*
	daylight	*UV light*	*Na_2CO_3*	*DSA*	
Isosalipurposide					
0.17	Yellow	Brown	Orange	Pinkish-brown	Isosalipurposide
0.32	Colourless	Dark-violet	Colourless	Yellow	Salipurposide (isomer)
0.44	Colourless	Colourless	Yellow	Pinkish-brown	Naringenin
0.55	Colourless	Colourless	Colourless	Pinkish-brown	Salipurposide (isomer)
0.83	Colourless	Blue	Colourless	Pink	Phenolcarboxylic acid
Phloridzin					
0.24	Colourless	Dark-violet	Colourless	Brown	Phloretin
0.42	Colourless	Colourless	Colourless	Brown	Phloretin
0.54	Colourless	Pink	Colourless	Brown	Phloridzin
0.64	Yellow	Yellow	Pink	Brown	Chalcone
0.75	Yellow	Yellow	Pink	Brown	Chalcone
0.83	Colourless	Colourless	Colourless	Violet	Phloretic acid

6.4 Biological Activity of Phloridzin and its Aglycone Phloretin

The accumulation of large amounts of phloridzin and isosalipurposide in plant tissues in autumn may be considered from the following: on the one hand, as the cause of gradual attenuation of growth processes, and, on the other, as a way of detoxication of aglycones.

Comparison of the action of phloridzin and its aglycone phloretin (IAA-induced) on the growth of wheat coleoptile segments (in per cent of the control) is shown in Table 6.4 and Figure 6.2

Table 6.4. Effects of Phloridzin and Phloretin on Wheat Coleoptile Growth.

Version	*Coleoptile segment growth*	*Version*	*Coleoptile segment growth*
Water (control)	100	Phloridzin + IAA	158
IAA, 7 mg/l	291	Phioretin, 2 x 10^{-4}M	100
Phloridzin, 6 x 10^{-4}M	105	Phloretin + IAA	180

Note: Phloridzin and phloretin are used in a concentration, which does not affect coleoptile growth.

The inhibiting effect of phloridzin is 2.5 to 3 times weaker than that of phloretin. If the growth of coleoptile segments is auxin-induced, the required concentration of phloretin will again be 3 times lower than that of phloridzin. It is remarkable that the inhibitor concentration required for inhibiting IAA-induced coleoptile growth is lower than the concentration for endogenous growth inhibition. Wheat coleoptile tissues decompose phloridzin to form phloretin, after which phloretin is slowly consumed from the incubation medium unlike apple tissues. However, no products of its

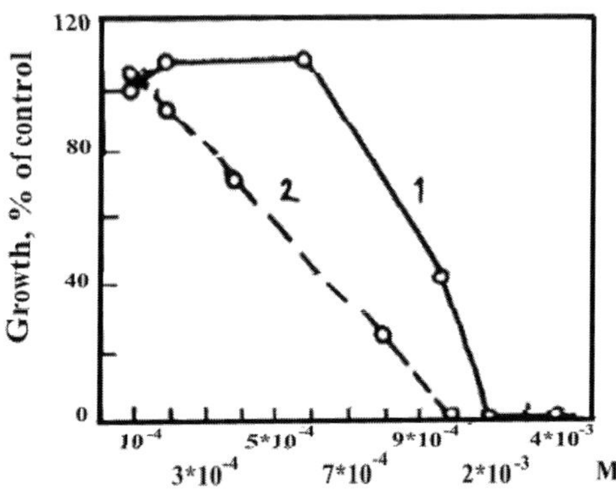

Figure 6.2. Effect of phloridzin and phloretin on growth of wheat coleoptiles.

decomposition can be found. Since phloridzin quickly changes to phloretin, it may be assumed that a factor inhibiting wheat coleoptile growth is actually phloretin, but not phloridzin, and, vice versa, glycosiding of phloretin is one of the ways of its detoxification. The decomposition of the natural growth inhibitors isosalipurposide and phloridzin is a physiological process preceding bud opening which is induced in buds when the plants emerge from winter dormancy. It is interesting that in this period neither willow stems nor apple tree stems exhibited inhibitor decomposition processes. Thus, decomposition of phenolic inhibitors seemed to be a process localized in organs with a high potential growth activity.

The ability of various plants organs to decompose isosalipurposide and phloridzin with formation of different substances is evidently associated with different metabolic pathways of these compounds in chlorophyll containing and chlorophyll-free tissues. Thus, contrary to decomposition of isosalipurposide by roots, its decomposition by leaves gave rise to such compounds as naringenin (flavanone), which could be regarded as a probable precursor of flavonols. The same authors pointed out that a still earlier reaction, which is chemical rather than enzymatic, was the reaction of ring closure and formation of two salipurposide isomers, and that deglycosiding and formation of the aglycone naringenin were already the second stage of this process. This latter reaction is of enzymatic nature, seemed to be catalyzed by a β-glycosidase type enzyme.

The chromatograms used for analyzing isosalipurposide transformation products did not reveal the yellow spots of flavonols sensitive to conventional reagents. From the above it follows that transformation of the chalcone isosalipurposide in the dark in vitro terminates at second stage. As to the subsequent reactions which involved formation of eriodictyol and luteolin occurred in the leaves of willow, they evidently took place only in vivo under light. The course of these reactions and the role of isosalipurposide as a possible precursor of flavonols may so far be discussed only hypothetically (Figure 6.3).

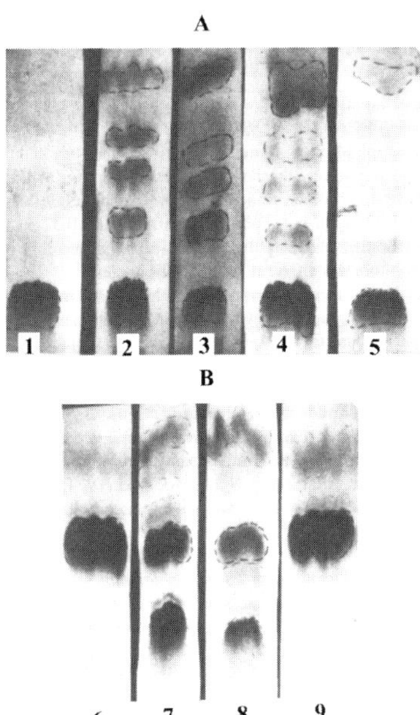

Figure 6.3 Decomposition of isosalipurposide (A) by segments of various organsof willow, and decomposition of phloridzin (B) by segments of various organs of apple-tree
1 – isosalipurposide incubated without tissue (control); 2 – buds; 3 – tips; 4 – leaves; 5 – roots; 6 – phloridzin incubated without tissue (control); 7 –tips; 8 – leaves; 9 – stem. Chromatography in 15% CH_3COOH solution.

Phloridzin as well as isosalipurposide were decomposed to form aglycone. Both phloridzin and phloretin produced yellow pigments. Since on chromatograms these pigments exhibit yellow fluorescence in UV light they might be aglycones. It is known that flavonoid glycosides are revealed as dark spots on chromatograms exposed to UV light. These yellow pigments are flavanones, since they do not react with AlC_3 and Na_2CO_3 like flavonols, which form bright yellow spots. At the same time similar to chalcones and aurones, these phloridzin transformation' products are yellow-coloured and turn orange-pink in the presence of Na_2CO_3 or, NH_4 OH.

Relatively easy transformations of isosalipurposide and phloridzin into compounds of other classes (flavanones, chalcones or aurones) were evidences of an important role of these products in the general metabolism of flavonoid compounds.

Isosalipurposide
2', 4', 6', 4-tetroxychalcone-
-2'-glucoside of chalconaringenin (chalcone)

1

Salipurposide
naringenin-5-glucoside
(Flavanone-glycoside)

2

Naringenin
(Flavanone)

3

Eriodictyol
(Flavanone)

4

Luteolin
(Flavone)

While decomposition of phenolic inhibitors has been adequately studied, the investigated much later ABA needs our special attention. Dorfling, 1987 presented the oxidation and glycosidation pathway of abscisic acid, which brings this inhibitor to the end of their inhibiting activity (Figure 6.4). Special interest is also ABA content in subtropical plants. Further discussion indicates specifics of these transformations.

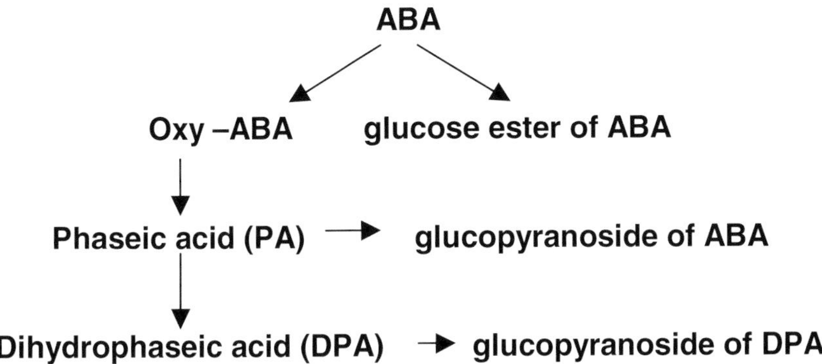

Figure 6.4 Catabolism of Abscisic acid (ABA).

6.5 Ethylene Evolution and Abscisic Acid Content in Citrus Plants.

Sub-tropical flora has its own specifics in ABA content and ethylene changes during the period of growth because citrus-plant growth is characterized by endogenous rhythm which manifests itself by several periods of increased growth activity over the course of the year. This rhythm considerably hinders a plant's ability to adapt to low above-zero and below-zero temperatures. Growth activation in autumn is especially undesirable as it prevents the normal transition to a state of relative dormancy.

The ethylene-releaser kamposan is known to shorten vegetative growth in autumn, thereby promoting plant adaptation to winter conditions . Different aspects of growth-regulation and of frost resistance in citrus plants have been widely studied, nevertheless, kamposan's effect on hormonal status in relation to growth and dormancy in citrus shoots, as well as on adaptation to low temperatures, has been insufficiently studied. In this respect, the study of the time-course of ethylene evolution and ABA content in different parts of lemon tree and tangerine tree shoots is very promising, as ethylene and ABA, interacting with other components of hormonal system, were involved both in growth-regulation and dormancy, as well as in adaptation to temperature changes (Chkhaiidze at all., 1993)

The purpose of this research is to study the effect of the ethylene-releaser kamposan on ethylene evolution and ABA content in shoots of lemon and tangerine trees during autumn and winter, so that we may have a better understanding regarding their adaptation to low temperatures (Figure 6.5).

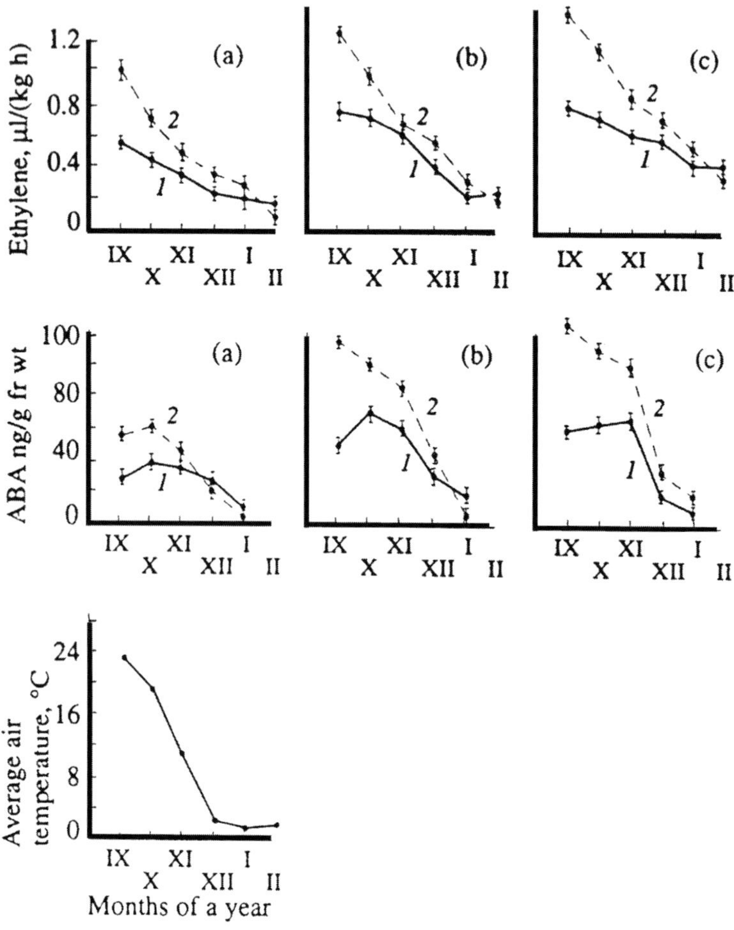

Figure 6.5. Time-course of ethylene evolution and ABA content in untreated (I) and kamposan-treated (2) leaves in autumn and winter period and the average air temperature. Unshin tangerinetree (a), lemon-tree cv. Mayer (b), and lemontree, cv. Gruzinskii (c).

Kamposan treatment enhanced ethylene evolution and increased ABA content in all actively growing shoots by the second and third days after treatment. Kamposan treatment resulted in a 13 to 14-fold increase of ethylene accumulation in the apical parts of shoots, and an 8 to 12-fold increase in leaves as compared to the corresponding organs of untreated plants. ABA content also increased by a factor of 15 to 16 in apical parts of shoots and by a factor of 4 to 10 in leaves. These changes in hormone levels depended on plant species and cultivars. The conclusion was made that the sharp increase in ethylene synthesis and ABA content in kamposan-treated plants led to the abscission of shoot apices and the cessation of visible growth approximately 10 to 15 days earlier than in the untreated plants (Table 6.5).

The time-course of ethylene emission and ABA accumulation in tangerine and lemon leaves took place in autumn, when plant adaptation to low temperatures occurred, and later in winter. Autumn-grown leaves were used for analysis because they were more adaptable in comparison to spring or year-old leaves.

Table 6.5. The effect of kamposan on ethylene synthesis and ABA content in tangerine and lemon tree.

Treatments	*Ethylene evolution*, μl/(kg h)*		*ABA content**, ng/g*	
	Apical parts of shoots (10 to 15 mm)	*Leaves (leaves 4 to 6)*	*Apical parts of shoots (10 to 15 mm)*	*Leaves (leaves 4 to 6)*
Unshiu tangerine tree				
None	72.0±14.2	45.4±6.5	111±8	83±9
Kamposan	907.4±12.7	350.0±9.0	1653±23	308±13
Lemon tree cv. Mayer				
None	79.0±9	43.3±6.2	154±17	91±14
Kamposan	1090.2±33.2	480.5±15.3	2359±20	652±10
Lemon tree cv. Gruzinskii				
None	80.4±14.0	50.5±5.0	165±15	93±2
Kamposan	1126.0±24.0	650.0±19.0	2554±30	1019±24

Notes: * Ethylene synthesis was measured on the second day after treatment.
** ABA content was measured on the third day after treatment.

With the arrival of low above-zero temperatures and the arrested visible growth of shoots (October and November), the intensity of ethylene synthesis in untreated leaves decreased. As compared to ethylene production in September, ethylene accumulation in November declined approximately 50% in tangerine leaves and 20% in lemon leaves, and continued to decrease gradually during January and February. Kamposan-induced enhancement of ethylene evolution was maintained at a high level from September (1.5 to 2 times higher than in the control leaves) until the end of December , and then decreased gradually through the end of January to the ethylene amount the same as in the control plants (Table 6.6).

ABA content in leaves of the control citrus plants slightly increased and was maintained at an enhanced level during November and December In February, it fell to a substantially lower level. Kamposan-treated leaves had higher ABA content in comparison to the control leaves through the end of December, then declined to the control-leaf level.

The evidence suggested that it is possible to shorten the autumn vegetation period of tangerine and lemon trees by treating them with the ethylene-releaser kamposan, apparently through an increase in ABA content and ethylene synthesis. Kamposan is known to enrich tissues with exogenous ethylene and to stimulate synthesis of endogenous ethylene. ABA, being an inhibitor of certain enzymes, affects the rate of ethylene formation . One might suggest first that the higher content of free ABA is the result of non-specific cell response to kamposan. which acts as a chemical stressor and, second, that ethylene induced an increase in cell ABA level. It was shown that kamposan decomposition in plants changed cell pH, destroyed chlorophyll in leaves, reduced photosynthesis, enhanced respiratory activity, and affected the extent of stomatal aperture closure in citrus plant leaves. These changes could affect intracellular energy balance. The sharp increase in ABA and ethylene content caused by kamposan could be one of the main causes of apical bud senescence and suppressed shoot growth.

Table 6.6. The Effect of Kamposan on Apex Abscission from Tangerine and Lemon Tree Shoots at Different Periods of Their Growth.

Treatments	*The number of shoots with apex*					
	Summer growth period			*Autumn growth period*		
	Before treatment	*After treatment*		*Before treatment*	*After treatment*	
		Day 5	*Day 20*		*Day 5*	*Day 20*
Unshiu tangerine tree						
None	50	50	7	50	50	5
Kamposan	50	8	0	50	10	0
Lemon tree cv. Mayer						
None	50	50	9	50	50	8
Kamposan	50	3	0	50	5	0
Lemon tree cv. Gruzinskii						
None	50	50	28	50	50	10
Kamposan	50	3	0	50	5	0

Kamposan had a prolonged effect. Even a single treatment affected ethylene production and ABA content in citrus plant leaves throughout autumn and part of winter. At low above-zero temperatures essential to the first stage of citrus plant hardening, we observed lower ethylene evolution and

ABA content. This trend was found both in control and kamposan -treated plants, the ethylene evolution by leaves of relatively frost-tolerant tangerine-trees being lower than in the leaves of frost-sensitive lemon-trees. These corroborate the evidence on the decrease in phytohormone content, including ABA in citrus plant leaves during autumn and winter when plants go into relative dormancy. This appeared to help plants better adapt to low temperatures.

The pattern of changes in ethylene synthesis and ABA content during cessation of growth (October and November) was different in kamposan-treated and untreated lemon trees. In particular, ABA content in the treated plants increased while it decreased in untreated plants. The pattern of changes in ethylene production and ABA content depend ed on temperature, plant species and cultivars as well as on the dose. Injuries sharply increased ABA content and ethylene production. Such phenomena were observed as a result of water deficiency, chemical or mechanical damage, ionizing radiation. Thus, ethylene production and ABA content in leaves and fruits of citrus plants were enhanced at harmful low temperatures resulting in elimination of the damaged organ. Ethylene synthesis by plant organs affected by low but not injuring temperatures did not damage the normal tissues. A similar phenomenon was observed when citrus plants were treated with kamposan.

Thus, the effect of kamposan was related to the changes in hormonal status of citrus plants, in particular, ethylene and ABA, which determined growth processes during the difficult autumn and winter months. Our data demonstrated that the ethylene releaser kamposan affected not only ethylene amounts but ABA content as well.

Thus effect of the ethylene releaser kamposan on ethylene amounts and abscisic acid content in the shoots of a tangerine tree and a lemon-tree were studied during autumn and winter. A sharp increase in ethylene emission and abscisic acid content was detected on days 2 and 3 after the kamposan treatment, followed by a gradual decrease in the production of hormones during autumn and winter. Adapting to low (above-zero) temperatures, the intensity of ethylene production and abscisic acid content in leaves decreased, these hormone levels were lower in tangerine trees than in a lemon trees. The pattern of kamposan-induced changes in phytohormone content depended on temperature as well as on the species and cultivars of the citrus plants.

6.6 Intact and Abscised Leaves and Their Properties.

In the autumn, the abscised leaves could play a role of the good substrate for the composting. At the same time leaves contained some inhibitors, the activity of which still existed in abscised leaves. The season activity of inhibitors from leaves of different trees were tested .

Preliminary investigations were concerned with the allelopathic potential of leaf litter from six tree species common to southeastern Oklahoma: Hackberry tree *(Celtis occidentalis*) and Water oak (Quercus nigra) to have the most inhibitory effects on the indicator species winter wheat (*Triticum aestivum L.* cv. Inna and Wakefield)(Horne, et. al, 1996). Hackberry and Water oak leaves were collected monthly from April through September 1996 and extracted with dH_2O (50 g dry leaves/L) at room temperature. The allelopathic effect of the leaf extracts on the germination and coleoptile length of winter wheat was evaluated. Three replications of 100 seeds of each wheat cultivar were incubated in Petri dishes of leaf extract or dH_2O in darkness at 24^0C.

Inhibition or stimulation of seed germination was expressed as a percentage of the controls (dH_2O only) after seven days after germination. Inhibition of seed germination by extracts from monthly collected hackberry leaves ranged from 26.5 % to 55.9% in cv. Inna and 42.9% to 79.4% in cv. Wakefield. Inhibition of seed germination by extracts from monthly collected water oak leaves ranged from 7.4% to 41.2% in cv. Inna and 14.3% to 57.2% in cv. Wakefield (Tables 6.7 and 6.8).

Inhibition of coleoptile growth by extracts from monthly collected hackberry leaves ranged from 72% to 81.2% in cv. Inna and 66.6% to 81.4% in cv. Wakefield. Inhibition of coleoptile growth by extracts from monthly collected water oak leaves ranged from 47.5% to 81.0% in cv. Inna and 42.6% to 83.5% in cv. Wakefield (Tables 6.7 and 6.8).

These results suggested that either seasonally mediated production of allelochemicals by hackberry and water oak or seasonal variability in leaf morphology affected extractions efficiency. The differences that took place in germination process between wheat cultivars also suggested the selectiveness of allelochemical activity.

Table 6.7 Percentage of Inhibition of Extract-Treated Wheat cv. Inna and Wakefield (compared to controls).

Month [a]	*Extract*			
	Hackberry		*Water Oak*	
	% Inhibition			
	Inna [b]	*Wakefield*	*Inna*	*Wakefield*
April	55.9	79.4	38.3	44.5
May	26.5	42.9	7.4	14.3
June	51.5	62.0	39.4	28.6
July	54.5	66.7	41.2	35.7
August	35.3	73.1	13.3	25.4
September	41.2	50.8	39.8	57.2

[a] Time of leaf collection

[b] Germination of Inna and Wakefield controls in dH_2O was 94% and 91%, respectively.

Table 6.8. Percentage of Inhibition of Coleoptile Growth of Extract-Treated Wheat cv. Inna and Wakefield Compared to Controls

Month [a]	*Extract*			
	Hackberry		*Water Oak*	
	% Inhibition			
	Inna [b]	*Wakefield*	*Inna*	*Wakefield*
April	81.2	81.4	81.0	83.5
May	72.0	66.6	53.2	42.6
June	74.8	73.5	47.5	42.8
July	76.3	68.4	67.8	71.4
August	76.1	80.3	65.6	61.4
September	72.0	71.4	62.1	52.6

[a] Time of leaf collection

[b] Coleoptile length of Inna and Wakefield controls in dH_2O averaged 46.0 and 36.4 mm, respectively.

The same experiments were done with intact and abscised leaves of elm, box elder, sumac and red maple. Bio-assay plants were used wheat seeds (WH) , clover (CL) , mustard (M). See tables 6.9 and 6.10 below.

Table 6.9. Effect of leaves extracts on the growth of some crop seedling (length of 5 tallest plants in cm).

	Spring			
	WH	CL	M	L
Elm	2	1.5	0.4	2
Box elm	0.7	0	0	0
Sumac	0.7	0	0	0
Maple	1	0	0	0
	Summer			
Elm	2	1.3	0.4	1.5
Box elm	1.9	0	0.6	0.9
Sumac	0	0	0	3
Maple	1.3	0	0	0
	Autumn			
Elm	5	1	0.2	1.5
Box elm	0.9	0	0	0.1
Sumac	3	0.5	0.7	0.7
Maple	1.6	0.3	0.1	0.8
	Abscised			
Elm	5.6	1.8	0.9	1
Box elm	3.4	0.9	0.9	0.8
Sumac	0.5	0	0	0
Maple	1	1	0	0.2

* WH – wheat; CL – clover; M – Mustard; L – lettuce;

Table 6.10 Lettuce Seed Germination Data (Abscised Leaf and Composted Leaf Water Extracts).

Concentration	*Abscised Leaves, %*	*Composted Leaves, %*
Control - Water	87	87
Concentrated	0	0.5
Diluted 1:2	20	50
Diluted 1:4	53	74
Diluted 1:8	65	83
Diluted 1:16	74	87

At the same time abscised leaves from red maple still contained various phenolics, which were observed by the paper chromatogram under UV-B light. The strongest inhibition was observed by some phenolic acids with Rf 0.50, and by coumarin derivative, Rf 0.69 (Table 6.11).

Table 6.11. Red Maple Abscised Leaves. (%% control).

Region	*Rf*	*Color in UV-B*	*Wheat*	*Clover*	*Mustard*	*Lettuce*
Control	-	-	94	99	98	96
Start	0	Dark Violet	53	76	96	81
2	38	Dark Violet	61	43	99	100
3	50	Violet	78	67	100	67
4	69	Bright Blue	43	84	60	67
5	88	Blue	60	87	94	67
6	94	Yellow	100	87	100	100

These experiments contributed to the search of the natural botanical herbicides, substances that could selectively inhibit the germination of the unwanted plant species (weeds).

CHAPTER 7

NATURAL GROWTH INHIBITORS AS ALLELOPATHOGENS AND BOTANICAL HERBICIDES

In the previous chapter we observed the inhibiting properties of leaves secondary substances and their transformation in plant tissues. Some of these products and their metabolites could play a role of allelopathogens or allelochemicals, which are sufficient in plant communities and ecosystems in general.

Plant - Producers of the Active Allelopathogens		
Secretion of plant metabolism's products	Conversion of secreted products by microbes to the stable allelopathic forms	Effects of the allelopathogens on other crops or weeds

Plants contain and secrete in the external medium significant amount of growth inhibiting substances. This chemical effect of plants on growth and development of other plants grown in their vicinity is called allelopathy. Leaves exudates from various willow species such as *Salix rubra, or Salix viminalis*, contain phenolic inhibitors like naringenin-derivative-isosalipurposide. For example, leaves and roots of apple tree contain phloridzin, which is a strong respiratory inhibitor. The roots and leaves of the wild plant *Nanaphyton* grown in semi-desert region of Mongolia, contain strong phenolic inhibitors. It was also observed that seeds during inhibition secrete allelochemicals. Tobacco seeds are the example of autotoxicity when leachings from the seeds (in this case abscisic acid) suppressed germination of the same tobacco seeds. Some of these leachates suppressed the imbibition and germination of seeds with different intensity. This demonstrates the selectivity of these natural botanical compounds similar to the selectivity of synthetic herbicides. There is sufficient data showing that phenolics and alkaloids play the role of selective agents. Secondary compounds can be modified in the transgenic plants and genetic mutants. Hence, molecular genetics is a tool, which helps to regulate the level of secondary metabolites in plants. The future search for natural botanical herbicides is essential in connection with ecological concerns aiming at limiting environmental pollution and eliminating the harm that had been caused by synthetic herbicides.

Let us discuss the ability of some root exudates to affect the germination of seeds of different crops: monocots and dicots (Table. 7.1 and 7.2).

It is important to know that only some phenolics are observed in the willow root exudates (1), which have no analogues in the roots (2) and leaves (3). These data show that botanical herbicides might be searched for and found among the common allelopathogens, though some substances could be retained by willow roots and some were excreted into the external medium.

Chromatography of these water exudates and the subsequent investigation of chromatogram in UV-B light showed that most

Table 7.1 Effect of root exudation on germination of crop seeds. (Non-concentrated exudates)

Variant	*% to tap water (control)*			
	wheat	*clover*	*lettuce*	*mustard*
Tap water	100	100	100	100
Spider plants (*Chlorophytum*) exudates	54	93	75	100
Willow (*Salix vitaminalis*) exudates	58	79	74	138
	Stem length (5 tallest plants, mm)			
Tap water	29	23	18	25
Spider plants (*Chlorophytum*) exudates	15	21	14	2
Willow (*Salix vitaminalis*) exudates	7	18	13.5	3.5

Table 7.2 Biological activity of willow root exudates after paper chromatography (Biological activity in % to control (water)).

		Clover		*Lettuce*	
Rf	*Colour in UV light*	*Germination*	*Stem length*	*Germination*	*Stem length*
0	Blue	91	76	90	64
0.14	Blue	94	68	98	58
0.3	Violet	86	80	93	76
0.5	Blue	56	52	71	76
0.67	Yellow	87	68	89	88
0.88	Yellow	52	56	63	64

of these substances are polyphenols such as coumarin or phenolic acids. The phenolic substances which are retained by the cells had different properties than those which were located in the root exudates. These data confirm the idea that the excreted substances had an allelopathic nature and were involved in the ecological relations between different plant species. Kefeli et al., (2001) showed that during the composting process water extracts contained many inhibiting substances that might form toxic exudates. Paper chromatography revealed the presence of phenolic acids and coumarins in these water extracts. The highest amount of these inhibitors was measured in abscised leaves of red maple (*Acer rictrurn L.*) that were collected in the fall. One gram of dry leaves were mixed with 20 ml of water that was used to prepare the extracts. The pH value of the solution was between 5.4 and 5.6. Inoculation of the water extracts with fungi imperfecti raised the pH to 6.0. Inoculated extracts that were incubated for a week at room temperature showed a further raise in pH to 7.2.

Concurrently, we observed that during composting the amount of phenolics was drastically reduced. The same prosecc of degradation was observed for ABA. ABA comimng through the roots inhibited the seedlings growth (Borisova, Bonavanture, 1992). Leaves could be a dood substrate for compost where cellulose, pectine, and other polimers could become a source of carbon (Kefeli, 2002) and phenolics (Macoskey et al., 2000). Some phenolics could play a role of herbicides or allelopathogens (Kefeli, 1999). It was observed that during the composting processes the phenolic composition of leaves changed remarcably and the most apparent changes were observed among flavonoids (Wellton et al., 2001). Seed germination tests were performed with these water extracts and pure water (control) on lettuce and wheat seeds. Measurements of germination rate and the length of the seedlings demonstrated that phenolics were observed to loose inhibiting properties after dilution, or after contact with fungi.

The whole process of allelopathogens formation in the environment is tightly connected with the secondary substances formation and plant biomass accumulation.

Many plant species increase their competitiveness by producing and releasing chemical attractants, stimulants, or inhibitors. Molisch first used the term allelopathy in 1937 to describe all chemical interactions, whether positive or negative, among organisms of all levels of complexity, including microorganisms. Currently, allelopathy refers to the detrimental effects of compounds released by higher plants of one species on the germination, growth, or development of plants of another species (Putman and Duke, 1978). This approach distinguishes allelopathy from plant competition and reinforces the different potential of organisms to utilize environmental resources (Willis, 1994).

Allelochemicals are present in virtually all plant tissues, including leaves, flowers, fruits, stems, roots, rhizomes, and seeds. Allelochemicals are released through volatilization, root exudation, leaching, and decomposition of plant residues (Rice, 1984). Allelochemicals are generally secondary compounds or by-products of primary metabolic processes (Figure 7.1). Allelochemicals also consist of acids, aldehydes, and cyanogenic glycosides, thiocyanates, lactones, coumarins, quinone, flavonoids, phenols, tannins, alkaloids, steroids, and others (Rice, 1984). Combinations of allelochemicals may act in an additive or synergistic manner (Einhellig, 1996). Once released, growing conditions, the physical, chemical, and biological properties of the soil could further alter allelochemical efficacy.

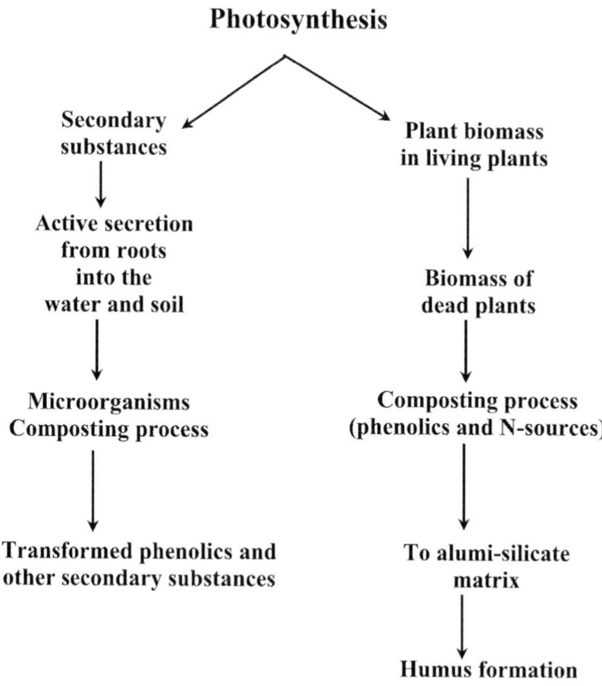

Figure 7.1. Secondary substances, plant biomass accumulation and humus formation during allelopathic effects.

Investigation of reduced herbaceous growth under several tree species and examination of extracts from leaf litter of several tree species showed an inhibitory effect on the germination and growth of brome grass. Similar observation was done by Yamamoto et al., 1999 investigating Barnyard grass.

A reduction in seedling growth of an indicator crop in the presence of extracts from a donor crop suggested allelopathic potential. Allelopathic potential is the degree of growth inhibiting activity of one plant on another and may differ among plant species, cultivars, or among plant parts of the same cultivar.

Local sources of plants with known allelopathic properties are common, and if the most effective plants could be identified, producers might have an inexpensive, biodegradable alternative to synthetic pesticides and herbicides. Leaves from these plant species, collected on the farm, could either be applied as a physical mulch or extracted and applied as a spray.

As a better understanding of plant allelopathy develops incorporation of this knowledge into no-till, minimum-till, intercropping, and rotational cropping systems as well as agroforestry systems may reduce the need for synthetic pesticides while reducing production expenses.

Preliminary investigations into the allelopathic potential of leaf litter from six tree species common to southeastern Oklahoma revealed hackberry (*Celtic occidentalis*) and water oak (*Quercus nigra*) to have the most inhibitory effects on the indicator species winter wheat (*Triticurn aestivum L. cv. Inna and Wakefield*). These two tree species were chosen for further investigation.

Hackberry (Figure 7.2) and water oak leaves were collected monthly from April through September 1996 and extracted with d. water (50 g dry leaves/L) at room temperature. The allelopathic effect of the leaf extracts on the germination and coleoptile length of winter wheat was evaluated. Three replications of 100 seeds of each wheat cultivar were incubated in Petri dishes of leaf extract or d. water in darkness at 24^0C.

Inhibition or stimulation of seed germination was expressed as a percentage of the controls (H_2O only) seven days after germination. Inhibition of seed germination by extracts from monthly collected hackberry leaves ranged from 26.5 % to 55.9% in cv. Inna and 42.9% to 79.4% in cv Wakefield. Inhibition of seed germination by extracts from monthly collected water oak leaves ranged from 7.4% to 41.2% in cv. Inna and 14.3% to 57.2% in cv. Wakefield. Inhibition of coleoptile growth by extracts from monthly collected hackberry leaves ranged from 72% to 81.2% in cv. Inna and 66.6% to 81.4% in cv. Wakefield. Inhibition of coleoptile growth by extracts from monthly collection was then studied. The monthly variation of the toxic activity of some woody plants was observed (Table 7.3, 7.4) (Horne et al, 1996; Kalevitch et al 1997).

These results suggested either mediated production of allelochemicals by hackberry and water oak or seasonally variability in leaf morphology effecting extraction efficiency (Figure 7.2). Differences in germination inhibition between wheat cultivars also suggested that some amount of allelochemical selectivity was occurring. The experiments with the hackberry leaves were continued with 4 biotests, which showed different sensitivity of seedlings of different species to leaves extracts. Leaves of hackberry were collected from the tree (TL) and after abscission (AL); both types of leaves were subjected to water extraction in the allelopathic chamber and than tested by mustard, clover and wheat test for seed germination. An allelopathic chamber consisted of two plastic trays-bigger and smaller. On the bigger tray we put envelopes with dry leaves, covered them by plate with filter paper, which slowly absorbed leaf extracts. On the plate we put seeds of tested plants, and after 2 days measured the effect of extracts on the process of germination. Later after 6 days the effect of extracts on the seedlings growth was evaluated. For separation leaf water extracts were run through paper chromatography in isoproranol-ammonia-water 10:1:1, and later analyzed in UV-B light. The extracts from TL contained more coumarin -like substances than AL.

So, various types of seeds, green and etiolated seedlings, roots and leaves were used for the search of substances with stable inhibiting and selective activity. Imbibed seeds and seedlings of 12 plant species from 4 families were investigated for their inhibiting and selective activity. Through biological testing, the selectivity of leaching from imbibed seeds of wheat and lentil was determined. Etiolated seedlings also secreted into the water inhibitors and other biologically active substances. The most pronounced effect was observed for the same plants: wheat and lentil, and less for pea, salad, fescue and other species. Seeds during imbibition excreted also ions, especially potassium. No definite correlation between growth inhibiting activity and ion concentration was observed. The pH of the leachates from seeds and roots wasn't much different. Leaves and roots could also be sources of biological allelopathogens.

Table 7.3. Percentage inhibition of extract-treated wheat cv. Inna and Wakefield compare to control

	Extract			
Month [a]	*Hackberry*		*Water Oak*	
	% Inhibition			
	Inna [b]	*Wakefield*	*Inna*	*Wakefield*
April	55.9	79.4	33.8	44.5
May	26.5	42.9	7.4	14.3
June	51.5	62.0	39.4	28.6
July	54.5	66.7	41.2	35.7
August	35.3	73.1	13.3	25.4
September	41.2	50.8	39.8	57.2

[a] Time of leaf collection

[b] Germination of Inna and Wakefield controls in dH_2O was 94% and 91%, respectively.

Table 7.4 Percentage inhibition of coleoptile growth of extract-treated wheat cv. Inna and Wakefield compare to control

Month [a]	*Extract*			
	Hackberry		*Water Oak*	
	% Inhibition			
	Inna [b]	*Wakefield*	*Inna*	*Wakefield*
April	81.2	81.4	81.0	83.5
May	72.0	66.6	53.2	42.6
June	74.8	73.5	47.5	42.8
July	76.3	68.4	67.8	71.4
August	76.1	80.3	65.6	61.4
September	72.0	41.4	62.1	52.6

[a] Time of leaf collection

[b] Coleoptile length of Inna and Wakefield controls in dH_2O average 46.0 and 36.4 mm, respectively.

Based upon our observation of growing plants in the Mongolian desert (Ulangom) we concluded that some plants like nanaphyton produced inhibitors which suppressed the growth of other species. Using biological test and chromatography we confirmed this hypothesis. Abscised leaves of some woody plants also form areas of unfavorable for growth of seedlings of grassy species. We collected the abscised leaves of various species. All leaves were used for water extraction. Extracts were analyzed for their inhibiting activity. Their influence on seeds of wheat, lentil, and bean, as well as

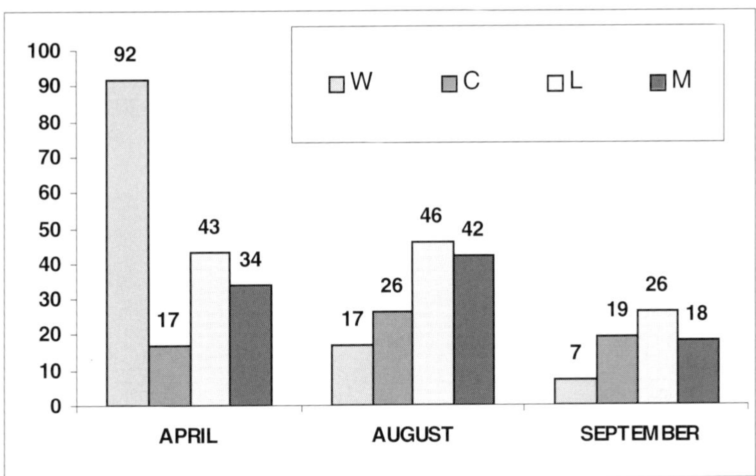

Figure 7.2 Effect of the Hackberry leaf age on its biological activity
Graduate water extraction in allelopathic chamber. 20 g of dry leaves in 140 mL of water.
Tested seeds:W - wheat, C -clover, L - lettuce, M - mustard.

roots and shoots of their seedlings was studied. Among these eight investigated species only the abscised leaves of the beech tree possessed strong inhibitory activity. The mixture of abscised needles of conifers strongly depressed the root growth and less the shoot growth of test seedlings. Mixture of leaves of leafy plants had less inhibitory effect. The role of abscised leaves as biological mulching was also discussed. Further investigation of botanical herbicides (BH) will be connected with the identification of their chemical nature (Table 7.5).

Table 7.5 Biological activity of plant root exudates and water extracts from the roots.

Variant	*Germination in % to control*			
	Wheat	*Clover*	*Mustard*	*Lettuce*
Tap water (control)	79	97	64	100
Papyrus exudates	59*	55*	59	83*
Willow exudates	66*	95	93	100
Spider plant exudates	100	99	100	94
Willow roots	42*	84	35*	61*
Spider plant roots	78	96	34*	98

Legend:
*valid inhibition of seeds germination in the comparison with control

Conditions of exudates collection and concentration:
- Papyrus exudates-1 month collection from roots of 4 plants. Starting volume-200 ml, concentration by evaporation up to 30 ml
-Willow exudates-1 month of collection from roots of 32 plants, Starting volume-1024 ml Concentration by evaporation up to 50 ml
-Spider plants exudates - 1 month of collection from roots of 20 plants, Starting volume-250 ml, concentration by evaporation up to 20 ml
Willow roots growing in water culture were collected from 10 plants, 5g, extracted by 50 ml of water, extract was concentrated up to 20 ml
Spider plant roots - 1 month growing in water, 20 plants, 5 g, extraction by 50 ml of water, evaporation up to 20 ml

Table 7.6 Effect of maple leaves extracts on the seed germination (in % to total amount of seed) 7 ml of tap water per Petri dish

Variant	*Wheat*	*Clever*	*Mustard*	*Lettuce*
Water	35.3	40.1	62.7	100
Sand 0.5 g	36.1	89.2	64.1	97.3
Sand 1.0 g	31.0	87.1	63.0	69.3
Sand 2.0 g	37.8	89.0	49.1	92.1
Leaves 0.5 g	27.4	38.2	13.0	25.6
Leaves 1.0 g	54.2	42.4	7.0	11.3
Leaves 2.0 g	50.0	32.0	14.4	8.3

In experiments with sand 0.5 g per 7 ml of water does not effect seed germination. On the other hand, maple leaves in dosage of 0.5 g / 7 ml of water inhibited seed germination of lettuce and mustard and showed less effect on the wheat. With increasing the dosages lettuce and mustard plants become more sensitive than wheat and clover (Table 7.6).

Some root exudates subjected to chromatography are of phenolic nature. Roots exudates from onion plants (*Allium cepa L.*) were collected in water during two weeks then evaporated and subjected to paper chromatography in isopropanol-ammonia-water 10:1:1. Three phenolic compounds of coumarin nature (bright violet color) were observed in UV-B light. Part of the chromatogram was subjected to wheat- clover test on seeds germination and root-stem length. One of the coumarin like substances (Rf 0.3) was able to inhibit the germination of wheat and clover seeds. This substance was supposed to play a chemical signaling role in allelopathic relations of plants. To check this idea we extracted onion roots by water and compared the substances from roots and from exudates by the same method of chromatography-UV-light investigations and biotests. In this case we did not observe any similarity in the phenolic substances composition. We suppose that onion roots synthesize special signaling substances which they secrete in the water medium.

Putman and Weston (1985) also brought issue of decomposing plant residues into allelopathogenic concept. These studies became particularly important now because of the trend toward reduced tillage agriculture, which by its nature purposely maintains plant residues on the soil surface. The toxic influences of straw or other highly carbonaceous plant residues were described in detail by Collison and Conn (1925). They concluded that two separate mechanisms were associated with the toxicity produced by plant residues. The first is toxic chemical agents (allelochemicals), which act quickly and are usually quickly inactivated by colloidal matter. The second involves the stimulation of microbial populations, which in turn immobilize much of the nitrogen, making it unavailable to higher plants.

McCalla and Duley (1948) reported that corn performed poorly under no tillage (stubble mulch) systems where sweet clover or wheat-straw mulch remained at the soil surface. Extracts of sweet clover and microbial products from decaying wheat were both found to be toxic to corn. Guenzi and McCalla (1962) extracted residues of several crops at the end of the summer and autoclaved half prior to bioassay. Although no autoclaved extracts were generally more toxic to germination, the autoclaved extracts were more inhibitory to seedling growth.

In later work Norstadt and McCalla (1963) identified a fungus (*Penicillium urticae*) that produced patulin, a strong inhibitor of seedling growth. Their work, and later studies by McCalla and Haskins (1964) clearly demonstrated that microorganisms as well as plant toxins contribute to the allelopathic potential of plants.

Patrick and Koch (1958) monitored the toxicity of decomposing plant residues using the respiration of tobacco seedlings as an assay. They found the greatest toxicity to occur when decomposition took place under saturated soil conditions. The stage of maturity of the plant residues also affected their toxicity. Residues from young plants were toxic immediately, whereas more mature plants required a longer period to release their toxic substances.

Allelopathy may adversely impact agricultural systems through various methods. Numerous weed species (now about 90) posses an alleged allelopathic potential. Although future studies may negate some of these accounts, it appears that sufficient evidence exists to implicate many. In several perennial species and particularly quackgrass, considerable chemical evidence has been accumulated. The negative impacts of crop residues on subsequent plantings were also well documented. These toxins may be leached from the plant residue, released upon its breakdown, or produced by microbes that use the plant material as a food source. Toxins from surface residues may affect crop performance in no-tillage systems. Therefore, more must be learned about their compatibilities. Replanting problems and auto toxicities result from the release of allelochemicals in many perennial-cropping systems. These often involve microbes that release toxins upon hydrolysis of conjugates into the plant tissue. Several species release allelochemicals that are inhibitory to nitrogen-fixing bacteria. Others apparently can regulate nitrogen availability through the effects on nitrification. Elroy Rice (1986) considered that most research projects in allelopathy have been designed in such a way that only the inhibitory results were considered significant in explaining the biological problem under investigation. Thus, any resulting stimulatory effects have been ignored, or mentioned only incidentally. Phytopathologists have been aware, however, of the significance of the chemical stimulation of growth and development of pathogens and parasitic plants by microorganisms and host plants. Hence, there have been many reports of allelopathic stimulation of fungi functions by other fungi or by plants, and many concerning stimulation of seed germination of parasitic plants by host plants. There have been numerous reports of allelopathic stimulation of algae by other algae in relation to bloom formation and phytoplankton succession. There have been at least a few reports of the chemical stimulation of each type of microorganism (bacteria, fungi, and algae) by the same and other types. Instances of allelopathic stimulation of plants by microorganisms have also been documented. The allelopathic growth stimulation of nonparasitic plants by plants has been grossly ignored, considering its potential economic importance. There have been several documented cases and a few instances were reported here about the remarkable growth stimulation of radish and downy brome grass by *Glechoma hederacea* (ground-ivy).

Decaying ground-ivy leaves (2 g/kg of soil) stimulated growth of downy brome shoots by as much as 770% and radish shoots by as much as 1064%. Downy brome root growth was stimulated up to 251% and radish root growth up to 1354%. Root exudates of ground-ivy significantly stimulated both shoot and root growth of radish. Volatiles from ground-ivy leaves did not significantly affect germination rate or radicle growth of the two test species.

The evidence indicates that ground-ivy is strongly allelopathic, at least some species. Both root exudates and decaying leaves of ground-ivy significantly stimulated growth of radish plants and decaying leaves significantly stimulated growth of downy brome. Additionally, root exudates of ground-ivy inhibited growth of downy brome. Volatile allelochemicals were apparently not involved; thus it appears that water-soluble compounds were responsible for the results. Rainfall leachates of the leaves were not tested for allelopathic activity but it is possible that such leachates could have significant effects.

The reduced stimulatory effect of decaying ground-ivy leaves in experiments is intriguing. The most important point is, of course, that there was still a highly significant growth stimulation of both test species in spite of numerous changes in the conditions. The greatest change in growth in the second experiment was the marked increase in control growth. The experimental changes in the control conditions were a slight increase in temperature; addition of N, P, and K; and addition of perlite. It is doubtful that the small temperature change or perlite was responsible for the increased control growth relative to that of the test. It is more likely that the added minerals were responsible for the difference, even though they were added to the test soil. Obviously, the large growth difference between control and test in experiment l was not caused by the addition of minerals in the leaf material of ground-ivy. Because only 2g of this material were added per kilogram of soil in the test and only a very small percentage of the leaf material consisted of minerals. This suggests that a chemical(s) in the ground-ivy might have increased the uptake of minerals by the test plants. Einelling (1986) presented his idea about the mechanisms and models of action of allelochemicals. Considerable speculation is still involved in any attempt to draw together the various interrelationships between allelochemical functions. Figure 7.3 suggests a possible sequence of events for the action of phenolic acids and closely related substances. Many probable interactions and the details of effects on each altered function have been omitted; in fact, currently there is almost no evidence to pinpoint specific biochemical and biophysical events responsible for these effects on plant functions. We suggested that interference wit cell membranes and interactions with phytohormones may be the two primary modes of action of these compounds, and these two need not to be mutually exclusive. Interference with specific protein functions is included at this level because several of these effects cannot be separated from membrane and hormone activity. Biodegradation of lignin can also accompanied by allelopathic affects (Lewis, Yamamoto, 1990)

Phenolic acids solubilize in membranes according to their lipid solubility, altering permeability and transmembrane transport. It is easy to relate disfunction of the plasma membrane and tonoplast to the failure of cells to maintain proper mineral nutrition and a favorable water balance. Likewise, efficiency in the energy systems of respiration and photosynthesis demands precise membrane organization, charge separation and the work of membrane-associated proteins. If phenolics partition into these membranes, perhaps binding to lipids, carbohydrates, or proteins, several mechanisms for interference are possible. These include various opportunities for disruption of electron flow and uncoupling of electron transport from phosphorylation. Allelochemical-induced membrane disfunction may also arise if the synthesis of membrane constituents is blocked, a possibility suggested because phenolics alter protein synthesis. Certainly several other biochemical and biophysical mechanisms for membrane interferences are possible.

The actions of several allelochemicals on hormone-induced growth and development responses might occur from alterations in hormone synthesis or reception sites, hormone inactivation, or combinations of these. Hypothetical mechanisms for the action of IAA, GA, cytokinins, and ABA all include potential actions at membranes. Hence it is possible that an allelochemical that affects membrane functions would interfere with these responses. Allelochemicals might also act by inducing the release of a secondary agents, like ethylene but this has not been yet investigated.

All the functions described above would not probably apply to a specific case of allelochemical inhibition, and certainly under the circumstances the modified factors would not have equal impact. For example, effects on photosynthesis would not apply to blocking seed germination, and hormone actions would not apply to inhibition of microorganisms.

The broadest effects of secondary compounds probably occur in the inhibition of seedling growth, where it is suspected to be the collective disruption of multiple plant functions that translates into disorganization and impairment of growth. Perhaps, this gives some rationale to the fact that seedling growth is typically more sensitive to allelochemical interference than seed germination (Figure 7.3).

The interaction model presented may be useful for evaluating the impact of nonphenolic allelochemicals. Because of their lipid solubility terpenes will absorb into membranes and loss of membrane integrity may be a starting point for their actions. One can only speculate that several other classes of allelochemicals must have primary effects on membrane functions. Much less we know about herbicidal effects of secretions from plant species. BH must possess inhibiting properties in the relation to the other plants or especially to their target organs and tissues. BH must be selective enough, that means that they can depress some species of plants and do not effect the others. They might be biologically safe and produce an effect at low dosages. They must be standard, i.e. have definite chemical properties probably most of BH could be recruited from allelopathogens, but they also can be isolated from various plant organs and tissues.

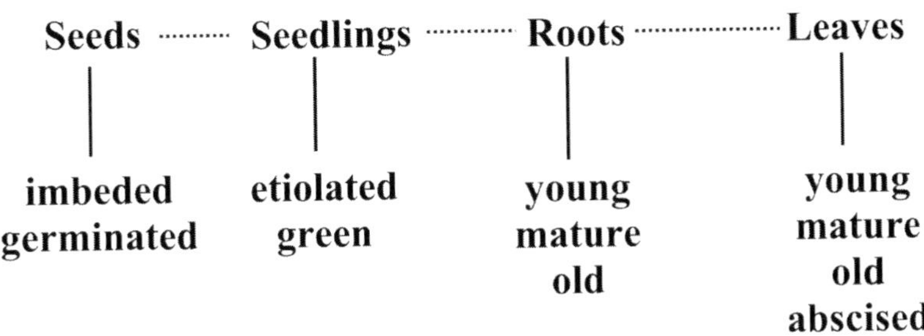

Figure 7.3 Scheme of sources of botanical herbicides.

It is certainly convincing that the new view on allelopathogens as botanical herbicides is tightly connected with the other plant physiological processes.

CHAPTER 8

NATURAL GROWTH INHIBITORS AND STRESS CONDITIONS.

8.1 General aspects of stress effects.

The concept of stress is based on the stress factor, dosage and cellular organization of the organism. The main ideas of different types of stresses are presented in "Experimental ecology" ed by Kefeli and Kudeiarov, 1991, (Horne et al., 1995; Kalevitch, Filimonova 1995).

It is important to take into consideration the dose of stress, which could induce different biological reactions (Table 8.1)

Table 8.1 Plant reaction and the dose of the ecological stress

	Dose		
Factor	*Low*	*Normal*	*High*
Temperature	Cold stress	Thermoperiodism Dormancy Stimulation of development	Heat stress Heat shock
Water	Anabiosis-diapause	Stimulation of development	Anoxic reaction
Light	Etiolation	Sequence of photo-biological reactions	Abnormalities Growth inhibition
Ions	Block of development	Normal root and air nutrition	Salinity Toxic effects
Gravity	Break of polarity	Optimal stem Root relations	Abnormal development

The interaction of stressors could make the stress situation more variable (Battagia, Brennan, 2000; Bassman et al., 2001). Some stressors like UV-C light could be at certain dosage neutral or slightly inhibiting agent of short time chilled wheat plants (Table 8.2 and 8.3). In general two factors can interact as synergists, antagonists or additive agents.

In the experiments of N. Kamacharya and V. Kefeli (Slippery Rock University), presented at California University of Pennsylvania, 1998, the data about combination of these factors were discussed.

Table 8.2 Use of UV-light as possible crio-protectors on imbibed wheat seeds and seedling growth (cm).

		Control	*Freezing 10 min*	*UV 40,000 µJ/m²*	*UV and Freezing 10 min*	*10 Freezing and UV*
Experiment I	Root	11.6	9.5	11.0	7.7	8.6
	Shoot	4.6	3.3	4.8	3.3	3.4
Experiment II	Root	47.7	4.4	45.5	12.8	0
	Shoot	25.1	2.1	24.7	6.1	0

Table 8.3 Use of UV-light as possible crio-protectors on germinating wheat seeds and seedlings growth (cm)

		Control	*Freezing 10 min*	*UV 40,000 µJ/m²*	*UV and Freezing 10 min*	*10 Freezing and UV*
Experiment I	Root	60.1	21.7	28.0	17.0	13.6
	Shoot	28.6	8.44	17.9	8.6	5.7
Experiment II	Root	52.0	11.4	16.8	12.9	**
	Shoot	33.1	4.6	16.7	12.3	**
Experiment III	Root	66.2	21.8	32.9	21.2	26.5
	Shoot	39.8	16.7	35.0	15.1	14.5
Experiment IV	Root	52.0	**	16.8	10.4	**
	Shoot	33.1	**	16.7	7.5	**

** *frozen*

8.2 Low temperature effects.

The stage and condition of plant or seedling, at which stress takes place is very important (Qun, Zeevaart, 1999; Ouellet et al., 2001). We observed that dry seeds of winter wheat Wakefield are not sensitive to freezing even after 3 h of treatment. However imbibed seeds became very sensitive (Figure 8.1).

The explanation of the cell death under certain stress conditions was explained in the book by Kefeli, Sidorenko, 1993 (schema presented in Figure 8.2).

In general the experiments with the combination of two stress factors were the following:

Wheat seeds of different stages: dry, imbibed, and one day germinated were taken. Seeds were distributed-10 seeds per Petri dish. The seeds were then exposed to freezing temperature (-20°C) for various duration. Three milliliters of distilled water was added to each Petri dishes. The dishes were then kept in light chamber with room temperature, 27°C. After a day of exposure of seeds to the freezing temperature, the seeds were taken out from the light chamber. The growth of root and shoot in each seed was measured in millimeters. The measurements were recorded in the laboratory notebook. The experiment was repeated three times for the conformation of the results (Figure 8.1).

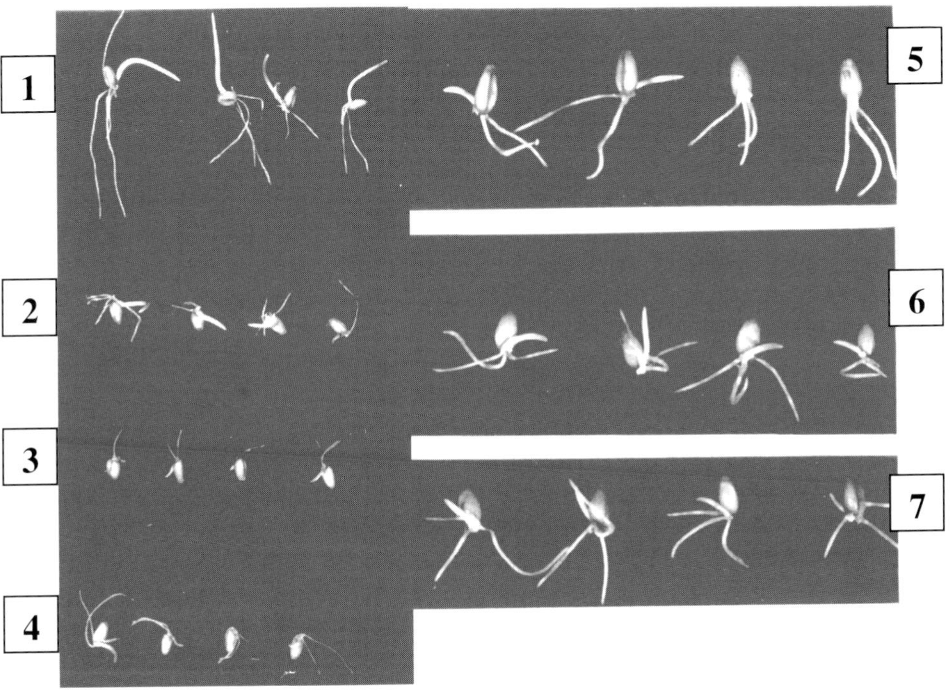

Figure 8.1 Stages of wheat seed germination and external treatment. Imbeded seeds and freezing(-21^{0}C).1 – control, 2 – 10 min freezing, 3 – 20 min freezing, 4 – 30 min freezing. Dry seeds and freezing(-21^{0}C).5 – control, 6 – 1 hour freezing, 7 – 3 hours freezing,

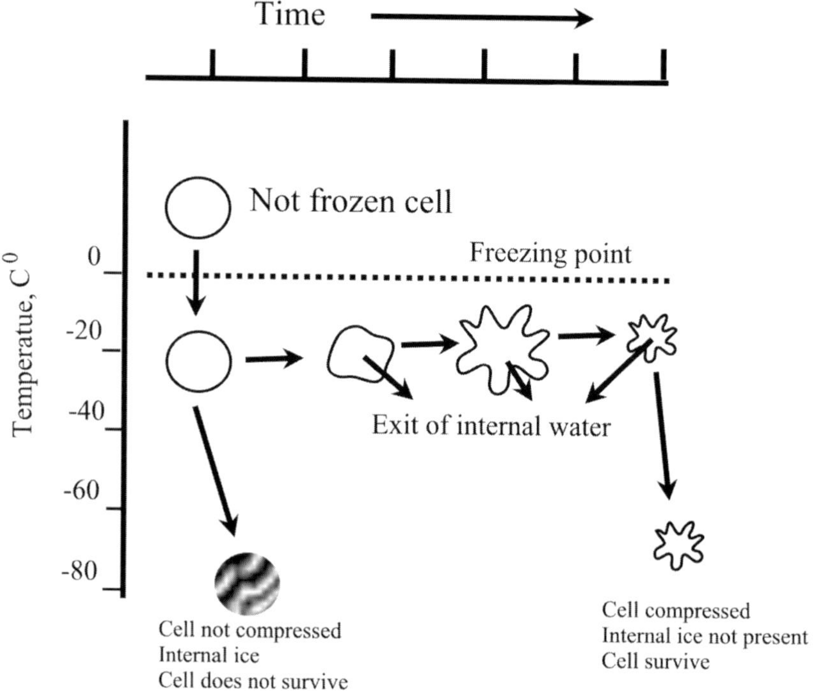

Figure 8.2 Influence of stress temperatures on cell

UV-light was used as a crio-protector in the second half of the experiment. The seeds of different stages were treated with freezing temperature for different duration time, and then exposed to 40,000 μJ of UV-light. At first it was done separately, and then as combination of two to observe the effect of UV-light as a crio-protector.

In general, the experiments dealt with germinated seeds of winter wheat Wakefield, which were subjected to freezing, for 10 min at –21°C. This treatment retarded more than 2 times root growth and more than 3 times shoot growth. UV-C with wavelength of 254 nm in the dose 40,000 μJ/sq cm decreased root growth of wheat seedlings 2 times and shoot growth - 1.5 times in comparison with control. Combination of treatments - freezing and UV reduced growth of seedlings more than each form of treatment separately. Germinated seeds are more sensitive to freezing and UV than dry or imbibed wheat seeds. Freezing activates ion secretion from the seeds, especially potassium, less sodium and calcium. (Table 8.4, 8.5)

Table 8.4 Ion concentration of the leaching as measured for different phases (mg/L)

Phases of freezing	*Variety*	Na^+	K^+	Ca^+
Imbided seeds	Control	2.55	49.6	7.0
	30 min	5.4	150.0	*
	1 h	3.8	116.3	*
	2 h	3.42	125.3	4.7
	3 h	2.36	116.7	4.0
Dry seeds	Control	4.6	8.4	7.6
	3 h	22.3	9.3	9.0
Imbided seeds	Control	2.44	12.8	8.4
	10 min	12.7	58.0	4.3
	15 min	20.3	119.0	4.0
	20 min	23.3	117.0	4.0
	30 min	24.7	115.0	4.0
Germinating seeds	Control	3.8	14.6	10.8
	10 min	14.0	66.7	8.0
	15 min	18.0	102.0	7.7
	20 min	19.0	93.7	6.0
	30 min	18.7	92.0	4.3

* - *Solution not enough to determine*

Table 8.5 Use of UV-light as crio-protectors on germinating seeds

		Control	*Freezing 10 min*	*UV 40,000 µJ*	*UV and Freezing*	*Freezing and UV*
Exp I	Roots	60.1	21.7	28.0	17.0	13.6
	Shoots	28.1	8.44	17.9	8.6	5.7
Exp II	Roots	52.0	11.4	16.8	12.9	**
	Shoots	33.1	4.6	16.7	12.3	**

** - *small, unable to measure*

In general, stress factor such as low temperature might affect the reactions of ion secretion and growth regulation (Figure 8.3). Stress induced the ion secretion as the transport through cellular barriers like tonoplast and plasmalemma. (Arshavski et al., 1992).

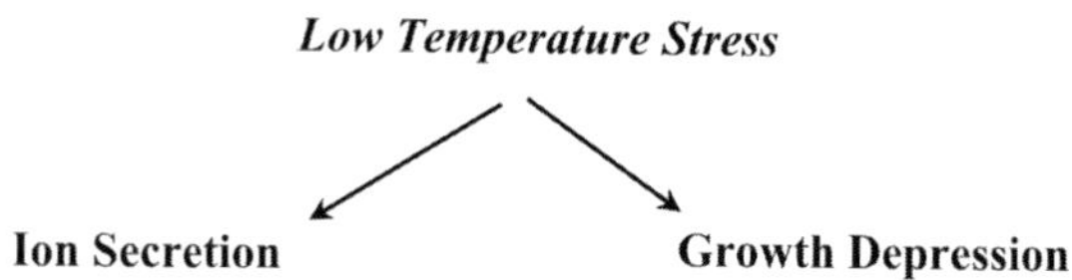

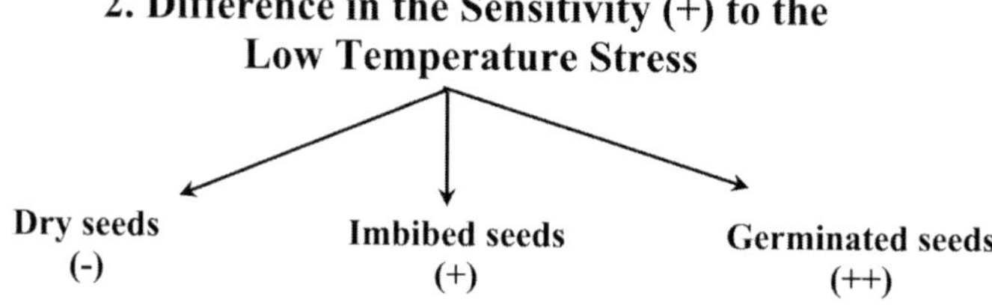

Figure 8.3 Pathways of cold stress

In our review (Kefeli et al., 1989) we described the increase of ABA and other growth inhibitors during such stress conditions. For example, water deficiency was accompanied by closure of stomata. Lehman and Vlasov, 1985 showed the change of ABA and its glucose ester during different types of stress conditions (Table 8.6)

Table 8.6 Tomato plants under stress conditions

Type of stress	*Length of Leaves (%)*	*ABA*		*ABA-glucoside*	
		ng/plant part			
		Leaves	*Roots*	*Leaves*	*Roots*
Control, water	100	1.48	0.54	50.7	46.4
Wilting	8	8.02	10.3	48.7	45.0
Heat shock	14	1.40	0.50	50.6	41.6
Cold stress	60	1.32	0.52	50.5	41.6

These data demonstrated that ABA in free form increased remarkably in leaves and roots during water deficiency but ABA - glycoside showed no change. The other types of investigated stress didn't affect the ABA level. High temperature induced stress proteins (heat shock) formation in the anucleated acatabularia cell (Kof, Kefeli, 1994). Thus stress effect nucleus-cytoplasm relationships. The role of inhibitors in this case is unclear.

8.3 Light as a factor of ontogeny and stress.

8.3.1 Exogenous and endogenous systems of photomorphogenesis and phototropism in plants.

Light is one of the most powerful factors regulating life processes in a plant organism during its entire ontogenesis. A mature seed and a seedling already possess systems, which are able to respond to light of various intensities and quality (photomorphogenesis). A seed germinates and the seedling is oriented towards a light source (phototropism). Autotrophic nutrition organs start to form (photosynthesis). Generative organs are initiated in the stem apices of a growing plant. As it is well known, many plant species are sensitive to the day length and respond to its changes by accelerating or retarding, and sometimes even by completely inhibiting the formation of regenerative organs (photoperiodism). It would be wrong to assume that photomorphogenesis, phototropism, photosynthesis and photoperiodism are just processes that follow one another in the ontogenesis of a plant and are not interrelated in any way. Most frequently, these processes are concomitant and interrelated. For this reason, singling out each particular process and defining its specific features involves a number of serious difficulties.

8.3.2 Photoreceptors

One of the very first stages in the effect of light on a biological model is the interaction of light with a photoreceptor localized in this model. This area in studying the light effect appears to be less known at present. The most general pattern of such an effect of tight may be schematically represented as follows:

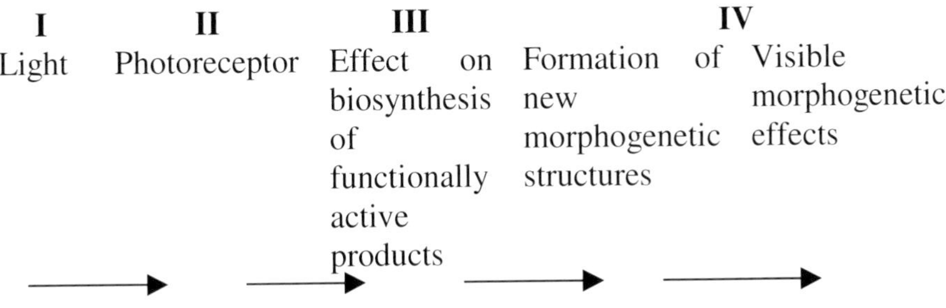

The photomorphogenetic effect of light is realized through the phytochrome-chromoproteid occurring in plants in two forms, namely, P660 and P 730 which can be converted into each other under light. The P 730 form is a physiologically active part of the phytochrome system, a sort of an effector molecule which controls morphogenetic processes in a growing plant. Apart from the phytochrome, a plant may contain some other photoreceptors, for example, those involved in the accomplishment of the action of the high-energy blue- far red light system. However, this problem has remained highly controversial up to now.

The phototropic response differs from photomorphogenesis in a number of characteristics. In the first place the difference between these photoprocesses is different photoreceptors. Unlike the photomorphogenetic receptor, the phototropic photoreceptor is sensitive to visible short-wavelength light. The active spectra of phototropism obtained for different plant objects are extremely similar. It is quite probable that this active spectrum points to the participation of flavoproteins (or flavins) acting as photoreceptors. Carotenoids constitute another group of substances which may act as photoreceptors in phototropism. The data describing the active spectrum of action of phototropism with maximum wavelength in the ultraviolet and blue spectrum regions have now become textbook material. Flavins are distributed equally in the cells of all tissues, while carotenoids are concentrated in some isolated sites. At the same time some carotene-deficient mutants, that contain a small amount of these pigments, still preserve their phototropic response. A hypothesis has been put forth about a double-pigment system of phototropism in which carotenoid acts as a leading photoreceptor (like chlorophyll in photosynthesis), and flavin is assigned the role of an associated pigment (like carotenoid in photosynthesis). Riboflavin absorbs energy in the UV region and transmits it to carotenoid. It is likely that carotenoid acts as a screening pigment, while riboflavin which is a photochemically active pigment (vitamin) is directly involved in the control of the auxin level. The fact that the photoreceptor system of phototropism controls the level of phytohormones involved in the curvature of a plant organ is now practically unquestionable (Table 8.7).

Although for most plants a phototropically active region is the blue light region, maize coleoptiles exhibit a predominant response to red light.

Table 8.7 Characteristics of processes of photomorphogenesis and phototropism

Characteristics	*Photomorphogenesis*		
	Low-energy reaction	*High-energy reaction*	*Phototropism*
Active spectra region; light, Spectrum (maxima) nm; Photoreceptor; Duration manifestation of photo effect (morphogenic effect	Red – far red; 660-730 Phytochrome; Short, min. Inhibition of hypocotil elongation, increase of cotyledons, differentiation of primary leaves, development of negative geotomic reaction in stem	Blue-far red 420-480; Phytochrome (?); Long, hours; Synthesis of anthocyans, inhibition of hypocotyl growth	Blue 340-470; Flavins, carotenoids; Short, min; Predominant cell extension on one side of photosensitive organ causing its curvature

Summing up the data available on the photoreceptors of photomorphogenesis and phototropism, it can be concluded that they are two independent photo biological systems with distinctly pronounced properties.

8.3.3 Light and phototropism

Whereas the problem of the photoreceptor for photomorphogenesis may be considered largely solved, it is only being stated for phototropism. At the same time, for photomorphogenesis the problem of a hormonal or inhibitor mediator is still highly controversial. While for phototropism this problem had practically been solved a long time ago, and it is only the details that are being tackled by the researchers now (Armin, Deng, 1996).

As it is known, the phototropic reaction of a plant is its asymmetric growth or curvature under the effect of unilateral light. The spectral region most active for phototropism is a blue light region in which carotenoids and flavins absorbed. Auxins are considered to be the effector system of phototropism.

A 10 sec. Exposure to UV light (254 nm) is sufficient for the oat coleoptile tip to make the first positive curvature towards the source of light; 10 min irradiation with blue light (436 nm) causes the second positive curvature - at the base of the coleoptile. These curvatures occur due to a lateral auxin transport which was demonstrated in classical and brilliant Thimann's experiments with ^{14}C-IAA applied to the tip of a coleoptile from an agar block. If the tip of the coleoptile was intact, polarization of the auxin transports was most markedly pronounced: 72-75 percent of the auxin supplied to the coleoptile was transported toward the dark side, while 28-25 percent remained in the light-exposed part. It is interesting that practically no

^{14}C-IAA was decomposed in the course of the experiment (3 hours). The role of the tip of the coleoptile in the distribution of auxin is obvious: with a tip the distribution of auxin between the light-exposed and dark parts was 35:65, while without a tip, this difference became smaller - 46:54 (Thimann, 1965, see the review of Kefeli, 1975).

So, the effector functions in phototropism are performed by auxin which was first shown by Cholodny and Went and later confirmed by Thimann. However, it would not be fair not to mention the objections that were raised against the auxin transport hypothesis. Some researchers while using high doses of ^{14}C-IAA did not observe any differences in its distribution in phototropism. Others related phototropism to changes in the activity of IAA-oxidase on two sides (light-exposed and dark) of a phototropically curving seedling. Finally, the third group of researchers gave the leading role not to the auxin, but to the inhibitor which synthesis is strongly light-induced, and which suppresses cell extension on the light-exposed side of the coleoptile (Galston, 1961).

There is also a version which seems to reconcile both hypotheses: the auxin and the inhibitor ones. According to this version (Naqui & Engvild, 1974), light causes partial decomposition of violaxanthin, one of the possible phototropism receptors, with formation of abscisic acid (ABA). The ABA produced inhibits the basipetal transport of auxin on the light-exposed side which results in an asymmetrical growth and positive phototropic curvature of the coleoptile. The phenolic inhibitor is assigned the role of an auxin inactivator in this process. This inhibitor might also act through the OIAA (oxidase indole-acetic acid) system as was suggested by Engelsma & Mayer (1965). Direct evidence was needed to verify the validity of the Cholodny and Went hypothesis and to demonstrate experimentally that in phototropism auxin is delivered from the apex predominantly in the lateral rather than in the basipetal direction. In other words, it was necessary to corroborate the results obtained by Thimann and to repeat the experiments in accordance with an extended program. This is exactly what was done by Shen-Miller (Shen et al., 1969), who irradiated etiolated oat seedlings with unilaterally directed blue and white light. Then, agar blocks with ^{14}C-IAA were immediately placed on the seedlings, in two hours the blocks were removed and the ^{14}C-IAA content was estimated in various parts of the coleoptile. It was found that light inhibited the basipetal transport of ^{14}C-IAA, the strongest photoinhibition being observed at the wavelengths of 350-480 nm. The maximum curvature of the coleoptile occurred also in this spectral region. Thimann also observed inhibition of the basipetal transport and translocation of ^{14}C-IAA in the lateral direction. These data helped solve the problem of a phototropism effector which is now part of the material presented in manuals and textbooks.

Recently, considerable attention has been paid to xanthoxin and abscisic acid as possible factors involved in regulation of the phototropism. By analogy with geotropism data presented by Wain (1977), it may be thought that the following chain of reactions occurs on the light-exposed side of a seedling with inhibited growth: violaxanthin under the effect of light is converted into xanthoxin and the latter changes into ABA which inhibits the cell-extension process.

Thus, the regulatory effect of light on the growth of plants can in general be defined by means of two processes - photomorphogenesis and phototropism. These two processes are similar in that they have common mechanisms of photoreception, the effector system, growth and photomorphogenetic changes. However, the receptors and effectors are different for phototropism and for photomorphogenesis. The schematic representation of the elements of the photomorphogenesis and phototropism systems in higher plants is given below.

	Photomorphogenesis	Phototropism
Nature of illumination	Diffuse (from all sides)	Unilateral
Active spectral region	Red – far red	Blue
Photoreceptors	Phytochrome	Caratenoids, flavins
Probable primary effector	Nucleic acida S (potentially active genes)	Auxins (lateral transport) or natural inhibitors
Change in seedling stem	Inhibition of elongation	Promotion of one-side growth
Secondary changes	Formation of stomata Vigorous leaf growth	Two –stage curvature (base and tip)

It should be pointed out that a number of aspects of the mechanisms of photomorphogenesis and phototropism still remain vague to the researchers. Thus, for example, the effectors of photomorphogenesis have not yet been studied adequately, the sequence of phytochrome-induced hormonal transformations is not quite clear; we do not know in-depth the photoreceptors of phototropism. Awaiting its solution is also the problem of interaction of the photomorphogenesis and phototropism processes both at the photoreception level, and at the level of the effector transmission of a light stimulus.

The role of phytohormones and natural inhibitors in the regulation of root growth has been studied much less profoundly than in case of stem growth. Thus, light quality, periodicity and orientation could be strong factors of development. The light intensity is also a very strong factor which could be a stress effector as it was shown by Kof and collaborators (1993).

8.3.4 Inhibiting effects of light intensity

There are many publications devoted to light as a stress factor for plants (Kefeli, Loznikova 1998; Kefeli et al., 1998). The response of leafletless and normal *Pisum sativum L.* phenotypes to changes in light intensity was studied according to ideas by Emil Nalborchick (Poland, personal discussion). The projection area of a leafless pea plant is reduced to a fraction of that observed for a normal pea plant. An increase in light intensity shifts this parameter insignificantly. Increasing light intensity inhibits the elongation of stems and tendrils and changes the total surface areas of stipules and especially leaflets. Increases in the specific leaf weight of leaflets and stipules correlate with the inhibition of stem elongation, which is determined by light intensity and plant genotype. Tall pea cultivars have the maximum daily dry weight gain irrespective of leaf type. In normal plants, all the plant organs contribute to biomass gain, whereas in leafless plants, biomass accumulation occurs predominantly due to stem and tendril. The light dependent curves of daily dry weight gain are not associated with the leaf type but are closely related to the height and genetic nature of the plant. Two types of cultivars were distinguished according to their sensitivity to radiation (Kof et al., 1993)

The plants were grown in special, the climatic chambers. Plants were developed in darkness or were irradiated by DKST-20 water-jacketed xenon bulbs (50, 230, and 350 W/m^2 PAR) with a 16 -h photoperiod and at a temperature of 18-22°C. Etiolated plants were grown under light-proof covers. At least 35 plants were used for each treatment (Protasova, Kefeli, 1982; Protasova et al., 1980)

During the experiments, the dynamics of stem elongation, total surface area of leaflets and stipules per plant, the dry weight of leaflets, stipules, and stems, and the SLW of leaflets and stipules were determined. The SLW was defined by the equation: SLW = m/s, where m is the weight of a leaflet or stipule (g) and s is the surface area of a leaflet or stipule (cm^2). The surface areas of leaflets or stipules were estimated by the weight method. In addition, we calculated the projection area (s) by measuring the distance from the stem to the uppermost point of a pair of leaflets for normal varieties or stipules for tendril varieties. This distance was taken as the radius of the circle (r) to calculate the areas of leaflet or stipule projections for one plant by the equation: $s = \eta r^2$. The projection area was also calculated by measuring a straightened tendril length taken for the radius. Both studies included leafletless vs. completely lacked leaflets. Instead of leaflets they formed well developed multibranched and twisted tendrils. At the same time these plants developed expanded stipules tightly embracing the stem. Thus, the projection area of the stipules, i.e., the area shaded by one 10-day-old leafletless plant grown at 230 W/m^2 PAR (irrespective of plan height), was several time smaller than the projector area of a normal plant (Figure 8.4a). This was true for 22-day-old and 40-day-old plants as well.

In accordance with the calculations of plant projection areas based on measuring leaf lengths up to the tips of straightened tendrils, the projection area was much larger in tendril varieties (Figure 8.4b). But because of the pronounced twisting of the tendrils, more compact canopy, and its high light-transmitting properties, the area shaded by one tendril variety plant was considerably smaller than that of a leaflet variety (normal) plant. Increased irradiation resulted in every case in decreased plant projection area calculated by summarizing leaf and tendril length (Figure 8.4b). However, the area shaded by the leaflets of normal varieties or by stipules of tendril varieties was not significantly influenced by the irradiation. (Figure 8.4a). The negligible decrease of the area shaded by one plant was observed (Figure 8.4a) only at high PAR (350 W/m^2).

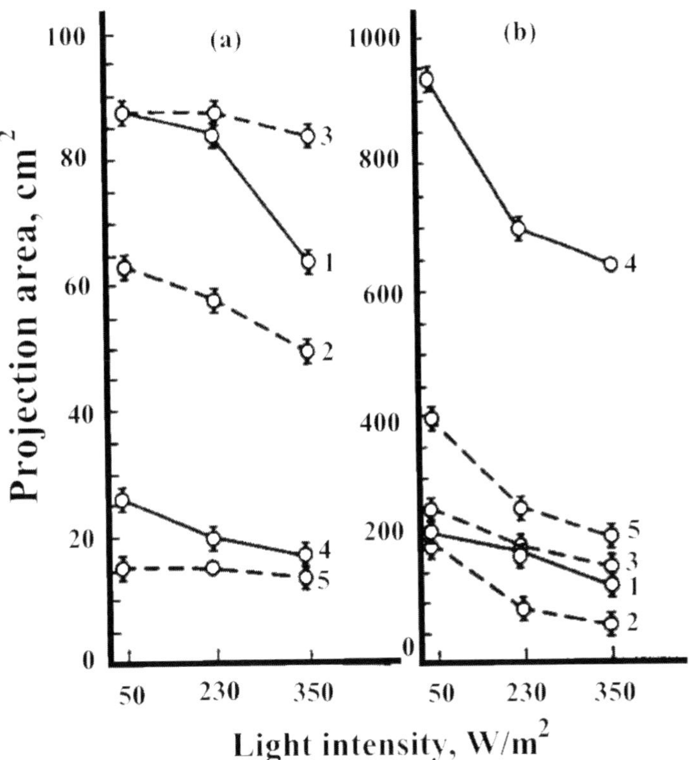

Figure 8.4 Effect of light intensity on the projection area of leaflets and stipules (a) or tendrils (b) in leaflet and tendril varieties of pea. 1 – Streletskii-31; 2 – Smaragd; 3 – Orlovchanin; 4 – Sprut; 5 – Nord.

8.3.5 Plant Height

Increased light intensity inhibited stem elongation in 5-day-old plants of all studied cvs. irrespective of their height and leaf type (Table 8.8). The inhibition of stem elongation was also observed for 10-, 2O-, and 4O day-old plants, i.e., for plants at the phases of 11 to 12, 15 to 16, and 20 to 21 nodes, respectively (Table 8.8)

Table 8.8 Characteristics of pea cultivars used in the experiments

Type of leaf	*Type of stem*	*Cultivar*
Leaflet (normal)	Tall	Streletskii-31
Leaflet (normal)	Short	Smaragd
Leaflet (normal)	Short	Orlovchanin
Leaflet	Tall	Sprut
Leaflet	Short	Nord

8.3.6 Surface Area of Leaflets and Stipules

The total surface area of leaflets and stipules was also influenced by light intensity. It should be noted that stipules in tendril varieties were greater; thus, their total surface area was much larger than that of the stipules in normal leaflet varieties (Figure 8.6). At the same time, the total area of stipules and leaflets in normal plants exceeded the total area of stipules in leafless varieties. Apparently, the greatly expanded stipules that considerably increased the photosynthesizing surface of such plants partially compensate for the lack of leaflets in tendril varieties (Figure 8.6)

The effect of light intensity on the surface area size of photosynthesizing organs (leaflets and stipules) varied depending on leaf type. In normal leaf cvs. The total leaflet surface area per plant and the summarized area of leaflets and stipules per plant increased considering the increase in irradiation from 50 to 230 W/m^2 and declined upon raising the light from 230 to 350 WIm2. Only the 10-day-old plants of tall-stem leaflet cv. Streletskii-31 and short-stem leaflet cv. Smaragd show no decline in the total area of stipules upon increasing irradiation to 350 W/m^2. Later (40-days-old plants), the pattern of light effect on the surface area of stipules and leaflets was similar for all cultivars (cv), though it was not so distinct for cv. Smaragd. (Figure 8.7b). It should be emphasized that the leaflet surface area changed to a greater extent than the stipule area. This may be a consequence of the angle at which light reached the plant: light falls at a right angle on the leaflets but at oblique angles on the stipules (Table 8.9).

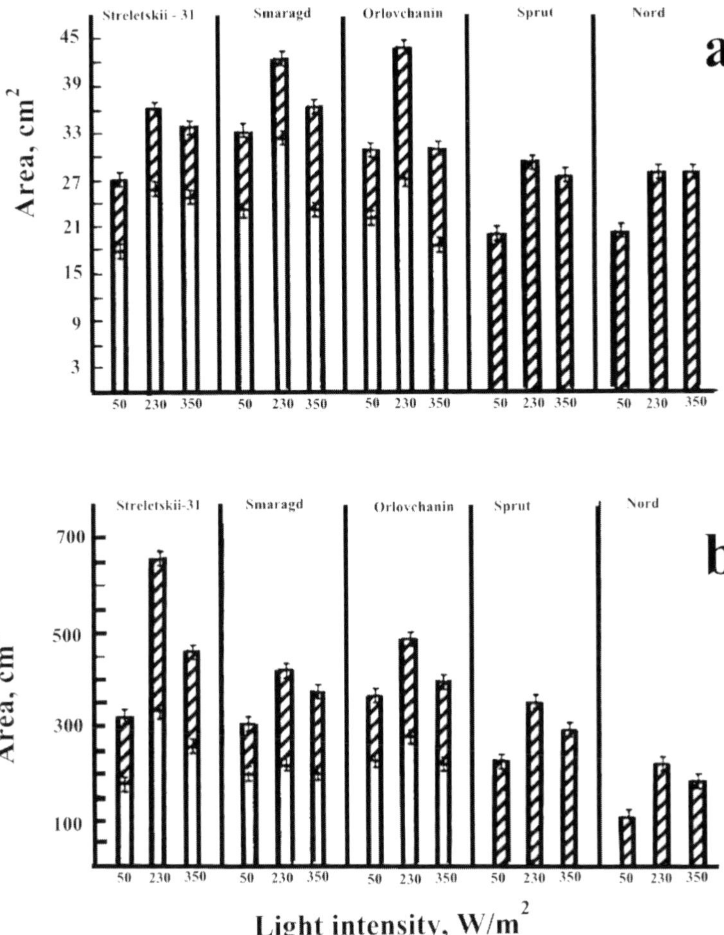

Figure 8.5 Effect of light intensity on the total surface area of leaflets (solid bars) and stipules (hatched bars) in 10-day-old(a) and 40-day-old (b)plants of different pea cultivars.

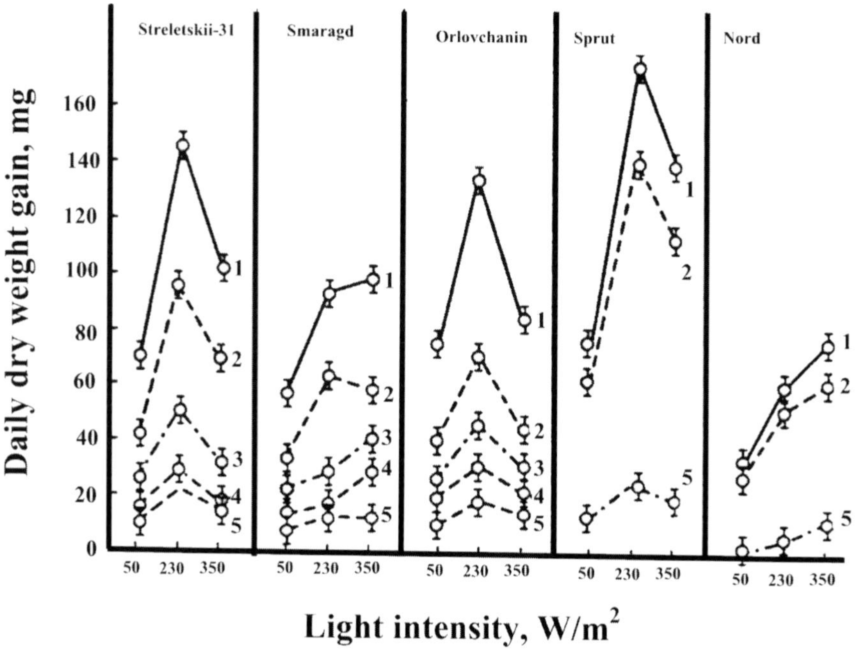

Figure 8.6 The dependence of daily dry weight gain on light intensity in different pea cultivars. 1 – total dry weight of the pea plants; 2 – dry weight of the stem; 3 – dry weight of leaflets and stipules; 4 – dry weight of leaflets; 5 – dry weight of stipules.

A much different pattern was observed for leafletless plants. Increasing irradiance from 50 to 230 W/m^2 enlarged the total surface area of stipules in 40-day-old plants (Figure 8.9). The subsequent increase of PAR to 350 W/m^2 had no appreciable effect on total stipule surface area per plant, which was the most obvious for the short-stem cv. Nord. This is also apparently affected by light angle. The total surface area of stipules in 40-days-old plants rose as the irradiance increased from 50 to 230 W/m^2 (Figure 8.6b) and then declined upon subsequent increase of light intensity from 230 to 350 W/m^2. Thus, light similarly affected the total photosynthesizing surface area of stipules in tendril varieties and stipules and leaflets in leaflet varieties in 40-days-old plants.

Effect of increasing light intensities on stem height in different pea cultivars expressed in Table 8.9.

Table 8.9 Effect of increasing light intensities on stem height in different pea cultivars.

Age, day	*Number of nodes*	*Stem height (cm), light intensity (W/m² PAR)*			
		0	50	230	350
		Streletskii-31			
5	5-6	25.6±0.1	8.6±0.7	5.1±0.5	3.3±0.2
10	11-12		26.7±1.8	15.5±1.3	9.3±0.4
20	15-16		52.9±1.5	29.1±1.0	22.3±0.8
40	20-21		145.7±1.9	96.2±2.1	89.6±1.5
		Smaragd			
5	5-6	18.0±0.3	4.2±0.3	3.7±0.1	2.7±0.2
10	11-12		12.4±0.7	9.5±0.2	7.4±0.3
20	15-16		29.9±0.9	17.3±0.9	13.5±0.3
40	20-21		77.6±1.3	64.0±1.6	54.0±1.7
		Orlovchanin			
5	5-6	11.9±0.1	5.3±0.1	4.16±0.1	3.2±0.2
10	11-12		14.0±0.5	10.0±0.1	7.3±0.3
20	15-16		26.0±0.8	20.6±1.3	14.7±0.4
40	20-21		82.4±1.3	81.4±1.3	61.0±0.6
		Sprut			
5	5-6	19.5±0.1	8.4±0.4	6.5±0.2	5.0±0.2
10	11-12		29.0±0.1	20.8±0.4	14.8±0.4
20	15-16		56.4±0.7	45.9±1.3	31.3±0.9
40	20-21		143.0±1.3	137.0±1.4	111.0±1.1
		Nord			
5	5-6	19.5±0.1	4.8±0.3	4.4±0.1	4.0±0.2
10	11-12		15.5±0.4	13.9±1.1	11.6±0.2
20	15-16		26.5±1.9	24.6±0.5	21.6±0.3
40	20-21		68.0±1.7	64.6±1.8	60.1±1.2

8.3.7 Specific Leaf Weight (SLW)

The short-stem cys. had higher values of SLW for leaflets and stipules (Tables 8.10 and 8.11).

Table 8.10 Effect of light intensity on SLW of leaflets in 10-day-old normal pea plant

Cv.	*SLW of leaflets (g/cm² x 10⁻³); light intensity (W/m² PAR)*		
	50	230	350
Streletskii-31*	9.56±0.86	12.19±0.14	12.57±0.10
Smaragd	10.85±0.34	12.80±0.26	14.05±0.14
Orlovchanin	13.53±0.17	14.80±0.18	15.62±0.37

*-pea variety

The increase of light intensity not only inhibited stem elongation but also enhanced the SLW values for the leaflets and stipules.

Table 8.11 Effect of light intensity on SLW of stipules in 10-day-old normal pea plant

Cv.	*SLW of stipules (g/cm^2 x 10^{-3}); light intensity (W/m^2 PAR)*		
	50	230	350
Streletskii-31	8.03±0.12	10.40±0.20	11.90±0.17
Smaragd	9.36±0.16	11.20±0.16	12.20±0.15
Orlovchanin	10.69±0.18	13.00±0.12	13.91±.0.61
Sprut	8.73±0.21	11.30±0.14	12.76±0.62
Nord	9.06±0.11	11.75±0.30	13.50±0.13

8.3.8 Daily Dry Weight Gain

Influenced by light intensity, the SLW and surface areas of photosynthesizing organs (leaflets and stipules) can effect the increase of plant biomass. The highest daily gain of dry weight was observed in the tall-stem leafletless cv. Sprut (Figure 8.7).

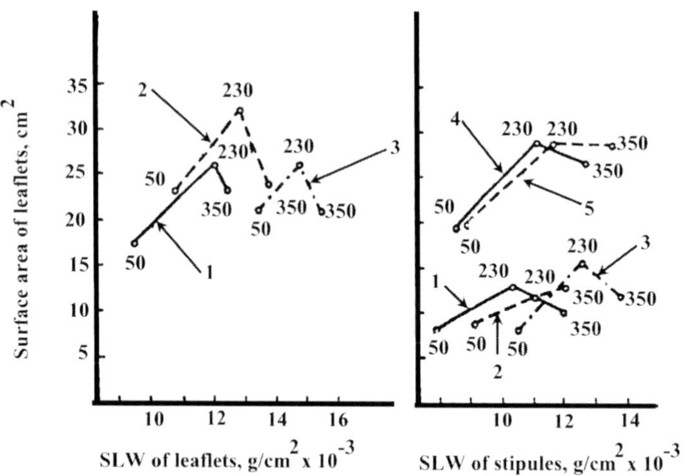

Figure 8.7 The correlation between SLW and photosynthesizing surface area of leaflets (a) and stipules (b) in leaflet (1,2,3) and tendril (4,5) pea varieties at different PAR intensities (designated in the figure). Designation of the varieties are as in Figure 8.6

The tall-stem leaflet Streletskii-31 occupied the intermediate position between the short-stem and Sprut cvs. The largest daily dry weight gain among short-stem plants was observed for cv. Orlovchanin. The main difference between normal and tendril plants is that in leaflet plants, the biomass gain proceeds because leaflet, stipule, stem, and tendril tissues accumulated (Figure 8.8), while dry weight gain in tendril cys. is mainly due to increasing the stem and tendril tissues (Figure 8.8).

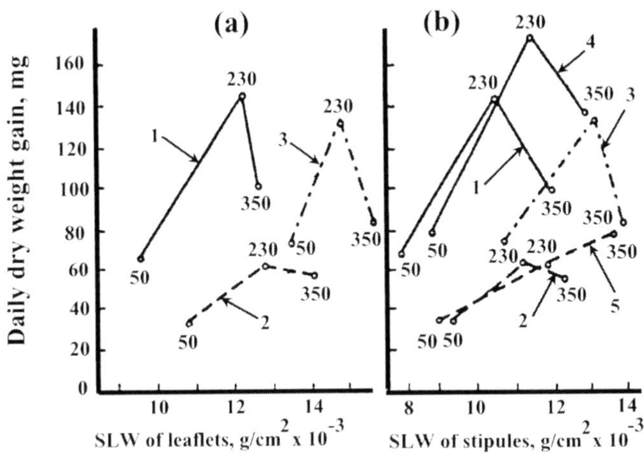

Figure 8.8 The correlation between daily dry weight gain in normal (1,2,3) and tendril (4,5) pea varieties and SLW of leaflets (a) and stipules (b) at different PAR intensities (designated in the figure). Designation of the varieties are as in Figure 8.6

This study of the effect of light on dry weight accumulation shows that under conditions of increasing light intensity from 50 to 230 W/m^2 PAR, the daily gain of dry weight per plant increased in cvs. Streletskii-31 (tall-stem, leaflet), Orlovchanin (short-stem, leaflet), and Sprut (tall-stem, tendril). The subsequent increase of light intensity to 350 W/m^2 lowered this index. At the same time, at a PAR intensity of 350 W/m^2 this maximum rate of dry weight accumulation in two short-stem cultivars, Smaragd (1eaflet) and Noid (tendril), was not reached.

This comparative study of the growth and morphogenetic processes that occur in pea genotypes characterized by contrasting leaf types revealed peculiarities in leafletless pea plants. First, it should be noted that the expanded stipules of young leafletlet plants partially compensate for the lack of leaflets. This is supported not only by data on the total surface area of leaflets or stipules but also by data on the ratio between the total surface area and the SLW for leaflet and for stipules at different PAR intensities. (Figure 8.7). The total surface area and SLW ratio for stipules in tendril varieties and for leaflets (but not stipules) in normal varieties had similar values. Later, when the spreading tendrils in leafletless plants acquired the function of photosynthesizing organs, the compensatory role of stipules in tendril varieties declined.

Furthermore, the differences between tendril and normal varieties manifested themselves in the drastic reduction of the tendril (af af genotype) plant projection area compared with that of normal (Af Af genotype) plants. This index is independent of plant height, which is genetically determined, and is only slightly sensitive to radiation. On the contrary, the highest values for the projection area calculated according to the length of the whole leaf together with its straightened tendrils were obtained for leafletless varieties with highly developed tendrils. The significant reduction of the plant projection area in all leaf type varieties in response to increasing light intensity indicates the high light sensitivity of the tendril elongation process.

Stem elongation in both tendril and normal pea varieties is also very sensitive to radiation. The extent of its inhibition increased with increasing PAR (Photosynthetically Active Radiation) intensity. The inhibition of stem elongation dependened on genotype or light intensity is not closely related to the surface area changes of any photosynthesizing organs but is accompanied by an increase in leaflet and stipule SLW (Specific Leaf Weight). In addition, the stipule SLW of leafless varieties is more sensitive to irradiance than the stipule SLW of normal varieties. This is especially interesting in connection with the idea that SLW is an integral tool for controlling plant architectonics and providing the plant with boundary surfaces to improve the external contacts for the best utilization of incident PAR. Moreover, these experiments show that the total leaf surface area and light distribution in the crop stand can be controlled by changing the SLW, which is tightly connected with PAR intensity and its profile.

According to the principle of maximum production, deduced mathematically and proven experimentally, photosynthesis and its energy equivalent. i.e., the specific growth rate (daily weight gain), are functions of the photosynthesizing surface area and plant organ thickness (SLW), allowing for maximum and optimum exchange with the environment.

Is this true for pea plants with different leaf types? Does this principle persist under varying PAR conditions, i.e., are the architectonics of pea plants with different leaf types adoptively expedient? The term adaptation refers to the ability to ensure maximum production under given environmental conditions or the ability to maintain maximum efficiency (daily weight gain), depending on the optimum interrelation between photosynthesis and respiration.

The maximum daily weight gain is characteristic of tall plants and is not associated with leaf type. In other words, the recessive mutant gene (*af* - *af* genotype) does not affect the biosynthetic potential of pea plants but changes the relative contribution of different organs and tissues to the daily weight gain. In leaflefless varieties, the weight gain proceeds mainly at the expense of stem and tendril tissues, but in normal plants, all plant organs contribute to it.

Two groups of cvs. that respond to light independently of leaf type were revealed. In the first group (Streletskii-31, Orlovchanin, and Sprut), the maximum daily weight gain was observed at medium PAR intensities (230 W/m^2); in the second group (Smaragd and Nord), the maximum weight was not attained at a high PAR intensity (350 W/m^2). This separation was supported by studying the light response curves for the dependency of daily weight gain on SLW (Figure 8.10a) and on the total surface area of leaflets and stipules (normal varieties) or just stipules (tendril varieties) (Figure 8.10b).

The daily dry weight gain in respect to the total surface area or to the SLW of leaflets and stipules was similar at equal PAR intensifies for all varieties within each group. Differences between the two groups of cvs. were also observed in the light curves for the dependence of total surface area on SLW of leaflets and stipules (Figure 8.9).

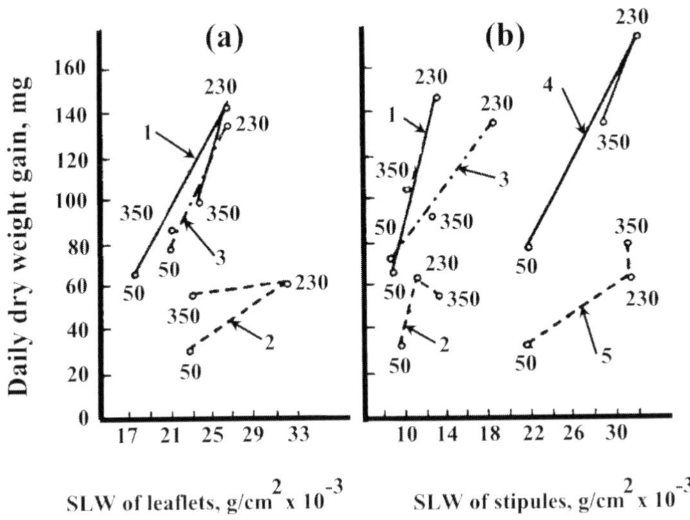

Figure 8.9 The correlation between daily dry weight gain in normal (1,2,3) and tendril (4,5) pea varieties and surface area of leaflets (a) and stipules (b) at different PAR intensities (designated in the figure). Designation of the varieties are as in Figure 8.6

These differences in all the cases were most pronounced for the stipules. Apparently, the specifics of the second group of cvs. (Smaragd, Nord) are genetic, supported by the lowest dry weight gain in these varieties.

It is quite evident that the sensitivity of dry weight gain to light intensity is independent of leaf type, which is determined by the recessive or dominant condition of the *Af* gene. However, this sensitivity is connected with plant height, which is also controlled by plant genotype; a similar pattern was observed earlier for stem elongation in plants differing in their height parameter.

Light intensity appears to affect various organs controlling pea plant productivity differently. This encourages studies of processes connected with photosynthesis in normal and leafless phenotypes. Lafond and Evans demonstrated that the activity of Rubisco, the key enzyme of photosynthesis; calculated on a weight basis, was almost equal in leaflets and tendrils. When calculated on a chlorophyll unit basis, the values for tendrils were much higher than for leaflets because the content of chlorophyll in tendrils is half the amount in leaves. At the same time, the Rubisco content amounts to about 50% of the soluble protein in green leaves. Because of the almost equal content of soluble proteins in tendrils and leaflets, Lafond and Evans concluded that $C0_2$ fixation in tendrils is more effective than in leaflets. Harvey and Goodwin made the same conclusion (see Kof et al., 1993). They showed that $C0_2$ fixation in tendrils was considerably higher when calculated per photosynthesizing surface unit The high level of tendril-plant canopy light-transmittance, especially during late periods of the growth, can provide more efficient photosynthetic utilization of light energy by the lower green mass (storeys) of the plants.

The high values of daily weight gain in tendril varieties can be explained by the photosynthetic activity of both expanded stipules and tendrils as well as by the high efficiency of light utilization in high and low plant storeys. The decrease in daily weight gain at increasing light intensity to 350 W/m^2 for most of the varieties is most likely explained by inhibited stem elongation, reduced total leaflet and stipule surface, and shortened tendrils, whose contribution to the photosynthetic potential of *afaf* plants is quite high.

The consistency of the relative rate of dry weight gain in pea varieties with different leaf types under similar conditions of PAR supports the idea of self-regulating formation of boundary surface structures, allowing for optimum exchange with the environment and maximum utilization of the incident light energy.

Our previous experiments with the high light intensity (Protasova et al, 1980, Protasova Kefeli, 1982,) showed that pea plants accumulate p-coumarate and quercetin-glucosilcoumarate during the stem inhibition.

Similar growth depression during UV-B light was accompanied by hormonal changes (Table 8.12).

Table 8.12 Effect of light intensity on stem length and quercetinglucosyl-coumarate content in pea plants.

Irradiance, erg(cm^{2}s)x10^3	*Plant height, cm*	*Quercetinglucosyl coumarate, mg/g dry wt*
0	42	2
25	25	7
200	13	20
300	8	26
420	5	36

As we assured ourselves, dwarf growth of the mutants or in plants exposed to high-intensity light resulted in the promotion of p-coumaric acid synthesis, which. in peas was found as glucoside or a QGC acyl compound.

p-Coumaric acid itself, as a natural hydroxylated substance, can activate IAA oxidase and IAA degradation will occur as the result. At the same time, the synthesis of IAA derivatives is effectively hindered in the presence of p-coumaric acid, and IAA level in the tissue is reduced. In addition. p-coumaric acid is able to suppress the gibberellin action, thus resulting in stem growth inhibition.

p-Coumaric acid accumulation as QGC may also be attributed to the fact that it represents the C_6-C_3-fragment of the lignin molecule. However, in response to external (light) or internal (mutation) factors. it is not utilized in lignification and exists in a mobile equilibrium between QGC and p-Coumaric acid.

Etiolation as the reaction of the plant on light deficiency is one of the main stress factor of plant. We investigated the process of de-etiolation in respect to hormonal-inhibitor relationships (Kefeli et al, 1993)

8.3.9 Growth regulators and light deficiency

The growth of seedling during de-etiolation depends on the intensity and spectral composition of the light. The participation of the phytochrome system in this process is confirmed by the reversibility of the effect and by the negative correlation between the growth rate and the content of the physiologically active form of phytochrome. It has been suggested that phytochrome control of growth processes is implemented with the participation of phytohormones, especially ethylene.

In certain cases exogenous ethylene can stimulate the action of light. For example in etiolated pea seedlings both light and ethylene slow down the elongation of the internodes. In other cases light and ethylene induce opposite reactions. Thus light induces an opening of the apical loop in pea seedlings; while exogenous ethylene blocks this process. The formation of endogenous ethylene depends on the conditions of illumination. In isolated green leaves light suppresses its evolution but at increased concentrations of CO_2 it stimulates this process.

Some data indicate that intact green plants formed virtually the same amount of ethylene both in light and in the darkness. In etiolated seedlings light either suppressed or, in a number of cases, enhanced the liberation of ethylene. Such contradictory experimental data maintain the interest in the problem of participation of ethylene in the photo- dependent processes. The present work is devoted to the question of the participation of ethylene in the photo-induced change of growth in pea seedlings grown under various conditions of illumination.

In the experiments we used seven -day pea seedlings (*Pisum sativum* L.) of the Konservnyi variety, grown in tap water under white light fluorescent lamps with a 16-h photoperiod and illumination 8 W/m^2 or in darkness. Beginning with the fourth day, the etiolated plants were irradiated daily for 5 mim with a red Philips lamp (France) through an interference filter with 650±5 nm. The illumination was 8 W/m^2. All the experiments were conducted at 24^0C.

Ethylene evolution by whole plants was determined by incubation plants for 30-60 min in hermetic glass chambers with a volume of 800 ml. The experiment was conducted 1 h after the last irradiation with red light. Green plants were incubated in white light ,etiolated plants and plants irradiated with red light were incubated in darkness.

The dynamics of the evolution of ethylene by the leaves and portions of the stem (third internode, including the apex and the loop) was determined after a single irradiation or whole etiolated seedlings with red light. After various time periods following irradiation, the leaves and stem were cut off and incubated in darkness in 15 ml glass chambers with tightly closed lids. To avoid distortions of the results on account of the formation of wound ethylene, the site of the cut was rinsed with distilled water and ethylene was determined no later that 30 min after the cutting. Arabidopsis plants were treated with UV-B light (280-320 nm), 6-8 W/m^2 and then were exposed to the white light. Ethylene was analyzed by gas-chromatograph with a flame-ionization It is known that the evolution of ethylene depends on the $C0_2$ and 0_2 concentration , therefore the level or these gases was monitored during the experiment. For this purpose. at the end of the exposure 1 ml of air was collected from each chamber with a syringe and analyzed by gas chromatography.

The CO_2 concentration in the vessels at the end of the experiment was 0.5-1.2 %. The isolation and purification of IAA from plant tissues was done by extraction and liquid chromatography. Indolepropionic acid (IPA) was added to the plant tissue before homogenization as an internal standard. The purified samples were methylated with diazomethane, and IAA was determined quantitatively by Hitachi M-70 chromato-mass spectrometer (Japan) in a system of detection according to three molecular weights: 130- indole; 180 - methyl ester of IAA, and 203-rnethyl ester of IAA. The column was 2m \ 3 mm, 3% 0V-l 7 on Gazchrom Q (80-100 mesh). The carrier gas was helium, velocity 40 ml/min, and the column temperature was programmed from 180 to 250^0 C, 10^0 C/min.

The tables and figures present the statistical means of three to four biological repetitions and their standard deviations. The analytical repetition was two to three times.

Seven-day seedlings grown under various conditions of illumination. differed substantially morphologically and produced ethylene differently (Table 8.13). The stem length of the green seedling growth on a 16-h photoperiod was one fifth as great and the weight of the leaves 5.5 times as great as in the etiolated seedlings. Even a 5 min daily red-light illumination of plants grown in darkness induced almost 30 % inhibition of stem growth and a threefold increase in weight of the leaves in comparison with non irradiated plants i.e., led to the change in growth processes characteristic of de-etiolation. Moreover, the production of ethylene by whole plants was increased by 27-28 % in comparison with the dark variant. The production of ethylene by green plants was twice as great as on a per seedlings basis as that by plants grown in darkness, and three times, as great on a per gram-of fresh weight basis.

Table 8.13 Evolution of ethylene, height of stem and weight of leaves of seedlings growth under various conditions of illumination.

Condition	*Height of stem, cm*	*Weight of leaves of one seedling*	*nl h^{-1} g^{-1} fresh weight*	*nl h^{-1} seedlings*
Darkness	11.6±0.9	11±2	0.72±0.11	0.55±0.07
Darkness + Red light (5 min daily)	8.3±0.4	36±6	0.91±0.04	0.70±0.03
White light (16 h daily)	1.9±0.4	81±10	2.43±0.38	1.33±0.21

Thus, in the process of de-etiolation the production of ethylene by whole plants was increased and in the process the growth of the leaves was enhanced while growth of the stern was inhibited. If the photoinduced changes in the growth of these organs was mediated by ethylene it should be assumed that the dynamic of the ethylene production after a single irradiation would be different in the leaves and stem.

Both the leaves and the apical portion of the stem (third internode, including the loop arid apex) liberated 50 % more ethylene during the first 30 min after a single irradiation of etiolated seedlings with red light in comparison with the leaves and stems of non- irradiated plants (Figure 8.10) in the leaves the production of the ethylene gradually decreased until the end of the experiment, but after 3.5 h it remain higher that in the dark variant. In the stem, ethylene production had already dropped to the level of the dark variant after 1 h. while by the end of the experiment the stems of the irradiated plants released 40% less ethylene than the stems of non-irradiated plants. These data show that in the organs whose growth is enhanced under the action of light, the production of ethylene increased, while in the organs whose growth is inhibited after irradiation, after a brief rise, a stable decrease in the production of ethylene was observed, down to lower level that in plants growing in total darkness.

In a number of studies it has been shown that IAA can be an inducer of ethylene formation (see the survey). Possibly in our experiments also. the changes in the concentration of ethylene under the action of light are associated with a change in the content of endogenous IAA. To test this hypothesis. we determined the content of IAA in the leaves and third (from the top) internode of seedlings growth under various conditions of illumination.

From Table 8.14 it is evident that the leaves of seedlings contain almost twice as much IAA calculated per gram of fresh weight in comparison with the leaves of etiolated seedlings and eight times as much, calculated per seedling. In the apical portion of the stem of one green seedling, there was one fourth as much IAA as in the apical portion of the stem of one etiolated seedling. Calculated per gram of crude weight of the stem, the IAA content was negligibly decreased

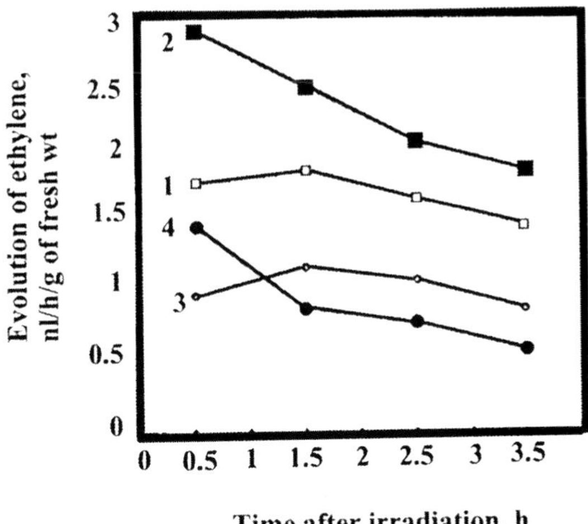

Figure 8.10 Evolution of ethylene of leaves (1,2) and the apical portion of the stem (3,4) of etiolated seedlings (1,3) and seedling irradiated once (5 min) with red light (2,4)

Table 8.14 Content of IAA in leaves an apical portion of the stem of seedlings growth under various conditions of illumination

Condition of illumination	*Content of IAA, ng*	
	Per gram of fresh wt	*Per seedling*
LEAVES		
Darkness	22.7±1.2	0.3±0.02
Darkness + Red light (5 min)	33.3±3.4	1.1±0.1
White light (16 h daily)	43.3±5.2	2.6±0.3
APICAL PORTION OF THE STEM		
Darkness	50.0±5.3	2.0±0.2
Darkness + Red light (5 min)	45.0±2.5	1.7±0.1
White light (16 h daily)	47.1±5.1	0.5±0.1

under the action of light. The maximum amount of IAA was detected in the fastest growing organs- green leaves and etiolated stems.

A 5 min-daily illumination of etiolated seedlings with red light induced an increase in the IAA content in the leaves and decrease in the IAA content in the apical portion of the stem in comparison with seedlings growth in total darkness. Thus, during de-etiolation the content of IAA in the leaves is increased and that in the third internode of the stem is decreased.

During de-etiolation, intensive growth of the stem. necessary to carry the photosynthesizing organs to the surface of the soil, gives way to growth and development of leaves. A brief illumination with red light is sufficient to induce changes in the growth processes characteristics of de-etiolation. Our experiments show that during illumination the content of IAA in the stem decreased but the amount in the leaves increased. After irradiation with red light there was either a movement of IAA out of the apical portion of the stem into the leaves or a synthesis of IAA in the leaves arid inactivation in the stem. Since IAA has attractant properties. The increase in its concentration in the leaves led to an increase in the attractant ability of the leaves and to their growth (table 8.14)

By activating ACC' (1 –aminocyclopropane- I- carboxylic acid) synthase. IAA can enhance the formation of ethylene in leaves, but formation of ethylene can occur from ACC in the leaves. It is known that the initiation of many physiological processes is accompanied by an increase in ethylene formation. In a number of studies it has been shown that ethylene can activate an alternative pathway of respiration and thereby participate in the enhancement of growth processes . Also, ethylene can accelerate sugar transport.,

The synthesis of ethylene by seven-day green, etiolated, and red light (650 nm) -irradiated pea seedlings was studied. The production of ethylene was the greatest in green seedlings; illumination of etiolated seedlings by red light for 5 min increased the formation of ethylene in the leaves but decreased it in the stems. It is suggested that the changes in the synthesis of ethylene are due to endogenous IAA, the content of which increased in the leaves during irradiation arid decreased in the apical portion of the stem.

8.3.10 UV-light as stress factor

As ecological factor UV light not only effect ion system of the plant but also affects phytohormones, like ethylene. (Jallilova et al 1993, Rakitina et al 1994).

The system of sensitivity to UV-B light depends on the genetic make-up of mutants and depends not only on physiological but also on morphological mechanisms. One of such mechanism is the wax formation. Enskat and co-authors, 2000 used scanning electron microscopy to show that sometimes investigation of wax crystals by scanning electron microscopy might have few problems. The freeze-embedding method may be helpful in cases where the leaves are unstable under the electron beam. In such cases, the transfer to stable carriers allows high-resolution SEM of wax crystals. Preliminary investigations indicate that the method may be of advantage for the investigation of wax films, especially very thin ones. These authors presented a new method for the isolation of wax crystals from plant surfaces.

The wax-covered plant surface, e.g., a piece of a leaf or fruit, is brought into contact with a preparation liquid, e.g., glycerol or triethylene glycol, and cooled to ca. - 100^{0}C. When the plant specimen is removed, the epicuticular wax remains embedded in the frozen liquid. After it warms up, the wax layer can be captured on appropriate carriers for further studies. This isolation method causes very little stress on the wax crystals; thus the shape and crystal structure are well preserved. In many cases it is possible, by choosing a preparation liquid with appropriate wettness, to isolate either the entire epicuticular wax layer or only discrete wax crystals without the underlying wax film. These crystals are well suited for electron diffraction studies by transmission electron microscopy and high resolution imaging by atomic force microscopy. The absence of intracuticular components and other impurities and the feasibility of the selective isolation of wax crystals enable improved chemical analysis and a more detailed study of their properties. John Bassman et al. (2001) consider, that there is an evidence that enhanced UV-B radiation (increased UV-B irradiance plus the spectral shift to shorter UV-B wavelengths at the ground) can decrease photosynthesis and growth in many plant species. However, there is considerable inter- and intraspecific variation in the magnitude and type of response to enhanced UV-B radiation. Angiosperms appear to be more susceptible to injury than gymnosperms because their epidermal cells are generally less effective at attenuating UV-B radiation (Day et al. 1993). Within angiosperms, variation in response to enhanced UV-B radiation may result from differences in their inherent leaf anatomy and shoot-growth pattern (Nagel et al. 1998; J. H. Bassman, G. E. Edwards, and R. Robberecht, unpublished manuscript). For example, trees with determinate or semi- determinate shoot-growth patterns may be more susceptible to injury than those with indeterminate shoot growth. This is because a particular complement of leaves would be ex-posed to enhanced UV-B radiation for a long period of time in the former, whereas in the latter, leaves are quickly sub-merged within the developing crown as new leaves are formed at the apex. Thus, a particular leaf experiences a constantly attenuating UV-B radiation environment in an in-determinate species. Submergence within the crown, and therefore attenuation of UV-B radiation, would occur faster in more rapidly growing species.

Species with faster growth rates frequently have high rates of net photosynthesis. Photosynthesis is a common target for injury by UV-B radiation because the high content of enzymes and lipids associated with carboxylation and the light-harvesting complexes are important UV-chromophores (Strid et al. 1994). Thus, species with high photosynthetic capacity may incur comparatively greater injury than those with lower photosynthetic capacities.

The concept of effects of UV-B light on photosynthesis in cotyledons of a resistant and susceptible species was a topic of research of Battaglia and Brennan (2000).

Elevated exposure to ultraviolet-B (UV-B) radiation influences the growth and development of higher plants in a variety of ways, with different species exhibiting varying degrees of resistance and susceptibility. The effects of relatively short-term high-intensity exposure to UVB upon photosynthetic C02 fixation were studied in cotyledons of cucumber (*Cucumis sativus* L. cv. Poinsett) and sunflower (*Helianthus annuus* L. cv. Cray Stripe). Treatment with 194 kJ m^{-2} of UVB radiation delivered over 16 h led to significantly reduced $C0_2$ fixation rates in cucumber; while sunflower showed no inhibition or a slight increase. Other photosynthetic parameters were studied in order to determine the basis for the relative susceptibility and resistance of these two species. The concentrations of both chlorophyll a and chlorophyll b were unchanged in response to UV-B treatment in cucumber but showed statistically significant increases in sunflower. Carotenoids did not change significantly in either species. Flavonoids (i.e., methanol extractable ultraviolet absorbing compounds) decreased in cucumber and were unchanged in sunflower. A decreased proportion of open stomata was observed in cucumber following UV-B treatment, while sunflower remained unchanged. Chlorophyll fluorescence analysis showed UV-B-induced decreases in the ratio of variable to maximum fluorescence under both dark-adapted and steady state photosynthesis conditions, and these decreases were more severe in cucumber than in sunflower. These results indicate that the differing abilities of cucumber and sunflower to withstand brief exposure to high-irradiance UVB are based on a combination of factors, including but not limited to the content of photosynthetic and UV-screening pigments, stomata function, light-harvesting efficiency, and electron transport capabilities (Bassman et al., 2001). In general UV-effect on plant cell depends on intensity and quality of UV.

Alex Kalevitch in his Ph.D thesis (1992) described the reaction of dwarf and tall pea seedlings on UV light, wavelength 250 nm and light intensity 20 watt/sq. M. All data are presented in %% to control, non irradiated plants (Table 8.15) (Kalevitch, 1991).

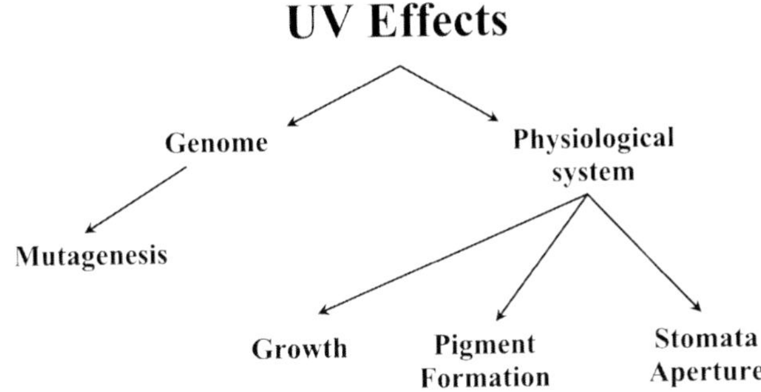

These data showed that even UV A light (365 nm) induced strong accumulation of waxes surrounding stomata. And UV-B and UV-C lights enhanced this process even more. Dusty (1996) considers the cuticular wax as testimony to its essential functional ecological factor.

Besides morphophysiological effects of UV-light it can effect DNA, inducing some genome processes (Post, Bettermiller, 1996)

Table 8.15 Effect of monochromatic UV-light on pea plants

Parameter	*% to control*	
	Pea cv	
	Dwarf cv. Prevoskhodny	*Tall Ranny*
First leaf square	23	36
Weight of the first leaf	40	37
Number of stomata per mm^2 on the 1st leaf	40	39
Stem length	90	97
Wax scale accumulation	Very high	Very high

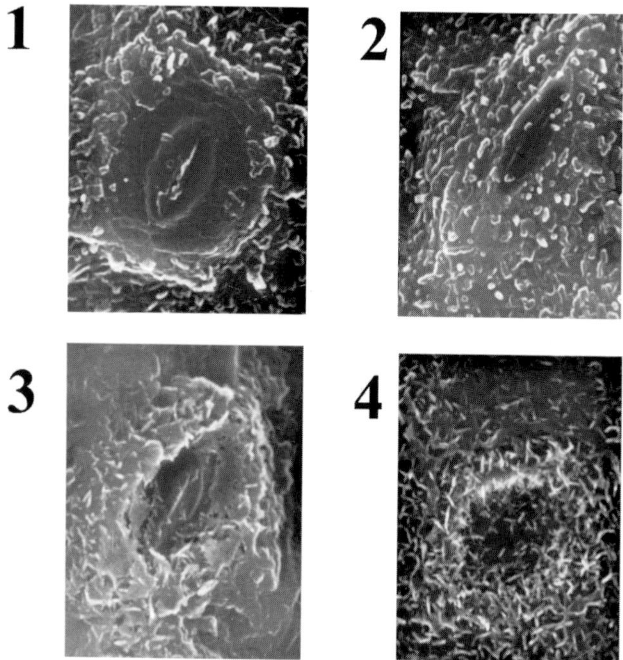

Figure 8.11 Effect of UV-light on wax formation of pea plants. 1- Control (no UV irradiation); 2 – UV-A irradiation; 3 – UV-B irradiation; 4 – UV-C irradiation.

Effects of UV-light on seedlings of various species showed different responses. (Kefeli et al , 2000). Some plants showed growth effects under the treatment by both types of UV-light (30, 000 J/sq.cm) like pea, bean and wheat, lettuce and arabidopsis seedlings are very sensitive: the growth of roots of irradiated arabidopsis seedlings was inhibited 50% if the dose 3000 J/sq.cm was applied. As a rule roots were more sensitive to UV effect, than shoots. One of the primary effect of UV-C on the seedlings was activation of potassium and sodium secretions. Calcium was more stable and didn't secrete actively. Dry seeds were more stable to UV stress, than germinated seeds or seedlings. Arabidopsis mutants with different sensitivity to UV-C were tested. Level of ethylene and ABA was the highest in the UV-B resistant mutants.

The special mechanisms of inhibition of growth induced by UV-B and UV-C lights will be discussed. Usually greening of radish seedlings didn't change the sensitivity of seedlings to the UV-C light (Figure 8.12). Pretreatment by day light of grown in the dark seedlings and/or after treatment exposure had not effect on their growth processes (Kefeli et al., 2000)

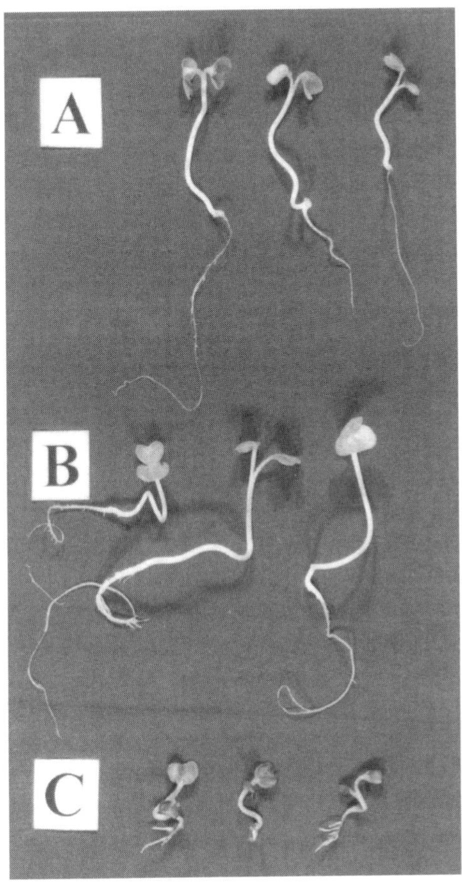

Figure 8.12 Radish seedlings, UV-C irradiation.
A – Control; B – 10000 μκJ/ M^{-2}, C – 30,000 μκJ/ M^{-2}

Low dosage of UV-C irradiation had no effect on the antocyanin formation in roots and shoots of corn (Figure 8.13). However, the high dose of UV blocked the growth and antocyanin formation , which actively forms in the corn seedlings under the white light (Touchnine and Kefeli, 1995).

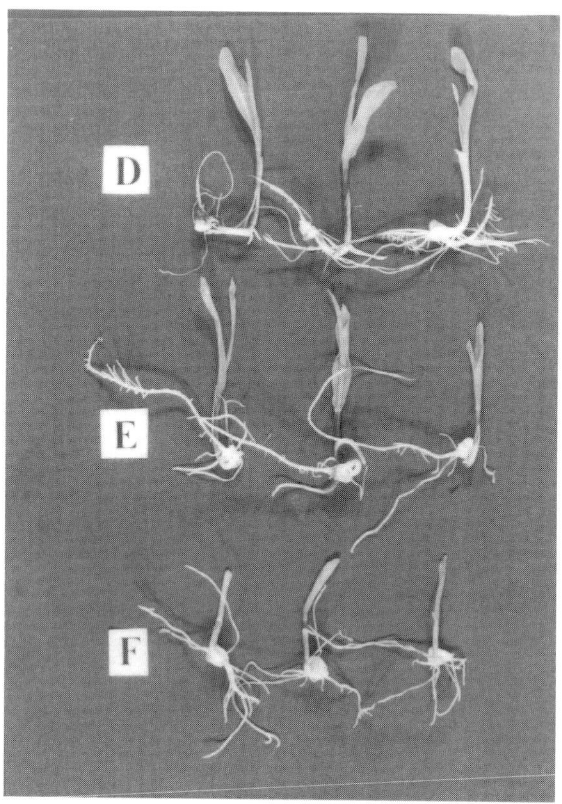

Figure8.13 Corn plants after UV-light irradiation.
D – Control; E – 10000 μκJ/ M^{-2}, F – 30,000 μκJ/ M^{-2}

8.3.11 Light and ethylene production

Genetical heterogenecity complicates interpreting studies of UV influence on plants. The collection of mutant lines of Arabidopsis makes it possible to select genetically related plants with a well-studied genotype. We chose from this collection mutant lines of Arabidopsis that differ in several genes and in their susceptibility to UV-B irradiance. We used these selected mutants to study the effect of UV-B light on growth and ethylene production (Jalilova et al., 1993; Rakitina et al., 1994). The main shoot in the irradiated *Enkheime* (chosen mutant strain) plants was 30% shorter than in the untreated plants (Table 8.16).

Table 8.16 Influence of UV-B radiation (60-min dose) on the length of the main shoot and the number of lateral shoots and pods at the end of the growth period (% of control plants)

Lines	*Main shoot length*	*Number of pods*	*Number of lateral shoots*
Enkheime (moderately sensitive)	70.1±3.1	78±2.8	78±3.1
E_6 glabra (sensitive)	51.0±2.3	73±2.1	81±3.6
Vc'er'gl'an (resistant)	98.4±0.6	85±3.2	86±2.0

UV-light suppressed the growth of sensitive E_6 glabra plants by almost 50%. At the same time, the shoot length of irradiated and non irradiated vc1er'gl'an plants did not differ appreciably, indicating plant resistance to UV-B concerning this trait (Rakitina et al., 1994)

In irradiated plants of the resistant line, the pod number was only 15 % less than in untreated plants, but 19-27% less in plants with moderate or low UV-B resistance. All three lines responded to irradiation with a 14 - 22% decrease in the number of lateral shoots (Table 8.16).

Figure 8.14 shows that the fresh weight of *the Enkheime* plants decreased 20% by the third day following 10-min UV-B irradiation. It then increased to the control level by the fifth day. The 30-rain irradiation resulted in a 20 - 33% fresh-weight loss for the *Enkheime and Et glabra* plants, with restoration to the control level occurring by the 11th to 12th day. Following 10- and 30-min irradiation, the fresh-weight of vc 'er 'gl 'an was only 8 - 12% lower than the control level. Higher doses of UV-B radiation showed a more distinct difference (Kefeli et al., 1993)

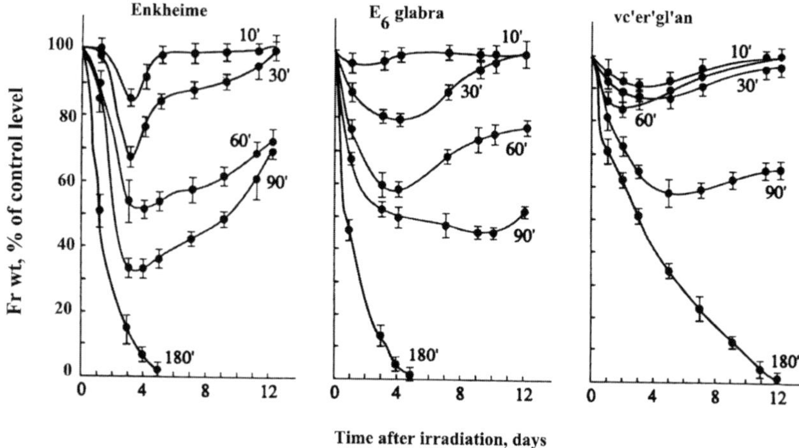

Figure 8.14 Fresh weight in plants irradiated with various doses of UV-B radiation (relative to control level, %). 10', 30', 60', 90', and 180' – irradiation doses in min.

between the lines. After 60-min and 90-min irradiation, the moderately sensitive Enkheime and sensitive E_6- glabra plants failed to restore the fresh weight even by the 12th day following treatment. On the contrary, in resistant vc 'er 'gl 'an plants irradiated for 60 min, the fresh weight achieved the control level by the sixth day. With the maximum UV-treatment (180 min), the fresh weight in Enkiteime and Et glabra plants decreased irreversibly by the fifth day, whereas in vc'er'gl'an plants, the same events occurred seven days later (Figure 8.14).

The 10-, 30-, and 60-min irradiations did not significantly change the dry weight of the resistant vc'er'gl'an plants, but the dry weight of plants with moderate and low resistance markedly declined following 30-min and, especially, 60- and 90-min irradiation (Figure 8.15).

Thus, UV radiation inhibited growth and impaired biomass production in Arabidopsis plants of all lines studied, with the effect dependent on the UV-radiation dose and plant susceptibility.

Without UV treatment, the plants of all lines produced rather small amounts of ethylene (about 0.5 nl/(g h)). However, when irradiated for a period as shown as 30 min, all plants had enhanced ethylene synthesis. Extending UV-exposure duration to 180 min resulted in a 10- to 16-fold rise in ethylene pro-duction, as compared to that evolved by control plants. Intensified ethylene evolution mainly occurred within the 2 to 4 h following irradiation of all lines, and then dropped drastically, approaching the control level within the next two days. However, the lines studied differed in the intensity and dynamics of ethylene production. Thus, ethylene production reached its maximum at a different time for each line. After 30- or 60-min irradiation, in resistant plants it occurred by the 4th h; in plants of moderate resistance, the maximum was between 2 and 4 h; in the sensitive E_6 glabra plants, the most active ethylene production occurred in less than 2 h. Under high radiation doses, the resistant plants significantly intensified their ethylene evolution. Following 90-min UV-B exposure, ethylene production by these plants was twice as high as that by the plants of the other two lines. After 180-min exposure, ethyl-ene production of resistant plants exceeded that of moderately resistant plants by 15% and that of highly sensitive ones by 70%. Thus, plants of the resistant line vc 'er 'gl 'an are characterized by more efficient ethyl-ene production than the Enicheime and E_6 glabra plants.

Short-term UV-B radiation impaired Arabidopsis plant growth, not only immediately following treatment at the early phases of development (see Figs. 8.14 and 8.15), but also at the end of the growth period (Table 8.16). The decrease in fresh weight just after irradiation was a result of water loss, which is especially evident by the second to fourth day following high radiation dose: compare (Figure 8.14 and 8.15). Transpiration disturbance caused by UV light be responsible for such a weight loss in our experiments, as well as in many others.

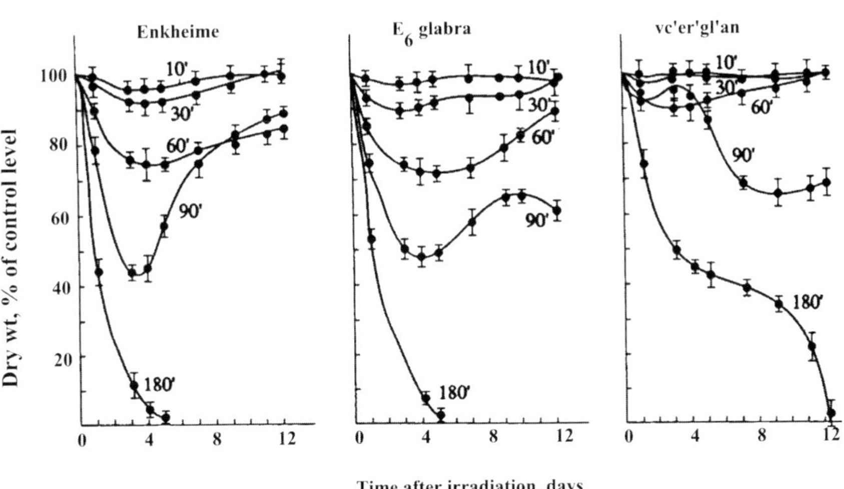

Figure 8.15 Dry weight in plants irradiated with various doses of UV-B radiation (relative to control level, %). 10', 30', 60', 90', and 180' – irradiation doses in min.

It is well known that various plant species vary greatly in their response to UV-B radiation. Our results demonstrate the intraspecies variations in plant susceptibility to UV-B. Thus, concerning growth responses to UV-B, we can regard the laboratory type *Enkheime* as intermediate between the resistant *vc'er'gl'an* and sensitive E_6*- glabra* lines.

It is suggested that flavonoids play some protective role against UV-B radiation, as their absorption maximum are located within the range of 280-320 nm, and UV-B radiation usually enhances flavonoid biosynthesis in the shoot epidermal layer. The protective role of flavonoids was demonstrated in experiments on irradiating a *Rosa damascena* cell culture. A highly resistant cell line by selecting cells surviving irradiation with sequentially increased UV doses. The flavonoid content of these plants was 14 times higher than that of the initial cells. In our experiments, the line E_6- glabra was selected with its very low flavonoid content caused by a disturbance in biosynthesis as a result of mutation. The high sensitivity of this line to UV-B radiation could be related to this trait.

Growth delay is considered as a protective response to UV-B radiation. This could account for the high resistance of vc 'er 'gl 'an in our experiments. These plants have narrow, thickened leaves, like some xerophytes. They actively produce biomass, but the main shoot length of these non-irradiated plants was shorter (19.1±0.4 cm) than that in *Enkheime* (23.6±1.2 cm) and *E6 glabra* (22.3±0.7 cm). Therefore, vc'er'gl'an plants tolerate higher radiation doses with less prominent damage.

Various factors (temperature, light, or water condi-tions) that induce stress responses enhance ethylene biosynthesis. Stress-induced ethylene can be synthesized in two different ways: either with the aid of a specific ethylene-producing enzyme, which is tightly bound to the membranes, or via the non-enzymatic transformation of 1-aminocyclopropan-1-carboxylic acid to ethylene. The latter can be triggered by various peroxides and free radicals. Photooxidation of membrane components, peroxide formation, and, as a consequence, free radical emer-gence usually goes along with the UV-induced damages. Therefore, both ways of ethylene syn-thesis in the investigated lines of Arabidopsis is quite possible.

In our experiments, UV radiation provoked an appreciable enhancement of ethylene emission in all three lines of Arabidopsis. Higher radiation doses corresponded with increased ethylene production, which was highest in the resistant line (Figure 8.16). Increased ethylene production appears to be a typical response to stress, and simultaneously, one of the traits of plant resistance to UV radiation.

To sum up, it is worth emphasizing that using genet-ically related lines of Arabidopsis allows us to assume that there are relationships between plant responses to UV-B and gene specificity unique to each line. The plant response to stress factors demonstrate the partici-pation of hormones, ethylene in particular in a plant genetically determined resistance to UV-B radiation.

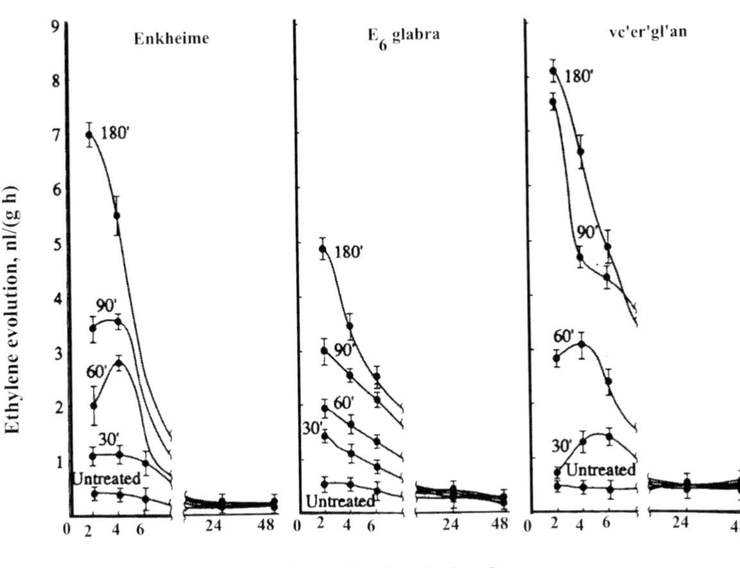

Figure 8.16 Ethylene evolution by plants irradiated with UV-B light.. 10', 30', 60', 90', and 180' – irradiation doses in min.

8.3.12 UV-B light and ABA-ethylene ratio

Predicted enhancement of solar MI-B (280-320 nm) radiation can have serious environmental impact on plant communities. Every two out of three tested plants showed some sensitivity to UV-B irradiation, with weeds being more resistant than agricultural crops. In natural ecosystems, UV stress can lead to the gradual substitution of sensitive species with resistant ones, thus reducing plant diversity. Therefore, clarifying the reasons for plant differential resistance to UV-B radiation stress has become the hot area in current research. The hormonal aspects of plant resistance, especially the role played by ABA and ethylene in stress responses, have attracted considerable interest.

Previously we demonstrated that ethylene evolution was accelerated with increasing UV-B dose and depended on plant resistance. The objectives of this work are to study the possible changes in the levels of another stress phytohormone, ABA, and to elucidate whether ABA and ethylene can interact during plant response to UV-B radiation stress. ABA content and ethylene evolution were monitored in Arabidopsis lines differing in their resistance to UV-B irradiation.

Three genetic lines of *Arabidopsis thaliana* (L.) Heunh. were used: wild-type Enkheime, the mutant E6 glabra susceptible to UV-B radiation, and the resistant mutant vc'er'gl'an. In some experiments the ABA-deficient mutant was used. The plants were grown in Petri dishes (49 plants per dish) in axenic agar mineral medium [5] in a growth chamber under illumination from LB-65 fluorescent lamps for 16 h per day and a temperature of 25^{0}C during the day and 22^{0}C at night. When the plants were 12 to 13 days old (the phase of the rosette) they were exposed to UV-B light for either 30 60, 90, or 180 min, which corresponded to the doses of 14.4;43.2; and 86.4 kJ/m^2. LE-30 erythema lamps with an emission maximum of 310 nm were used as a source of UV-B radiation.

Plant survival was expressed as the percentage of seedlings surviving the UV-B treatment out of the total number of treated seedlings. Nonviable plants lost their turgor, were discolored, and much of their cells were subject to lysis.

The content of free ABA was determined by gas-liquid chromatography. The shoots of 500 - 600 seedlings (1.7-2.0 g) were grounded in 4-5 ml of cooled 90% methanol. The homogenate was filtered through a GF/C glass fiber filter (Whatman, Great Britain) and reduced to a water residue, which. was purified according to the method of Dorffling and Tietz [6]. The purified extract was evaporated to dryness, dissolved in a small volume of methanol, and additionally purified by thin layer chromatography ilufol UV-254 Silica Gel plates using a solvent mixture of chloroform:methanol, and water (70:12 :0.5). ABA in the samples was methylated with an ether solution of diazomethane at 0^{0}C. Diazomethane was obtained by adding 2 ml of 45% KOH and 8 ml of ethyl ether to 1 g of N-nitroso-N-methylurea at 0^{0}C. The ABA content was determined using gas chromatography equipped with electron capture detector. The column (3 x 10^{-3}X 2.5 m) was packed with ChromosorbW-HP (100-120 mesh) coated with 3% Silicon OV-17 (Serva, Germany). The temperature of the detector column, and injection chamber was 350, 225, and 250^{0}C, respectively. ^{3}H-ABA (69 Ci/mmole) at a concentration of 1pg/ul was added to the fixed material as an internal standard. This concentration was too low to be detected by the electron capture detector, but high enough to estimate the absolute radioactivity in a Rack-beta Spectral-1219 scintillation counter Wallac LKB, Sweden. In order to estimate ethylene evolution, the plants were grown in vials (10 - 12 plants per flask). Following plant exposure to UV-B, the flasks were sealed and exposed to light for 2 h for ethylene accumulation. The measurements were performed 2,4,6, and 24 h after treatment usinga Chrom-4 gas chromatograph. The procedure of ethylene estimation has been described earlier.

The tables and figures represent the means and their standard deviations from two to four experiments performed in two to six replicates.

All of the plants survived when they were exposed to UV-B light for 30 or 60 min. Following a 90-min exposure, only part of the plants survived: about half of the plants of the wild line Entheime, 79% of the resistant line vc'er'gl'an, and only 43% of the susceptible line E_6 glabra (Table 1.7). Longer treatment (180 rnin) resulted in the death of all the Enkireime and E_6 glabra plants, whereas 18% of the vc'er'gl'an plants survived. These facts further demonstrated that the vc'er'gl'an line was the most resistant to UV-B radiation stress.

Table 8.17 Plant survival after UV-B irradiation (% of survived plants)

Line	*Exposure, min*	
	90	*180*
Enkheime	53±1	0
E_6 glabra	43±3	0
vc'er'gl'an	79±3	18

In the untreated plants, cis-ABA content in the wild-type Enkheime was 2 - 2.5 times lower than in the mutant lines (Table 8.18). Two hours after 90 min of irradiation, ABA content increased 2.1 and 1.6 times in the wild type and susceptible mutant, respectively. Four hours after the treatment, the ABA content dropped in both genotypes and increased again after 24 h. UV-B radiation did not affect the ABA content in the resistant mutant vc'er'gl'an during the first four hours, and only after 24 h did the ABA content increased 1.5 times.

After 180 min of irradiation, ABA accumulated in all three genotypes and to a greater extent than after 90 min of irradiation. However, the time-course of ABA accumulation differed in different lines. In the plants of the wild type, the ABA content increased 3.5 times in two hours and 5.6 times in 24 h. In the plants of the susceptible mutant ABA content increased 1.8 times 2 h after irradiation, although it slightly decreased after 24 h. In the irradiated resistant vc'er'gl'an plants, the rate of ABA accumulation was higher than in the untreated plants: 2 h after the treat-ment, the ABA content was by 30%, 4 h by 80%, and 24 h 150% higher than in the untreated plants. Leung and Giraudat, 1998 think that ABA plays a major role in adaptation to abiotic environmental stress.

Table 8.18 The effect of duration of UV-B irradiation on the content of cis-ABA in plants

Line	*Exposure, min*	*The content of cis-ABA (ng/g fr wt) following exposure to UV-B light, measured after*		
		2 h	*4 h*	*24 h*
Enkheime	0	4.3±0.5	3.0±0.6	3.8±0.4
	90	9.1±0.1	6.3±0.8	8.6±0.5
	180	14.9±2.4	8.1±0.4	21.4±1.2
E_6 glabra	0	7.1±0.1	7.7±0.4	5.9±0.6
	90	11.7±0.6	8.5±0.3	9.2±0.1
	180	13.0±0.5	14.4±0.2	10.1±0.4
vc'er'gl'an	0	9.5±0.8	8.0v0.4	10.4±0.3
	90	9.4±0.2	8.1±0.5	15.5±0.7
	180	12.6±0.5	14.6±0.1	26.2±5.3

It is interesting to notice that in our experiments trans-ABA was detected only in irradiated plants and at a significantly lower level than cis-ABA (Tables 8.18 and 8.19). After 90 min of irradiation, trans-ABA was found in three treatments and, after 180 min of exposure, in eight out of nine treatments. Evidently, UV-B radiation stimulated trans-ABA formation in the plants.

Table 8.19 The effect of duration of UV-B irradiation on the content of trans-ABA in plants

Line	*Exposure, min*	*The content of trans-ABA (ng/g fr wt) following exposure to UV-B light, measured after*		
		2 h	*4 h*	*24 h*
Enkheime	90	1.9±0.1	0	2.1±0.3
	180	1.8±0.1	1.0±0.1	8.0±0.2
E_6 glabra	90	1.7±0.2	0	0
	180	3.0±0.2	1.9±0.1	2.4±0.1
vc'er'gl'an	90	0	0	0
	180	1.9±0.1	0	4.4±0.5

We have previously detailed the time-course of ethylene distribution in the plants of three Arabidopsis lines for 48 h following UV-B irradiation (3). Ethylene synthesis in these plants as related to the duration of UV-B treatment is presented in Figure 8.17. Ethylene evolution enhanced with an increasing dose of absorbed UV radiation and depended on plant resistance. Non-irradiated plants of all lines evolved about 0.5 ml of ethylene per g of wt per hour. After 180 min of irradiation, ethylene accumulation in the susceptible line E_6 glabra accelerated nine times, 14 times in the initial line Enkireime, and 17 times in the resistant line vc'er'gl'an.

Ethylene evolution in the ABA-deficient Arabidopsis mutant was 4.8 times higher than in the plants of other lines (Figure 8.17). The highest rate of ethylene synthesis occurred two hours after the 180-min irradiation of the ABA-deficient mutant. The rate decreased during the ubsequent 24 h until it reached the level typical for the untreated plants (Figs. 8.17 and 8.18). After the 90-min irradiation, the rate of ethylene evolution was four times slower as compared to the 180-min irradiation. It decreased also during the following 24 h. The 30-min and 60-min exposures only slightly affected ethylene production from the plants of this line.

Earlier we demonstrated that the lines Enkheime, E_6 glabra, and vc'er'gl'an significantly differed in their resistance to the UV-B radiation stress as judged by stem length, biomass production, and number of pods. Our new data on plant survival after UV-B irradiation (Table 8.19) is consistent with the results of our earlier research: E6 glabra is a susceptible line, vc'er'gl'an is a resistant line, and Enkheime is an intermediate genotype.

Resistance to stress is often related to plant capacity for ABA accumulation. In addition, plants that are tolerant to desiccation or air pollutants usually contain more ABA under normal conditions as well. Rice varieties with higher ABA levels were damaged by ozone stress. One could expect that UV-B radiation would also induce ABA accumulation, as during the first three days following irradiation, when the plants lost some water, and ABA was known to be involved in regulating the tissue water regime. However, UV-B irradiation of the *Rumex patientia* leaves did not result in ABA accumulation. In our experiments, the higher UV-B doses increased the ABA content in all the tested lines of *Arabidopsis* (Table 8.19), with the higher doses being more effective. In both the untreated and irradiated plants, we observed some fluctuations in the ABA content. Similar fluctuations were noted in the leaves and panicles of pearl millet and in the needles of spruce. In the E_6 giabra line, which was susceptible to stress, the ABA content peaked four hours after the 180 min irradiation and then declined. In the resistant line and wild-type plants, the ABA content increased steadily, and, one day after irradiation, it was almost twice as high as in the susceptible line.

Intense sunlight is known to induce cis-ABA isomerization into trans-ABA in the plants. Exists the possibility of cis-ABA isomerization as affected by UV-B radiation. They concluded that such isomerization was possible only to a small extent.

In our experiments, the trans-ABA content in the plants was very low (5 - 10 times lower than the cis-ABA level), but it clearly grew with increasing doses of absorbed ultraviolet radiation (Table 8.19). High doses of UV-B light could induce isomerization just as it occurred in vitro.

Ethylene production is a more rapid response to UV-B radiation stress than ABA accumulation. According to our results, ethylene evolution peaked during the first hours, whereas the maximum ABA level was achieved only one day after irradiation. Ethylene production enhanced with increasing doses of UV-B radiation in all three tested lines and achieved its maximum values in the resistant line vc'er'gl'an (Figure 8.17).

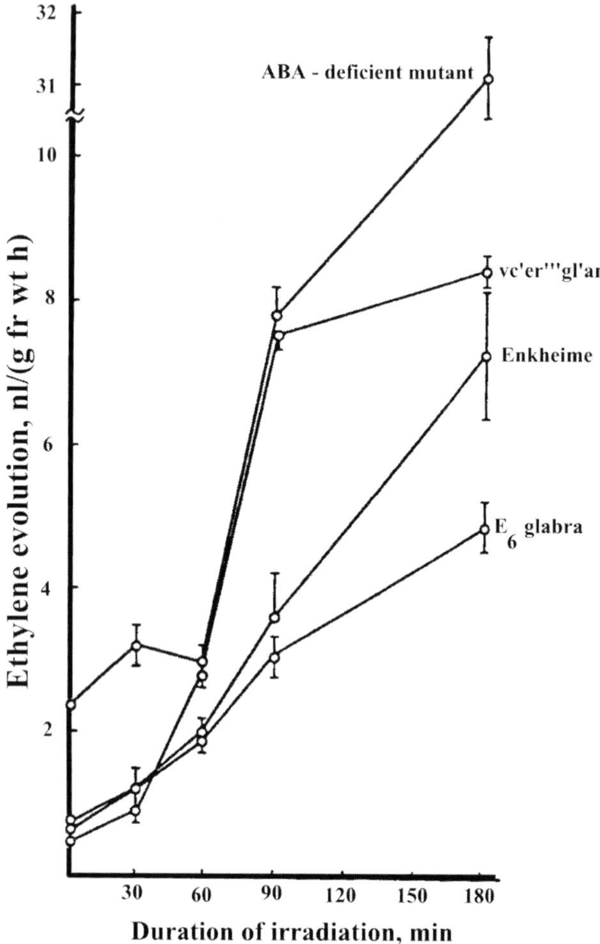

Figure 8.16 The average rate of ethylene evolution from the plants of different lines in 2 h after UV-radiation.

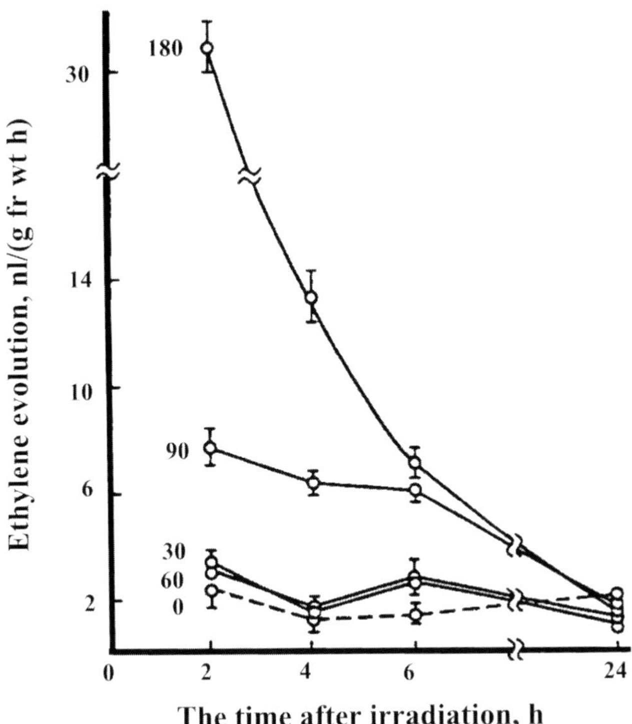

Figure 8.18 The kinetics of ethylene evolution from ABA-deficient mutant in the day following UV-B irradiation. The figures0, 30, 60, 90, and 180 indicate the duration of irradiation (min).

Thus, the detrimental effects of UV were accompanied by the accumulation of two stress-induced phytohormones, ABA and ethylene. Their content depended on the dose of UV-B radiation and on plant resistance. The higher was the resistance of the line, the more ABA and ethylene was detected in the plants.

The problem of interactions between ABA and ethylene is far from clear, although some authors have discussed possible ways of such interactions. We have attempted to approach the problem using an ABA-deficient Arabidopsis mutant. In both non treated plants and plants irradiated with the maximum UV-B dose, ethylene production was higher than in the plants with the normal

ABA content (Figure 8.17). However, the low endogenous ABA content did not affect the time-course of stress-induced ethylene evolution (Figure 8.18), which was similar to the changes that we described earlier for other Arabidopsis lines (3). Ethylene evolution was enhanced during the first two to four hours after irradi-ation, then slowed, and finally reached the level of the untreated plants after 24 h. Our results are similar to those obtained by Tal et al (19) on the *wilty* mutant of tomato with a reduced ABA content. Ethylene synthesis by this mutant was significantly faster than by normal plants under both sufficient and deficient water supply.

The presented results testify that plant resistance to UV-B radiation is genetically determined, as the related lines of Arabidopsis, differing only in several genes, responded quite differently to a stress factor. These mutant genes are responsible for the higher resistance of the *vc'er'gl'an* line as compared to the susceptible E_6 *glabra* line and the wild-*type Enkheime*.

The presented results concerning the effect of stress conditions on growth and inhibitors level showed that some factor like water deficiency, UV-light effects are in the good correlation with the level of growth inhibitors and phytohormones but some like reaction on low temperatures are fare from good correlation. Most of the presented data concerned with the short time stress effects, however there are long time adaptation reactions (adaptations to unfavorable factors) as well as complex effect of some stress factors in one time-general adaptation syndrome. These processes and the relation of growth inhibitors needs the further investigation.

CHAPTER 9

NATURAL GROWTH INHIBITORS AND PHYTOHORMONES IN THE INTACT PLANTS AND ISOLATED CELLS, ORGANS AND TISSUES

Natural growth inhibitors were discovered in plant tissue as far as the 1920's. However, unlike auxins, they were for a long time considered separate from growth as admixtures interfering with detection of phytohormones.

A number of problems pertaining to the mechanism of action of growth inhibitors have not yet been solved. Thus, for example, we do not know whether, the accumulation of natural inhibitors in plant tissues is a factor preceding growth inhibition or a process that occurs simultaneously with the latter. The same problem awaits its solution with respect to phytohormones. We do not yet have at our disposal direct techniques which could be used to separate experimentally the synthesis of natural inhibitors and plant growth. There is only indirect evidence that both processes are closely interrelated. The nature of this relationship needs further investigation.

Individual forms of growth processes are regulated not only by changes in natural inhibitor concentrations, but also by the level of natural phytohormones. It would appear that there exists in a plant a specific equilibrium between stimulating and inhibiting substances, with different forms of growth regulated by a balance of appropriate phytohormones and growth inhibitors, which is observed in the primitive forms even in one cell organism – *Acetabularia* (Bonotto et al., 1979).

In view of the above, the growth and dormancy of plants can be regarded as a system controlled by a ratio between various groups of growth stimulators and inhibitors. According to this concept, each form of the growth process is regulated by a specific 'level' of these compounds, the quantities of which undergo changes. The groups of stimulators and inhibitors act alternately which ensures the dominance of one of these factors in regulation of various growth processes. Thus, bud opening or seed germination is regulated by the ratio between phytohormones (gibberellins, auxins) and growth inhibitors, with phytohormones dominating and growth inhibitors contained in minimum amounts. The formation and growth of roots are mainly dependent on the ratio between auxins and gibberellins, auxins being the dominating factor. Stem growth depends on still another pair: gibberellins inhibitors; here, similar to the previous process, growth inhibitors act as coordinating factors which compensate for the stimulating effect of phytohormones. If not for the retarding action of growth inhibitors, the growth of a stem and a root could be 'disorganized' as happens in the darkness, in etiolated shoots or when roots are affected by auxin-producing pathogenic organisms.

Other forms of growth - enlargement of leaves, growth and ripening of seeds are also controlled by a ratio in the content of appropriate phytohormones and growth inhibitors. When growth is retarded, the content of phytohormones drops abruptly, while the activity of natural growth inhibitors is promoted. Therefore, such processes as leaf abscission, and the inception of dormancy in seeds and buds take place in the practically complete absence of phytohormones. The phytohormone - inhibitor system may even be inherent in regulatory mechanisms controlling generative development (Table 9.1)

The study of a broad range of concentrations of natural inhibitors in their action on the growth of phytohormone-sensitive biotests and comparison of this action with the activity of auxins, gibberellins and cytokinins showed that, contray to phytohormones, none of the tested concentrations of natural growth inhibitors could display hormonal effects, i.e. strongly stimulate the growth of bio-assays. The same bio-assays helped establish that natural inhibitors affect growth in a more subtle way than such synthetic poisons as 2, 4-dinitrophenol; therefore, the retarding effect of natural inhibitors does not increase as abruptly as that of poisons with an increase in the concentration strength. This feature of natural inhibitors may be due to the fact that they accumulate in plant tissues gradually and the plant passes slowly from the active growth phase into dormancy.

Naturally it is not only the interaction of inhibitors with auxins that makes growth such a complex process. Natural growth inhibitors were able to reduce also the effect of other phytohormones (gibberellins and cytokinins) on the growth of specific bio-assays (Sarapuu & Kefeli, 1968).

Further studies of Wareing, Aspin, Ziegler and other researchers confirmed the validity of our statement that growth inhibitors are compounds which depress various growth processes rather than functions of individual phytohormones.

In general, the retarding effect of natural inhibitors on plant growth may be schematically represented as follows (Figure 9.1)(Table 9.1).

Neither phenolic nor terpenoid growth inhibitors depress primary reactions specific for the action of each phytohormone; rather they affect general metabolic processes, which are indispensable for any form of growth. Such general metabolic reactions include the synthesis of nucleic acids and proteins, ATP formation and other processes the production of a phytohormone precursor.

We cannot rule out the possibility of one growth inhibitor 'switching off' the synthesis of several phytohormones. Its action might be realized through cascade repression which is a more complex version of the scheme developed by Jacob and Mano and applicable mainly for microbes.

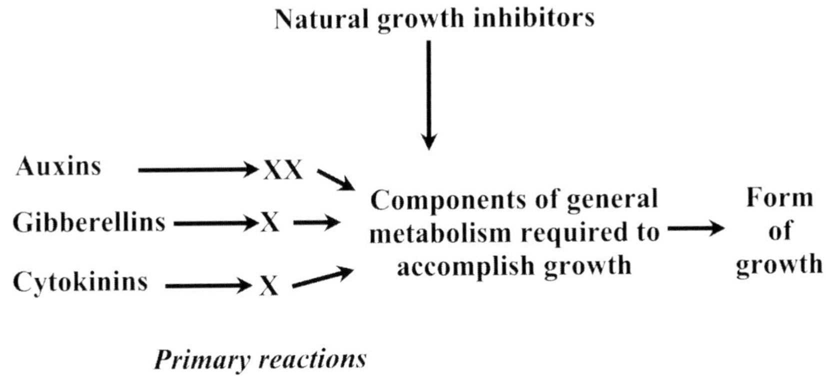

Figure 9.1 Interaction of natural growth inhibitors and phytohormones

Table 9.1 Balance of phytohormones and inhibitors during plant development

Dormant seeds and buds			
Gibberellins	Auxins	Cytokinins	Inhibitors
(-)	(-)		(+++)
Bud breaking or seed germination			
Gibberellins	Auxins	Cytokinins	Inhibitors
(++)	(++)	(++)	(-)
Initiation and growth of root			
Gibberellins	Auxins	Cytokinins	Inhibitors
	(+++)		(+)
Stem growth			
Gibberellins	Auxins	Cytokinins	Inhibitors
(++)			(+)
Leaf growth			
Gibberellins	Auxins	Cytokinins	Inhibitors
	(++)	(++)	(+)
Formation of generative organs, seed germination			
Gibberellins	Auxins	Cytokinins	Inhibitors
(++)	(++)	(++)	(+)
Leaf abscission			
Gibberellins	Auxins	Cytokinins	Inhibitors
	(-)		(++)

In compliance with this principle, cascade repression becomes possible if a structure gene controlled by the operator of the first system forms a substance acting as a repressor for the operator of the second system. A natural growth inhibitor accumulating in cytoplasm in autumn could act as such a repressor.

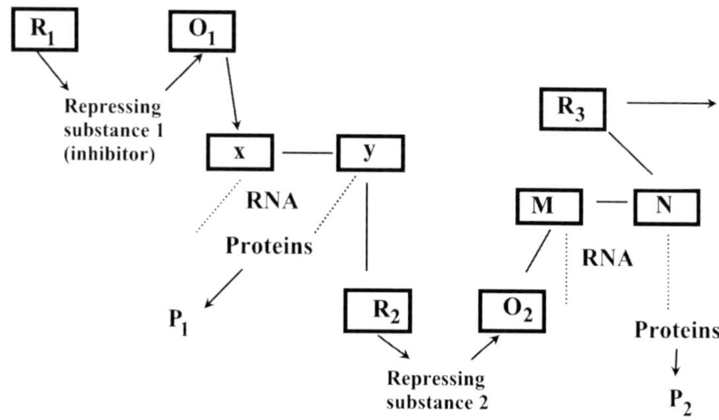

R1 and R2 are regulatory genes, O1 and O2 - genes-operators x, y, M and N -structural genes of cystrones.

It is precisely in this period that the content of natural inhibitors grows considerably and the phytohormone function is attenuated. In the autumn dormancy period, plants often exhibit no response whatsoever to the treatment with phytohormones.

Considering this scheme as applied to morphogenetic processes that occur in plants in spring (opening of buds, seed germination, differentiation in tissue cultures) it may be assumed that canceling out of cascade repression will open new synthetic pathways, in the first place, biosynthesis of phytohormones. This 'cascade induction' of biosynthesis which occurs during activation of growth processes may be caused by an exogenous factor called an artificial evocator.

An artificial evocator may be a chemical agent of specific (herbicide, poison) or even non-specific nature (ester, alcohol). Such agents are frequently employed for inducing termination of dormancy in buds of perennial plants. An exogenous evocator may also be a physical factor, such as low or high temperature, light. The primary function of such an evocator.

A

1. Self-inhibition

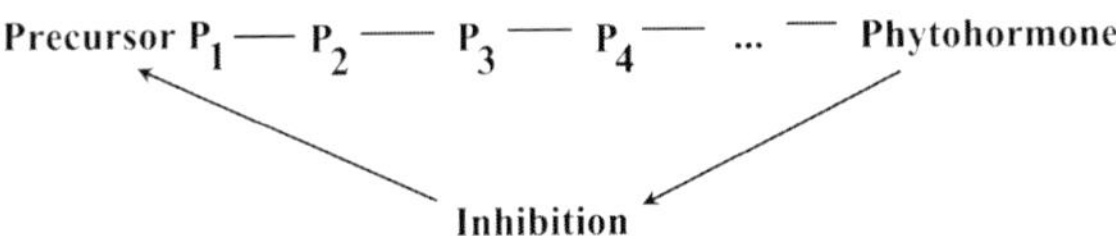

2. Inhibition by means of natural inhibitor $P_1 \rightarrow P_2 \rightarrow P_3 \rightarrow \cdots$ Phytohormone

Growth inhibitor

B

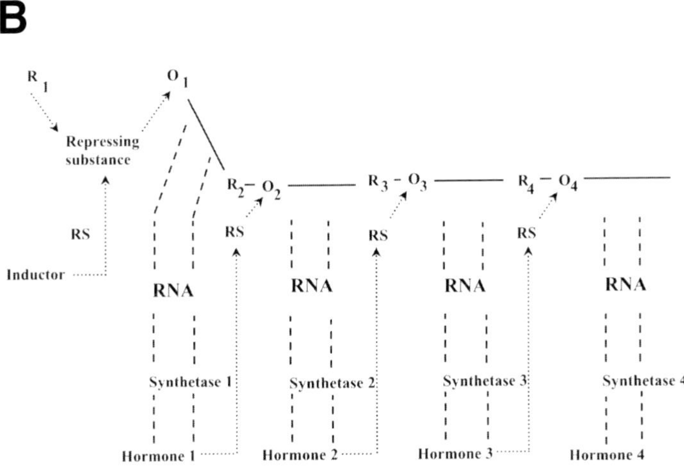

Figure 9.2 Cascade form of hormones biosynthesis

seems to interfere with the primary repressor bond which breaks to enable the synthesis of a number of endogenous inductors - phytohormones.

An exogenous or endogenous inductor neutralizes the effect of repressor R1 which represses the function of the operator (6) controlling the synthesis of hormone 1. Hormone I performs the function of an inductor, eliminates the inhi-biting effect of repressor R2 and induces the synthesis of hormone 2, which, in its turn, induces the synthesis of hormone 3 by acting upon repressor R3. The enzymes synthesizing hormones are called here synthetases.(Figure 9.2).

Each phytohormone is synthesized in succession by an enzyme called synthetase. The cascade induction of phytohormones may be taken as a basis for interpreting a rather common phenomenon when several phytohormones are formed simultaneously in objects with active growth processes (germinating seeds, a vigorously growing stem, breaking buds, sprouting eyes of tubers and others). According to the cascade induction scheme, the synthesis of each phytohormone is controlled by a separate cystrone or, more generally, a genetic block. A question now arises: Can we assume that hormone synthesis is controlled by only one 'hormone' cystrone rather than by several blocks? In this case an inductor or evocator induces at once the synthesis of all the three hormones by blocking the repressor substance. This assumption seems to us hardly probable, because sometimes only one phytohormone is present in growth systems, for example, auxin, while other phytohormones are either absent or contained in minimum amounts. Then the triggering of the repressed biosynthesis is not always simultaneous which means, that it may most likely be controlled not by one but by several linked structure genes (cystrones).

Obviously, the above systems of cascade repression and phytohormone biosynthesis induction are directly affected by external climatic factors which seem to be at the beginning of the regulatory process chain.

The basic point which determines the beginning of the inhibitory action of phenolic inhibitors on plant growth may be their accumulation in a plant cell. A more detailed perspective to this problem will be given below. Our experiments have shown (Kefeli, 1978) that accumulation of growth inhibitors in the autumn buds of woody plants is associated with an abrupt reduction in their decomposition processes. Autumn willow buds decompose the phenolic inhibitor isosalipurposide several times weaker than spring buds. On the one hand, it may be thought that the inhibitor accumulated in large quantities depresses, through an ATP formation mechanism, the synthesis of enzymes which oxidize the inhibitor, in other words, nonspecific inhibition of enzymatic decomposition of the inhibitor is effected with the aid of the inhibitor itself. On the other hand, introducing negative feedback, the phenolic inhibitor may retard its own synthesis which results in the equilibrium state characteristic of growth inhibition in the autumn period. To be able to effect inhibition through ATP synthesis, the phenolic inhibitor must first accumulate in a cell in microquantities while abscisic acid acts in low concentrations at the RNA synthesis level and represses thereby the activity of the entire growth system.

Generally speaking, the action of natural inhibitors on the activity of phytohormones and on growth processes in plants takes place at different levels: at the levels of synthesis, functioning and inactivation. Natural growth inhibitors may be regarded as products positioned at the key points of biosynthetic pathways which often come to be the final products of the metabolic systems and have a relatively long life in a biosynthetic system (as compared with unstable intermediate compounds commonly found in metabolic processes). These compounds may act as repressors of many metabolic processes.

The existence of metabolic blocks with branching biosynthetic pathways is all the more probable because two functionally reverse factors (phytohormones and growth inhibitors) have common metabolic precursors. These precursors are shikimic and chorismic acids for indole auxins and phenolic inhibitors, and mevalonic acid and its derivatives for gibberellins and abscisic acid.

Such metabolic precursors and metabolic blocks which branch off into the phytohormone and inhibitor formation pathways can be used for interpreting growth promotion and inhibition periods. Thus, active synthesis of auxins and gibberellins (i.e. the function of one branch) occurs in a period when growth processes are promoted; inhibition of this synthesis takes place at the inception of dormancy. Intensified production of phenolic compounds, including phenolic inhibitors, as well as of abscisic acid (i.e. the function of the other metabolic branch) is observed with the cessation of growth and at that time dominates over the phytohormone synthesis.

The alternation of the active synthesis of phytohormones and that of natural growth inhibitors could also be regulated at later stages of their biosynthesis. For instance, phenolic growth inhibitors can retard the production of physiologically active indoles from tryptophan. Finally, natural growth inhibitors can limit the activity of phytohormones after they have been produced. i.e. during their functioning.

Evidently, such interaction is essential for coordination of growth processes. The content of inhibitors in etiolated plants is known to be practically negligible. In an etiolated plant extension of axial organs predominates, while leaves and lateral shoots are underdeveloped. Light induces the synthesis of natural inhibitors in other words, it controls the activity of metabolic precursors. The production of these compounds is due to activation of the chloroplast functions.

Natural growth inhibitors formed in a greening plant seem to affect the phytohormone level and the stronger retard the growth of the stem the higher the light intensity. The growth of leaf blades is activated in this case. Consequently, apart from determining the start of the dormancy period natural growth inhibitors may be factors of correlative regulation of endogenous growth which balance the action of hormones so as to prevent their hyperfunction resulting in excessive growth of shoots, formation of knots and other signs of disorganized growth. Such disturbances in the life activity of plants are observed when they are infected with pathogenic agents, in particular, certain microorganisms which can synthesize large amounts of phytohormones These substances produce almost no effect on the growth of producing agents themselves, do not undergo noticeable oxidation or binding, i.e. they are in a free form; however, getting into the tissues of a higher plant they interfere with the normal course of growth processes and cause various pathologic changes in the plant.

These changes might be expressed in root nodules induced by *Rhizobium* (Figure 9.3) or by *Agrobacterium tumifaciens* (Figure 9.4). In these nodules or tumors increases the concentration of phytohormones cytokinins

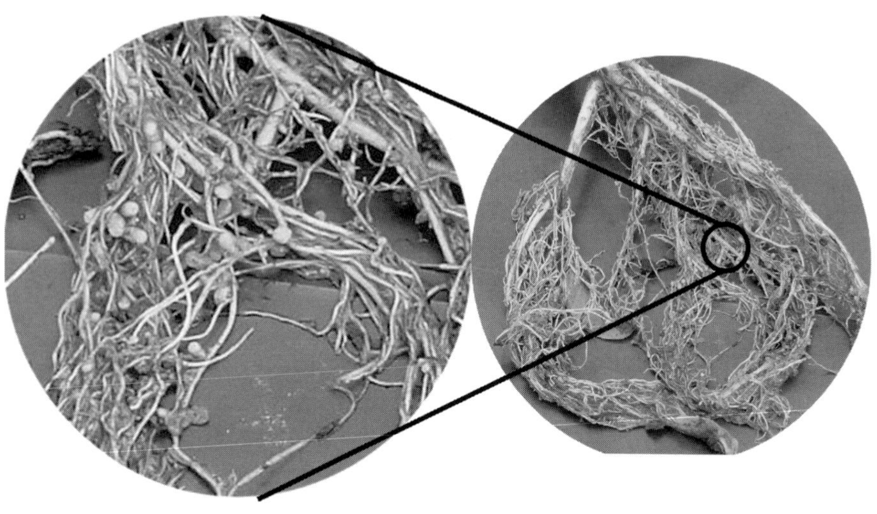

Figure 9.3 Nodules grown on clover roots (Rhizobium), nodules as center of high level of auxins and cytokinins.

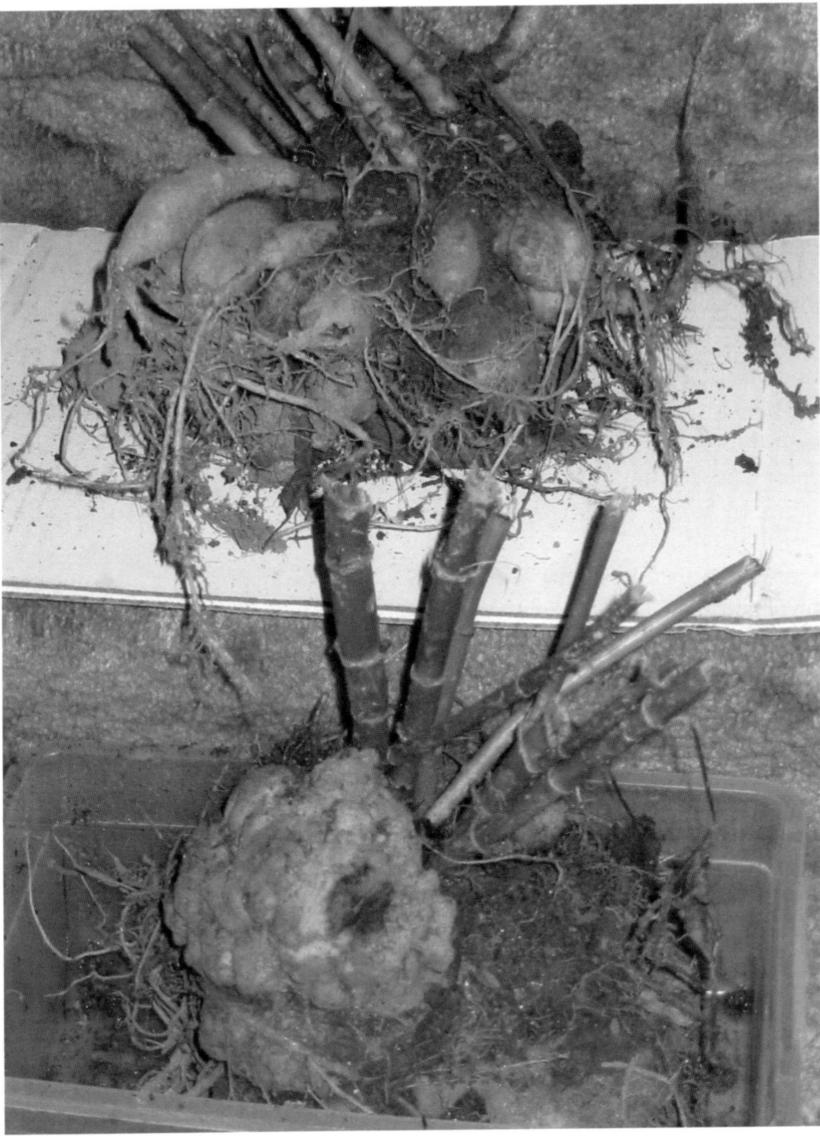

Figure 9.4 Tumor(lover part) induced by Agrobacteriym on Dahlie roots.Tumor as centera of high level of auxins and cytokinins. Upper part is nornal Dahlie roots.

and gibberellins (Chilton, 2001; Long, 2001; Ziemienowicz et al., 2001). Beside bacteria the fungi are also able to produce phytohormones and inhibitors.

With this object in view we have selected two groups of objects: *Taphrina sadebeckii* and *Taphirina epiphylla* fungi which induce pathologic growth in higher plants, and the hypocotyls and cultures of cabbage tissues. The *Taphrina* fungi were incubated on the Rollen-Tom medium with L-tryptophan (1 g/l) and without it at room temperature during 10 days. The concentration of indole auxins in the medium was estimated calorimetrically after treating the medium with the Salkowski reagent. Chromatographic analysis has shown that the mixture of indole auxins comprised mainly IAA and indoleacetonitrile (Figure 9.5), and that the IAA content increased appreciably in the tryptophan medium to reach 23 mg/l by the 10th day (Drakina & Kefeli, 1967; Kefeli, 1968).

The cabbage tissue culture and cabbage hypocotyls produced much less auxins than the fungi. Auxins were contained in the tissue itself and were released into the medium. Growing of the leaf blade callus on a medium with L-tryptophan during these days had a highly accelerating effect on the formation of auxins (Figure 9.5) both in the medium and in the tissue (Kefeli, Komizerko et al., 1972).

Apart from the substance with R_f IAA, the chromatograms also revealed the presence of a compound with indoleacetonitrile R_f whose content also increased after the addition of tryptophan into the medium. Accumulation of the substance with R_f IAA and measurement of its UV spectrum confirmed its identity The cabbage callus converted up to 0.1% of the introduced tryptophan into auxin: the amount of tryptophan converted into auxin in the fungus culture was 3 to 4%. The IAA content in the fungus culture was so high that it could be calorimetrically detected by the Salkowsky reagent, whereas in the tissue culture the substance with Rf IAA was identifiable only by means of coleoptile segments. The same technique was used to detect IAA in non-sterile leaves of Savoy cabbage (Kutacek & Kefeli, 1968). The above results suggest the following conclusion: a fungus and a higher plant, when placed in sterile culture conditions, can synthesize auxins, this ability being greatly promoted by the addition of tryptophan into the medium.

The tissues of a higher plant convert into IAA 20-30 times less tryptophan than a fungus. The factors causing this low activity of a higher plant are to be considered below.

Despite the fact that sterile tissues of a higher plant were able to convert tryptophan into IAA, it was necessary to be cautious with regard to a series of experiments conducted with non-sterile tissues in order to exclude the effect of bacterial auxins. Phenolics as signals play an inducting role in cohabitation of plant and pathogen (Lynn, Chang, 1990).

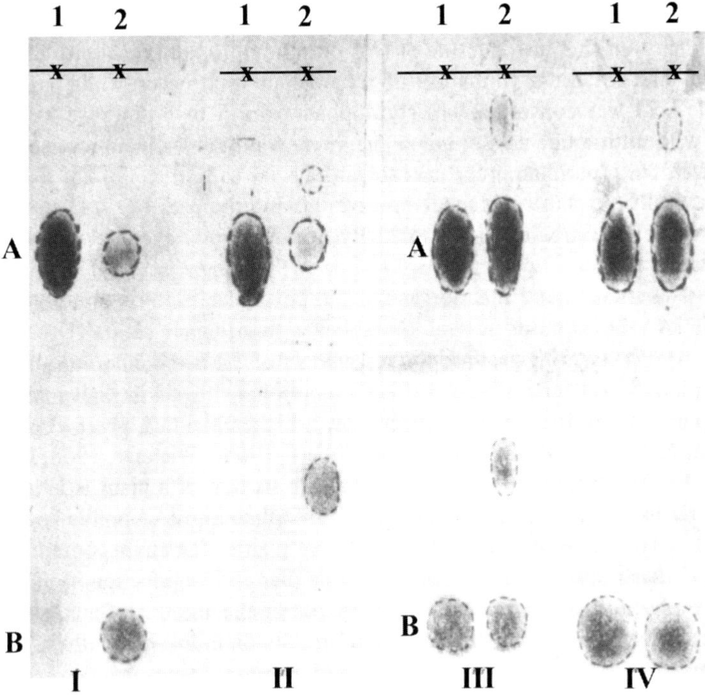

Figure 9.5 Chromatograms of extracts from culture medium of Taphina fungus developed with Salkowski and Ehrlich reagents (Drakina and Kefeli, 1967). I – medium without tryptophan (Salkowski reagent); II – medium without tryptophan (Ehrlich reagent); III – medium with (Salkowski reagent); IV – medium with (Ehrlich reagent); a – IAA; b – indoleacetonitrile; 1 – standards; 2 – indoles of extract from growth medium.

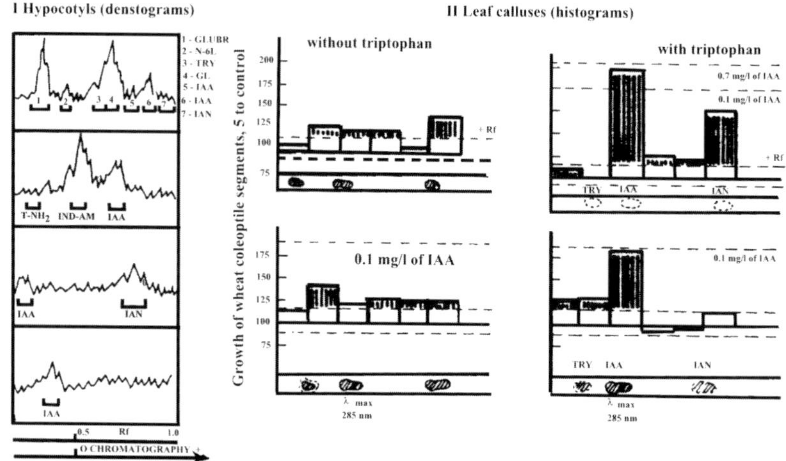

Figure 9.6 Production of indolyl-3-acetic acid and indoleacetonitrile in cabbage hypocotyl segments (I) and callus (II).
A – primary chromatogram (in mixture n-butanol-CH_3COOH-H_2O, 4:1:2); B – electrophoregram of zone 6; C – chromatogram of zone 7 (in mixture triechloroethylene – CH_2COOH, 100:2); D – chromatogram of zone 7 (in mixture isopropanol-25% NH_4OH-H_2O, 10:1:1).For histograms II, III: separation on paper in mixture mixture isopropanol-25% NH_4OH-H_2O, 10:1:1)

Dr. M. Kalevitch (Filimonova), 1992 showed that Botrytis cinerea can produce cytokinins of the zeatin group, predominantly zeatin-ribosine, as determined by HPLC. During the submerged cultivation the content of cytokinins increased remarkebly. These cytokinins were used as a hormonal component in Murasige - Skoog-media for tomato tissue culture cultivation or for the stimulation of *Spirodella* growth. Beside cytokinins the same fungi produces ABA up to 600 mg/l for *Botrytis cinerea* (Table 9.2, 9.3)

Table 9.2 Content of n-butanol soluble cytokinins in medium of fungi Botrytis anthophia (chromatogram region Rf 0.1 with the highest activity)

Weeks of fungi growth			
	4	7	8
Biological activity (%%)	183	150	144
Content mg/equiv. Kinetin/l	0.18	0.08	0.07

The level of ABA in the medium of *Botrytis* fungi is presented in the table 9.3

Table 9.3 Abscisic acid content in the medium of different Botrytis fungi

	Botrytis species								
	B. cinerea			B. allii			B. anthophila		
	4	6	8	1	2	3	4	7	8
Content of ABA mkg/l	291	513	593	35	36.2	63	0.2	9.3	9.3
Content of ABA mkg/g	75	78	107	2.1	6.0	17.1	0.02	1.7	0.9

In the nature the interaction of plants and microbial pathogens take place on the base of hormonal- inhibitor exchange. The forms of exchange could be not only restricted metabolites but also alterations in plant genome (Klee, Estelle, 1991; Somerville, 2001). Efficient vector systems for alien gene introduction and expression in the plant genome were produced on the basis of *Agrobacterium* plasmids. The various effects of plant specific expression of *Agrobacrerium tumefaciens* isopentenyl transferase gene (ipt or gene 4 T-DNA) have been discussed. We have used transgenic tobacco plants carrying the ipt gene from *A. tumefaciens* as a model to study changes in phytohormonal status (Makarova et al., 1997, 1999)

Transgenic plants carrying isopentenyl transferase (ipt) gene and normal tobacco plants (*Nicotiana tabacum L.)* were analyzed to compare their phytohormone status. Total Cytokinin (zeatin, zeatin riboside, isopentenyladertine and isopentenyladenosine) level and free IAA content were always higher in shoots regenerated from transgenic culture although the concentrations were lower in roots. In transgenic plants, IAA-oxidase activity was lower and the concentration of its protectant chlorogenic acid was increased. Transgenic plants also contained lower concentrations of ABA. It was also shown (Rezzonico et al., 1998) that ABA inhibits the induction of 1,3-glucanases the enzymes which have been implicated in stress responses.

These data correspond to the idea of Tamagnone et al. 1998 who investigated the intermediates in phenolic acid metabolism in the leaves of transgenic plants. They observe several complex phenotypic changes are induced when the transcription factor AmMYB308 is over expressed in transgenic tobacco plants. The primary effect of this transcription factor is to inhibit phenolic acid metabolism. In the plants that were produced two morphological features were prominent: abnormal leaf palisade development and induction of premature cell death in mature leaves. Evidence from the analysis of these transgenic plants suggests that both changes resulted from the lack of phenolic intermediates.

The pathway of phenolic acid metabolism in plants requires the initial steps of general phenylpropanoid metabolism and provides the precursors for lignin biosynthesis. However, intermediates in the pathway and derivatives of these intermediates are ubiquitous in plants and accumulate to significant levels in tissues that do not synthesize lignin. The function of these intermediates, if any, is not yet clear, although several different roles have been proposed.

Phenolic compounds, particularly hydroxycinnamates, are present at significant levels in plant cell walls, where they may act as molecular bridges (Kefeli et al., 1983). For example, in grasses, 4-coumaric acid and ferulic acid link lignin to polysaccharide polymers, such as glucuronoarabinoxylan, through labile ester and/or ether bonds (Jung et al., 1993; Wallace and Fry, 1994). In the *Chenopodiaceae*, hydroxycinnamic acids are found in all cell walls attached to pectic polysaccharides (Wallace and Fry, 1994). Pectins may also be cross-linked to other pectins or to other noncellulosic polysaccharides through ester linkages with dimerized hydroxycinnamic acids, such as diferulic acid (Fry, 1986. Parr et al., 1996; Waldron et al., 1997). Although the importance of this cross-linking role has been difficult to evaluate (Fry, 1983), there is evidence that cross-linking of matrix po1ysaccharides through diferulic acid bridges in oats plays a protective role, because it increases in an incompatible interaction with the crown rust pathogen to provide a barrier to pathogen invasion (Ikegawa et al., 1996). In our experiments (Makarova et al., 1997) we also observed some abnormalities in tobacco transgenic plants.

Regenerated transgenic plants carrying an active foreign gene for phytohormone synthesis, in particular the ipt gene that encodes isopentenyl transferase, have become a classic model system to study the growth and development of higher plants (Pirus, Yan, 1990)

Gene expression and the balance of endogenous hormones are known to interact, and the phytohormone balance affects plant morphogenesis and resistance to stress factors (Kefeli et al., 1990, Makarova et al., 1996a)

The present study was focused on producing ipt-transgenic tobacco plants and the investigation of their morphogenic and regenerative potentials. We addressed the question of how the expression of a single active foreign gene could affect the regeneration of the entire plant and the sensitivity to growth regulators in the cultured cells derived from ipt-transgenic plants.

Portions of the wild-type and transgenic plants were used for callus tissue production (Figure 9.7). Entire stems of young plants (three-to five-week-old) or the middle parts of the stems of the older plants were cut into 5-mm segments. Explants were placed into 100-ml flasks (five segments per flask, six to eight flasks in each treatment) or in 20 x 200 mm tubes (one explant per tube, 20 - 30 tubes in each treatment) on MS^- or MS containing various combinations of auxins (IAA, 2,4-D) and cytokinins (BA, kinetin). The average medium volume was 30 ml per flask and 10 - 15 ml per tube.

After they had been exposed to darkness for three weeks, we assessed the frequencies of callus formation in the wild-type and transgenic explants, calculating the proportion of explants that produced callus tissue out of the total number of explants placed on a given medium.

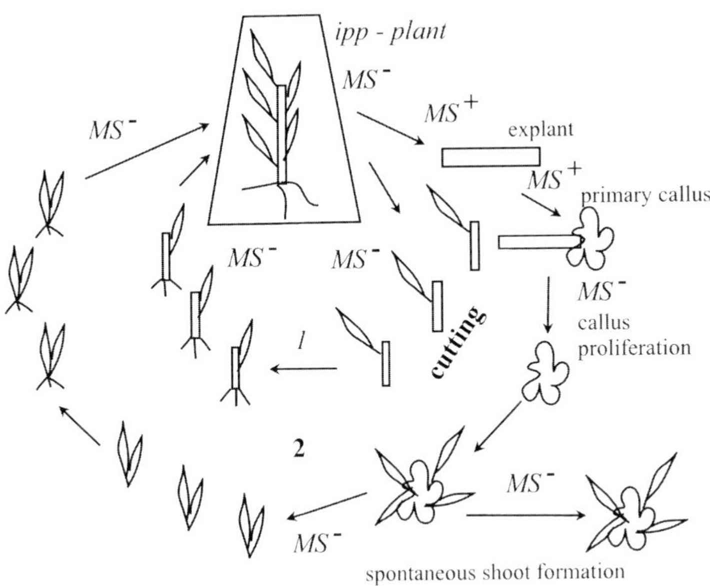

Figure 9.7 Scheme of tobacco plant regeneration from (1) cuttings and (2) callus tissue.

Abbreviations: BA-benzyladenine; cyt-callus – callus derived from an ipt plants; ipt – plant – a regenerant carrying the ipt gene for cytokinin synthesis; MS^+ - complete Marashige and Skoog nutrition medium; MS^- - hormone-free Marashige and Skoog nutrition medium; NPT – neomycin phosphotranspherase; PCR – polymerase chain reaction.
Abbreviations: Z – zeatin; ZR - zeatin riboside; IPA –isopentenyladenine; IPAs - isopenteniladenossine; ipi-plant - a regenerant carrying the ipi gene for cytokinin synthesis.

Primary callus tissues derived from wild-type and transgenic explants were subdivided into small pieces (150 – 200 mg) that were placed into tubes (10-20 tubes in each treatment) on MS^- or MS containing various combinations of auxins and cytokinins. Calli of both types were kept at 26^0C under continuous illumination or in darkness for at least half a year with subculturing every 21-30 days. At the end of each subculturing cycle, growth and morphogenic potential was assessed by weighing calli and by counting the proportion of morphogenic calli out of their total number.

Microshoots grown from wild-type and transgenic callus tissues were planted in flasks with MS and propagated by cutting after 6, 8, and 10 weeks.

In regenerated plants, we counted the number of leaves and roots and measured the height of shoots.

Since MS supplemented with 1 mg/l 2,4-D and 0.1 mg/l kinetin provided for active callus formation on both types of explants and further rapid proliferation of callus tissue, this medium was used in subsequent experiments as a complete nutrient medium (MS^+).

Thus, we used two types of MS: (1) MS^- (hormone-free) for growing regenerated tobacco plants, both wild-ype and transgenic, and for the maintenance of the collection of calli derived from transgenic plants (cyt-callus), and (2) MS^+ (containing 1 mg/l 2,4-D) and 0.1 mg/l kinetin) for the induction of callus formation on explants of both types and for the proliferation of the cultured wild-type cells.

Usually, plants had erected firm stems of medium thickness, with long internodes and large leaf blades; the plant root system was well developed. After six weeks of culturing, the leaves of wild-type regenerated plants turned yellow and fell off.

Regenerating shoots formed by ipt plants varied in their morphology. Most commonly, easily rooting and normally developing plantlets were formed. Extremely rare were rosette plantlets lacking roots. And almost in each subculture, there were observed several tubes with numerous microshoots and shootlike structures, sometimes covered with callus tissues, which were developed on a single explant the right tube.

During the first five subcultures, the stem cuttings derived from secondary regenerants developed roots three-five days earlier than the wild-type regenerants. Later, rhizogenesis became somewhat less active. During the first subcultures, the ipt plants were also ahead of wild-type ones by their rates of stem and leaf growth. After subculturing for four to six weeks, the transgenic tobacco shoots turned into well-developed plants, with some characteristic traits.

At the age of 7-10 weeks, the transformed plants grown in 1-1 flasks had shortened compact shoots with a great number of leaves, both large, normal ones and numerous small leaflets along the stem. Roots were well developed, although their total mass was lower than in the wild-type plants (Table 9.3). Transgenic plants had a longer juvenile period: their leaves yellowed later, and the capacity of their stem explants to form calli declined as well. New shoots were formed in the axis of almost all leaves of the ipt plants. After detachment from maternal plants, these shoots easily rooted and developed well.

9.1 Callus production by wild-type and transgenic regenerants

In wild-type plants, the highest frequency of callus production was observed in the explants from young plants. The explants from plants cultured for more than seven weeks formed calli quite rarely. High frequencies of callus formation were observed in the wild-type explants when various concentrations of 2,4-D and kinetin were added to the medium, independent of hormone ratio. Replacement of 2,4-D with IAA or NAA usually inhibited callus initiation. Initiation did not occur on the hormone-free medium (Table 9.4)

Transgenic explants formed callus tissue only when 1 mg/l 2,4-D and 0.1 mg/l kinetin were added to the medium. As in wild-type explants, the replacement of 2,4-D with 1AA was not stimulatory. Only one of the combinations of IAA and kinetin resulted in callus formation on some of the explants. When other combinations of various concentrations of auxins and cytokinins were added, stem explants swelled, split, and extensive necrosis occurred.

The ipt-explants maintained their capacity for callus initiation longer than the wild-type explants. Some decrease in the frequency of callus formation and in the rate of callus growth was observed only in old transgenic explants (10 weeks or more; data not shown). The areas of dedifferentiated tissue produced at the distal parts of the explants darkened and did not proliferate.

Table 9.4 Growth of wild-type and regenerated tobacco plants on MS$^-$

Type of regenerant	*Age, weeks*	*Shoot height, mm*	*No. of leaves*	*Fr wt per leaf, g*	*Root fr wt, g*
Wild-type	6-7	87.8±20.1	17.4±1.2	0.16±0.03	1.66±0.25
	10-11	161.0±20.7	15.5±1.3	0.28±0.03	1.49±0.29
Cys-transgenic	6-7	115.3±20.9	29.0±2.0	0.22±0.02	0.94±0.19
	10-11	128.2±25.1	27.6±1.2	0.34±0.03	0.90±0.10

Thus, we observed auxotrophic formation of the primary callus tissue (cyt-callus), susceptible to the hormonal composition of the nutrient medium, and a high frequency of callus production by ipt-explants of various ages.

9.2 Proliferation of Subcultured Calli

Since both types of regenerated plants, wild-type and transgenic, manifested the capacity for primary callus formation on MS containing 1 mg/l 2,4-D and

0.1 mg/l kinetin, and these callus tissues rapidly grew on this nutrient medium. we selected this particular medium for further investigation of callus proliferation.

During long-term subculturing, some differences in the sensitivity to the hormonal composition of the medium were found between the wild-type and transgenic calli.

Table 9.5 Effect of 2,4 – D, kinetin, and light on the growth of wild-type and cyt-callus

Callus	*Madium*	*Callus fr wt, g/tube*	
		light	*darkness*
Wild-type	MS^+	4.5±0.9	4.3±1.0
	MS^-	0.30±0.02	0.21±0.03
Transgenic Cys -	MS^+	5.3±1.7	4.8±1.4
	MS^-	4.4±0.9	4.1±1.0

Table 9.5 presents the data on callus proliferation during three-week-long culturing on the complete (MS^+) and hormone-free (MS^-) media in darkness and in the light. Cyt-callus proliferated at a similar rate on the hormone-free and hormone-containing media, in the light and in darkness. The growth of the wild-type callus ceased rapidly on the hormone-free medium regardless of illumination conditions. The loose and yellow wild-type callus, grown on the hormone-containing medium, had a high growth index. Callus density depended on the hormonal composition of the medium (data not shown).

Cyt-callus developed into a tissue of mixed embryo-genic-nonembryogenic type, homogeneous in density. In the light, the callus had a globular structure and was bright green. The globules consisted of numerous shootlike structures varying in size from faintly visible to 5 mm; when transferred to MS^-, each of them easily rooted and developed. These regenerants were propagated by cuttings. The morphogenic potential of cyt-callus grown on hormone-free medium could be improved by adding kinetin.

In cyt-callus grown in darkness, the proportion of nonembryogenic tissue was relatively small. By recurrent isolation from mixed cyt-callus tissue. nonembryogenic fragments that proliferated rapidly on MS^- were selected. Their high growth potential did not depend on the hormonal composition of the medium. After sub-culturing, this tissue rapidly proliferated and formed microshoots and shootlike structures. It is worth noting that microshoots developed during two-three subsequent subculturings in darkness. Sometimes. nonviable structures such as shoots were formed on the cyt-callus proliferating in darkness on hormone-free medium.

Morphogenesis of cultured plant tissues depends on their genetic organization. When plants regenerate from long-cultured wild-type tobacco tissues, they are usually morphologically modified: branched stems, deformed leaves with spots on them, etc.. Such morphological modifications could be the result of mutation or result from some genetic changes. The formation of numerous leaf and shoot deformations indicate a disturbance in growth correlations. One of the possible causes could be an elevated kinetin level.

The insertion of the active bacterial ipt gene into the genome of a higher plant and the expression of this gene result primarily in an increase in the level of endogenous cytokinins (the paper concerning this problem is currently in press).

Substantial experimental data have accumulated concerning the physiological properties of ipt plants and regenerants derived from tumor tissues. Thus, tissue cultures, initiated on ipt-explants, acquire prototrophy, i.e..· independence of the presence of phytohormones in the medium (Table 9.4). Wild-type calli acquire hormone-independence only as a result of site-directed mutagenesis and cell selection or habituation.

Expression of the bacterial gene for cytokinin synthesis influenced the growth and development of regenerated plants. When transgenic tobacco tissues were cultured in the light, we observed variations in the shoot shape. In darkness, cyt-callus sometimes produced abnormal shootlike structures with a reduced developmental cycle.

It is a well-known fact that malformations often result from disturbances in the hormonal balance. Tumor proliferation is determined by a specific ratio of auxins to cytokinins; vitrificated and rootless shoots arise due to an excess of cytokinins.

The basic difficulty in the production of ipt-transgenie plants is poor root formation. With improved methods of gene engineering, a new technique for the insertion of the necessary gene into the plant genome was developed and this difficulty was overcome.

Using a binary system comprising a "mini-Ti" vector plasmid and an integrating Ri plasmid, we succeeded in the in vitro regeneration of ipt plants from individual infected cells.

By now, collections of normally developing ipt plants (alfalfa, potato, tomato, etc.) are maintained in several laboratories.

Investigations of cytokinins, as well as other phytohormones, by the methods of gene engineering follow two basic lines; (I) the identification of genes expressed under cytokinin control, and (2) the production of transgenic plants with a cytokinin level modified by transfering the ipt-gene from T-DNA of *Agrobacterium.*

Earlier in the Piruzyan laboratory (Institute of Molecular Genetics, RAS), a new vector class was developed to help gene integration into the plant genome. We studied in detail the morphology of ipr-transgenic plants and their capacity for regeneration. It turned out that the expression of the ipt bacterial gene affected several growth parameters in tobacco regenerants. Thus, such plants produced roots several days earlier than the wild-type plants. During subculturing for several years, this capacity to form a well-developed root system was maintained (an inherited trait). although the roots of transformed plants grew more slowly than the wild-type plants (Table 9.6). Enhanced shoot initiation (numerous axillary shoots and shootlike structures on callus tissue) resulted in the production of plants with an increased number of shortened internodes.

In ipt-transgenic explants callus initiation occurred only in the presence of hormonal supplements to nutrient medium. This callus tissue was highly morphogenic and proliferated without exogenous hormones (Figure 9.10).

The model of transgenic plant with the artificially increasing level of one of hormones let to elucidate following questions-

-the effect of changed hormonal balance on the morphological properties of plants

- the effect of one dominating hormone on the level of the other hormones

-the influence of hormonal level on the formation of inhibitors

The investigation of growth regulation by treating plants with exogenous hormones had some serious limitations. Transgenic plants with changed levels of hormones or changed sensitivity to them allow one to elucidate in greater detail the contribution of individual hormones to the maintenance of plant integrity. However, even in this case. the contribution of individual hormones cannot be evaluated unambiguously.

Inserting the ipt gene of *Agrobacterium tumefaciens* encoding isopentenyltransferase (a key enzyme in cytokinin biosynthesis) into the plant nuclear genome results in cytokinin superproduction and, consequently, in disturbances in plant morphogenesis and regeneration. However, diverse physiological and genetic manipulations have helped to select transgenic plants without serious disturbances to their development. Nevertheless, the physiological behavior of a plant containing an active foreign gene for phytohormone synthesis, the ipt gene in particular, could be unpredictable.

The regeneration and morphological traits of transformed tobacco plants in vitro after insertion of the ipt gene from *Agrobacterium tumefaciens* and also the development of the capacity of the differentiated tissues of transformed tobacco to grow in hormone-free medium. The goal of this work was to study the patterns of endogenous phytohormones in wild-type and ipt-transgenic tobacco plants during various periods following transformation.

Our research described the phytohormone production by tobacco ipt-regenerants in vitro.

The contents of cytokinins (zeatin, zeatin riboside, isopentenyladenine, and isopentenyladenosine); IAA, and some indole derivatives in leaves and roots and also ABA in leaves were compared in wild-type and ipt-transgenic tobacco (*Nicotiana tabacum L.*) plants in vitro. Wild-type tobacco plants contained 1.5 times more cytokinins in their roots than in their leaves; both organs contained predominantly the derivatives of iso-pentenyladenine. The ipt-plants of various ages (6-8 month and 18-24 months) contained twice the amount of cytokinins as the wild-type plants; relatively higher amounts of zeatin and zeatin riboside and also free IAA were presented in the leaves than in the roots. Transgenic plants showed a decreased activity of IAA oxidase and an increased level of chlorogenic acid. ABA content in the ipt-plants was somewhat lower. Co-chromatography with tryptophan, indole-3-acetamide, indole-3-lactic acid, and IAA-aspartate demonstrated quantitative differences in the amounts of these compounds in the wild-type and transformed plants. The leaves of ipt-plants contained less tryptophan and IAA-aspartate and small amounts of indole-3-acetamide; only traces of all tested indole derivatives were found in the roots of ipt-plants. The ratio of free cytokinins to auxins, established in ipt-plants, may determine the previously described basic morphological features and the capacity for regeneration of transgenic plants.

Table 9.6 indicates that a major portion of cytokinins in the wild-type plants, comprising about 63% of total plant cytokinins, was found in plant roots. This percentage was similar in the plantlets maintained in culture for

Table 9.6 Content of free cytokinins in wild-type and transformed tobacco plants.

Plant form	*Exp. Series**	*Organ*	*Cytokinins, ng/g fr wt*				
			Z	*ZR*	*IPA*	*IPA's*	*Total*
Wild-type	1	Leaves	35.5	22.5	42.5	55.0	155.0
		Roots	50.0	35.0	80.0	100.0	265.0
	2	Leaves	37.5	22.5	80.0	35.0	175.0
		Roots	1000.0	37.5	94.0	76.0	307.0
Ipt-Regenerant	1	Leaves	308.0	210.0	52.5	50.0	620.0
		Roots	17.5	22.5	85.	97.5	222.5
	2	Leaves	120.0	115.0	182.0	107.0	524.
		Roots	87.5	42.5	73.0	35.0	238.0

* Z – zeatin;
ZR – zeatin-riboside;
IPA – isopentenyl-adenin;
IPA's – isopentenyl-adenines.

6-8 and 18-24 months after transformation, although the total amount of cytokinins slightly increased A comparison of individual cytokinins showed that the wild-type plants contained mainly IAA derivatives; during long subculturing, the amount of IAA in leaves of the wild-type plants somewhat increased, and the amount of IAAs decreased.

On the contrary, the roots of ipt-regenerants contained only 25% of total cytokinins, also mostly IPA derivatives. Leaves of the transformed plants contained more zeatin and its derivatives. The total amount of cytokinins was higher in the transformed plants and did not depend on the time period after transformation (Table 9.6).

Free IAA was concentrated mainly in the leaves of both wild-type and transformed plants (Table 9.7). Leaves of the transformed plants contained three times more and roots three times less free IAA than the corresponding organs of the wild-type plants.

It is worthy of notice that both types of plants synthesized similar physiologically active phenolic compounds (tryptophan, indole-3-acetamide, indole-3-lactic acid, and PA-aspartate) However, in leaves of the wild-type plants, the level of tryptophan was 10 times higher than PA; the corresponding ratio was only two in transgenic plants. Thus, the level of IAA precursor in the transgenic plants was relatively low, and the level of PA was very high.

Table 9.8 shows that transformation depressed the activity of PA oxidase in leaves and roots.

Leaf extracts from both types of plants contained chlorogenic acid, an auxin protector (the inhibitor of IAA oxidase). Its level was higher in the transgenic plants than in the wild-type plants (0.44±0.02 and 0.12±0.01 mg/g fresh weight, respectively).

The level of ABA was lower in the transgenic plants (45.5±1.8 as compared to 75.7 ± 4.5 mg/g fr wt in wild-type plants).

It is well-known that plant regeneration from plant tissues depends on the balance of endogenous phytohormones, and considerable changes in phytohormone ratios result in the formation of abnormal, usually nonviable plant forms.

Rhizogenesis is a prerequisite for plant regeneration. The capacities for root formation and root growth in vitro depend much on the ratio between endogenous auxins and cytokinins. Predominant auxin accumulation can trigger root formation; in other cases, auxins repress root growth. Some evidence indicates that IAA can activate rhizogenesis in leaf cuttings of kidney bean plants by stimulating cytokinin formation.

Cytokinins play some role in rhizogenesis, although it is not always clear what their particular contribution is to this process. Endogenous cytokinin was found in the basal part of tobacco cuttings at the moment of root utilization. Cytokinins are also known as inhibitors of adventitious root formation. Sometimes they inhibit root elongation (supposedly via their effect on the IAA level).

The insertion of an active ipt gene of *Agrobacterium tumefaciens* encoding isopentenyl transferase into the plant genome and its expression in the cells of transgenic regenerants results in the superproduction of cytokinins and, as a rule, auxins, as is supported by our data (Table 9.6 and 9.7)

The ability to produce various regenerated phenotypes and the morphological and physiological characteristics of transgenic calli primarily depend on the ratio between endogenous cytokinins and auxins. Normal rooting of cuttings, evidently, occurs only at a definite hormonal balance. Thus, when phytohormone concentrations increased by 3-10 times, as occurred in our experiments, the cuttings readily formed roots. However, when these concentrations increased by 50-100 times, neoplastic (tumor) growth was observed .

The special method of insertion of the foreign ipt gene and the selection of ipt-regenerants at the stage of microshoots allowed us to produce plant phenotypes that retained the capacity for root formation.

Earlier, we characterized the growth of these regenerants in comparison with that of the wild-type regenerants. For six to seven weeks after transformation, the ipt-plants were ahead of the wild-type plants in their stem height (115 versus 88 mm) and leaf weight (220 versus 160 mg), but their root system was poorly developed (0.94 versus 1.66 g). However, after longer subculturing (up to 10-11 weeks), the main stem of the ipt-plants usually ceased to elongate, whereas the wild-type plants continued to grow.

Thus, the transgenic tobacco plants used in this work had shortened stems, actively induced and rapidly growing axillary buds, a great number of leaves with a total weight higher than in the wild-type plants, and a poorly developed root system, which, however, was initiated earlier These characteristics indicate some disturbances in growth correlations. in this study, we analyzed the patterns of cytokinins and auxins in the leaves and roots of such regenerants. We observed several specific traits related to the introduction of the ipt gene into the tobacco genome. The wild-type plants contained 1.5 times more cytokinins in the roots (the natural site of their synthesis) than in the leaves. Both organs contained more IPA derivatives; the ratio between the two types of cytokinins was similar in the leaves and roots. The analysis of cytokinin level in leaves of the ipt-plants showed that their total content was higher than in the wild-type plants, mainly due to the cytokinin of zeatin type; the content of IPA derivatives was much lower in both leaves and roots. A most profound difference was observed after five to six subculturings following transformation. Later (after 18-24 months), the difference between the contents of Z and ZR disappeared, and their total concentration decreased substantially, although it was still higher than IPA derivatives.

Introduction of the ipt gene into the tobacco genome resulted in accumulation of not only cytokinins, but also auxins. In fact, the concentration of free IAA in the ipt-regenerants was approximately twice as high as in the wild-type plants. In both types of regenerants, leaves contained much higher levels of IAA than roots. It is interesting that the roots of transformed plants contained especially low concentrations of IAA (Table 9.7).

Table 9.7 Content of free IAA in wild-type and transformed tobacco plants

Plant form	*Organ*	*IAA, ng/g fr wt*
Wild-type	Leaves	40.0
	Roots	13.3
ipt-Regenerant	Leaves	112.7
	Roots	5.0

Note: Plants were analyzed 6-8 months after transformation.

Table 9.8 Activity of IAA oxidase in wild-type and transformed tobacco plants

Plant form	*Organ*	*IAA, ng/g fr wt*
Wild-type	Leaves	34.4±2.2
	Roots	62.1±15.2
ipt-Regenerant	Leaves	24.1±3.9
	Roots	32.2±8.1

The concentration of free auxin in the cell is related to the changes in the rates of its synthesis, degradation, and/or conjugation. A comparison between the levels of various indole derivatives (tryptophan, indole-3-acetamide, indolc-3-lactic acid, and IAA-aspartate) and IAA in the leaves of the wild-type and transformed regenerants showed that the increase in the level of free IAA in the ipt-plants was accompanied by a corresponding decrease in the level of its precursor (Figure 9.8), indicating the possibility of accelerated auxin synthesis.

The rate of auxin degradation by IAA oxidase can also contribute to changes in hormone content in the cell. The data found by other researchers concerning the correlation between the activity of IAA oxidase and the auxin level are rather ambiguous. Some researchers reported an inverse dependence between plant growth and the activity of IAA oxidase. On the other hand, there is some information about a direct correlation between the activity of this enzyme and tobacco growth in vivo and in vitro. The high activity of IAA oxidase is believed to reflect intense auxin metabolism. Tobacco plant transformation by insertion of the ipt gene resulted in a substantial decrease in the activity of IAA oxidase in both leaves and roots. In the case of roots, this decrease was correlated with a lower level of IAA (Table 9.7 and 9.8) and retarded root growth. Similar dependencies were observed earlier in etiolated pea seedlings.

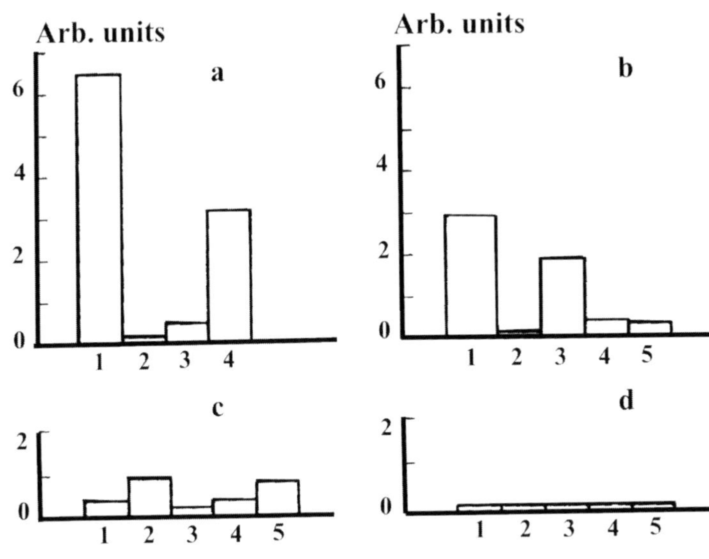

Figure 9.8 The level of indole derivatives in (a, b) leaves and (c,d) roots of (a,c) wild-type and (b,d) transformed tobacco plants. The level of indole compounds is expressed in arbitrary units (areas of corresponding peaks per union of fresh matter). (1) Tryptophan; (2) indole-3-lactic acid; (3) IAA; (4) IAA – aspartate; (5) indole-3-acetamide.

The low activity of IAA oxidase in transformed tissues could be determined by the increased level of chlorogenic acid, an inhibitor of enzymes degrading IAA. We observed a substantially higher level of chlorogenic acid in the transformed tobacco leaves, where the activity of IAA oxidase was reduced.

In our experiments, ABA level was somewhat reduced in the transformed shoots. A parallel can be drawn to the decreased level of endogenous ABA in the wheat plant treated with benzyladenine.

Thus, the insertion of the ipt gene into the tobacco genome changed the balance of phytohormones in the transgenic plants (regenerants). The level of free IAA and the total content of cytokinins in the leaves of the ipt-plants were higher than in the wild-type plants by three and four times, respectively. In contrast, the roots of the ipt-plants contained a lower amount of phytohormones, especially IAA (by 2.7 times). The ratio of cytokinins to IAA remained 1.5 to 2 times higher in both organs. When calculated per whole plant, the levels of both cytokinins and IAA approximately doubled after transformation, and the ratio of cytokinins to IAA was approximately the same in both wild-type and transformed regenerants (Table 9.9)

Table 9.9 Ratio of cytokinins to IAA in wild-type and transformed tobacco plants

Plant form	Organ	(Z+ZR)/IAA	IPA+APAs/IAA	Total cytokinins/IAA
Wild-type	Leaves	1.43	2.44	3.88
ipt - Regenerant		4.60	0.91	5.51
Wild-type	Roots	6.39	13.53	19.52
ipt - Regenerant		8.00	36.50	44.50
Wild-type	Whole plant	2.67	5.20	7.87
ipt - Regenerant		4.74	2.42	7.16

It follows that plant regeneration from tobacco stem cuttings in vitro is determined by a definite ratio of cytokinins to IAA. This ratio can be maintained by accelerated auxin accumulation in response to the increased level of endogenous cytokinins, as in the case of the ipt-gene introduction into the tobacco genome (Makarova et al., 1990).

On the other band, in the transgenic plants with a genetically determined level of IAA, cytokinin concentration can change in such a way that the development of regenerants occurs as in the initial plant forms (Butenco et al., 1979)

In our opinion, the correlation between the levels of cytokinins and chlorogenic acid is of certain interest. In fact, the ipt-regenerants contained a much higher amount of chlorogenic acid, in accordance with the low activity of IAA oxidase (Karanova, Makarova, Kefeli, 1979; Kutachek et al., 1981).

Thus, the insertion of an active bacterial gene for cytokinin synthesis into the tobacco genome resulted in an increase in the levels of both cytokinins and IAA. At the same time, the ratio between these hormones remained essentially constant. Apparently, it was this constant ratio that determined the capacity for successful plan regeneration. On the other hand, the observed changes in auxin metabolism (a decreased activity of IAA oxidase and an increased level of chlorogenic acid, an inhibitor of auxin-degrading enzymes) could determine the peculiar phenotypic traits of regenerants that we observed earlier: increased capacity of transgenic tissues for callusogenesis, increased capacity of cultured callus for organogenesis (the formation of shoots and shoot-like structures), and independence of transgenic plant growth in vitro of exogenous phytohormones.

The growth inhibitors biosynthesis and hormonal-inhibitor balance can be effected by this as well (Tamagnone et al., 1998). Karanova and co-authors (1999) showed that both lines of *Dioscorea* have similar activity of IAA-oxidase, but the mutant lines possess the higher ability to conjugate the excess of indolic auxins precursors-antranilate and tryptophan (Kutachek et al., 1981; Butenko et al., 1979; Kefeli et al., 1979).

The most pronounced was the lost of the sensitivity of cells to auxin (A-) or its precursor tryptophan, in the comparisson of the control, non treated by mutagen cells (A+) (Figure 9.9)

These two types of experiments with transgenic tobacco plants and with mutated *Dioscorea* cells showed that the change of hormonal levels in plant cells could also be accompanied by morphological changes or even by the development of the mechanisms of hormonal receptions.

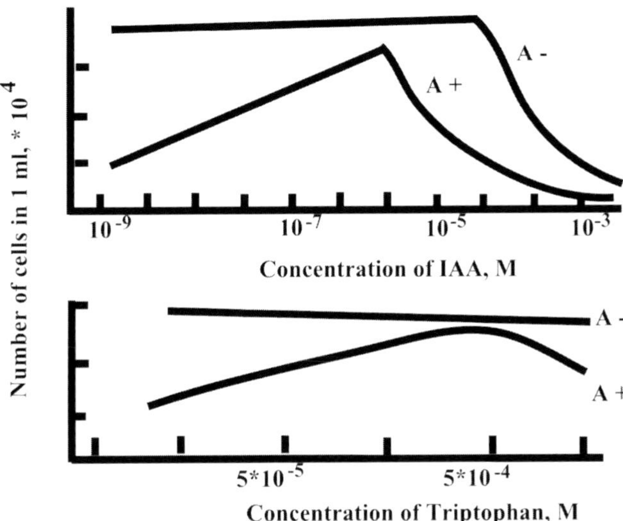

Figure 9.9 Sensitivity of Dioscorea cells to auxin-IAA and its precursor tryptophane. A – mutant cells; A+ control – non mutated cells.

The effect of the excess of cytokinins on the accumulation of phenolic inhibitors and related substances were investigated by authors back in 1985. We selected for the research cytokinin sensitive objects tissue culture strains of tobacco and soy as well as moss *Funaria hygrometrica*. In these objects by means of two dimensional chromatography (Figure 9.10 and 9.11) and HPLC were identified phenolic acids and flavonoids. Tobacco cuts contained scopoletin, p-coumaric and caffeic acids. Kinetin in the concentration 10 mkg/ml activate dramatically the concentration of chlorogenic acid in the callus (Figure 9.12) and isoflavone daidzein in soy callus (Figure 9.13). About 7 % of phenolic acids incorporated in the vascular elements of the tobacco callus where as daidzein in the soya callus subjected to glycosidation. Callus models do not allow considering the cellular process of a certain sequence in the differentiation (Filonova, 1985)

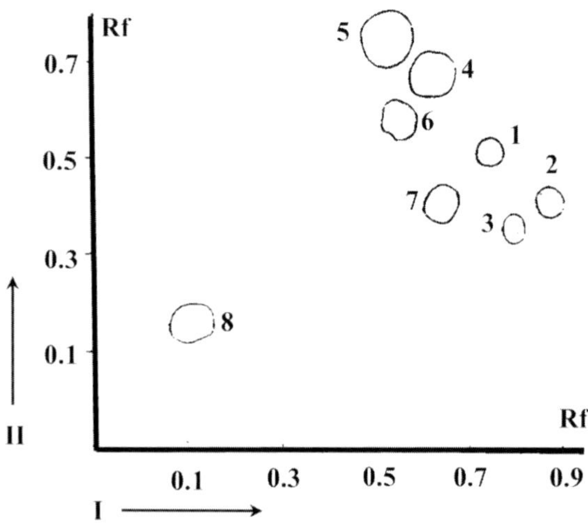

Figure 9.10 Two-dimensional chromatogram of the phenolic compounds of tobacco callus. l-n-butanol-acetic acid-water 4: 1:5 11-2% acetic acid
l-scopoletin 2-p-coumaric acid 3-caffeic acid 4-S-chlorogenic acids 6-S- non identified phenolics

So, it was selected the protonema of *Funaria* as an object for differentiation. The kinetin after 8 days of *Funaria* protonema growth induced in the target cell the formation of branch as a result of the asymmetric mitosis. After 48-50 hours the target cells transformed to buds and in 72 hours bud brings gametophores. Moss protonema contains two types of phenolic substances (Figure 9.14) the substances 1-4 exist in the protonema during its growth and the substance 5 appears only after 48 hours exposure protonema with kinetin . Hydrolysis of this substance with the subsequent HPLC chromatography allowed to identify this substance as cinnamic acid. This acid accumulated in the period, which precedes the branching of caulonema of the moss. The integrative reactions in higher plants are less clear, but the probable role of lectin/hormone ratio might exist (Lozhnikova et al., 1985).

Thus the process of differentiation of the tissues in calluses or in cell chains of moss is accompanied by accumulation of phenolic acids or their precursors. The corn root could be the good model for the investigation of anthocyanins formation in the root (Arapetjan , 1985, Touchnin , Kefeli 1995) anthocyanin appears only after the irradiation of the root in the light (Table 9.10). The zone of its location is the region of differentiation of the root. ABA and IAA inhibited the root elongation and simultaneous1y the formation of antocyanins was depressed. In general it is possible to conclude that transition from the process of cell elongation to cell differentiation is accompanied by the accumulation of phenolic substances and their precursors, but this process could be a variation in the isolated cells (Grotewold et al., 1998).

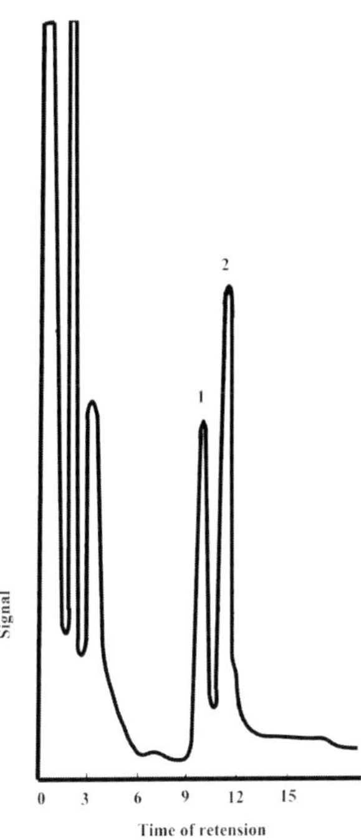

Figure 9.11 Gas chromatography of phenolic compounds of tobacco callus using liquid phase Apiezon L.1-neochlorogenic acid 2- chlorogenic acid(in form of tri-methyl-sillil esters)

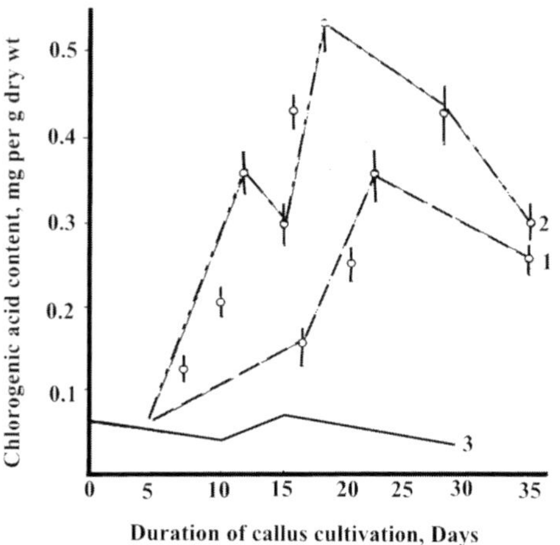

Figure 9.12 Content of chlorogenic acid in durind the tobacco callus growth A-Amount of chlorogenic acid/mg g of callus dry mass B-duration of callus cultivation , days

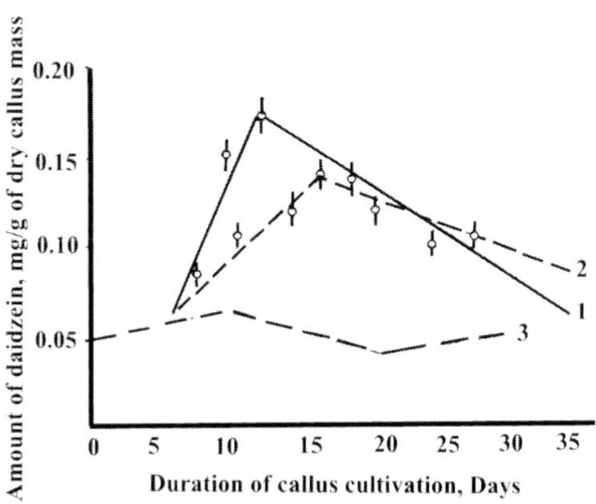

Figure 9.13 Accumulation of daidzein in soya callus

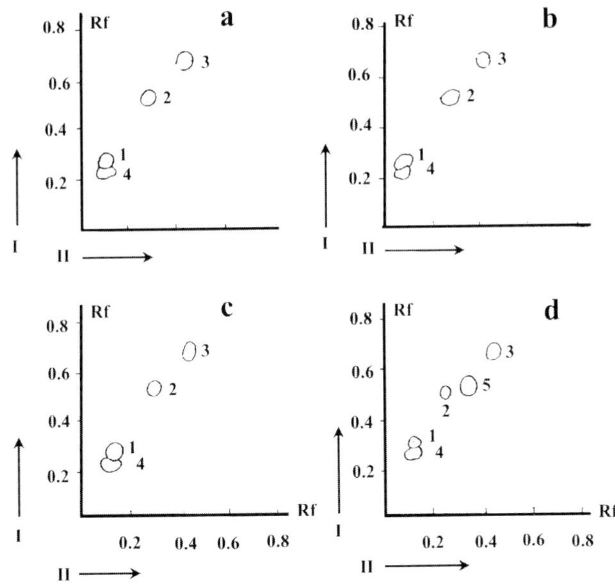

Figure 9.14 Two dimensional chromatograms of the methanol extract of Funaria protonema l-n-butanol-acetic acid-water 3:1:1 11-15% acetic acid Application of kinetin a- after 3 hours b-after 9 hours, c-after I8 hours d- after 48 hours

Table 9.10 Effect of indol-3-acetic acid and abscisic acid on the growth and the length of the anthocyanin part of the corn seedling root (mm)

Variant	*Root length, mm*		*Lan*
	Whole root (L)	*Anthocianin part (La)*	*Ratio La/L*
Control	104±7	79±9	0.76
ABA $5*10^{-6}$M	72±4	56±5	0.79
ABA $2.5*10^{-5}$M	64±2	37±3	0.59
ABA $5*10^{-5}$M	58±3	36±3	0.62
Control	120±9	83±12	0.70
IAA$5*10^{-6}$M	79±8	67±6	0.84
IAA $5*10^{-5}$M	70±5	53±6	0.75

Investigation of the models with the different types of organization demonstrates that natural growth inhibitors and phytohormones are connected in their concentrations-increase the level of cytokinins in tobacco callus leads to decrease of the level of ABA. Differentiation of the cells is connected with the increase of phenolic substances and the formation of the ones. Bacteria and fungi produce also plant hormones and inhibitors, which may play a role of factors of co evolution and pathogenesis. The non-tissue form of the growth of *Funaria* also connected with the certain level of phenolics, the concentration and amount of which increase during the differentiation. Disturbance in the balance of hormones and inhibitors in plant tissues leads to the break of growth correlations, and to the formation of the abnormal plant organs.

CHAPTER 10

NATURAL GROWTH INHIBITORS AND BIOTESTS

Isolated plant organs and tissues became models with the higher sensitivity to phytohormones and natural growth inhibitors. Usually intact plants and seedlings are less sensitive to natural growth inhibitors, but the process of seeds germination on the contrary is very sensitive. The intact plants are more common models for effects of gibberellins (Figure 10.1) or ethylene. It might depend on the metabolic processes which are depressed by natural growth inhibitors. In the previous chapters the growth and dormancy of plants were considered in terms of the relationships between hormonal and inhibiting factors. Induction and inhibition of growth processes: were explained by differences in the rates of synthesis and decomposition of phytohormones and natural growth inhibitors, on the assumption that phytohormones and inhibitors function in a plant in close interaction rather than as independent systems. The nature and main features of such interaction will be the subject of this chapter.

Earlier it was pointed out that natural growth inhibitors lack antihormonal specificity. This circumstance suggested that the action of the inhibitors does not affect only the initial reactions specific for each hormone, but extends to more general metabolic routes, such as, for example, the synthesis of nucleic acids, protein or ATP.

As we know, these metabolic routes are inhibited by specific synthetic inhibitors (antibiotics, antimetabolites, poisons). It seemed of interest to compare the action of these compounds which we shall sometimes call metabolic inhibitors, as distinct from natural inhibitors, with the effect of natural inhibitors on the endogenous growth of the bio-assays and on phytohormone-induced growth of plants grown on different soil types (Kefeli, 2002).

It would also be worthwhile to analyze the nature of interaction of natural inhibitors and phytohormones when they simultaneously affect various forms of growth (Kefeli et al., 2001).

10.1 Action of Metabolic Inhibitors and Antibiotics on Growth in Bio-Assays

Researchers when dealing with growth inhibitors usually try to select such concentrations as to make the action of the inhibitor selective. An inhibitor will lose its specificity if its concentration exceeds a certain limit or if the duration of the experiment is too long. When taken in very high concentrations, inhibitors depress many synthetic processes at once through retarding growth due to their excessive amounts in a cell. For estimating the retarding effect of inhibitors, the following criteria are usually used: minimum effective concentration (C_{min}) i.e. an inhibitor concentration which causes significant growth inhibition; a concentration of the inhibitor which causes 50% depression of growth (C_{50}); and an inhibitor concentration causing destruction of the object (CL).

Figure 10.1 Arabidopsis plants as test-models: 1 – control; 2 – GA_3 – 10 mg/l treatment.

These criteria are largely indirect. The specificity of inhibitor action may also be estimated directly with the aid of biochemical analysis of the process inhibited and by neutralizing the effect of the inhibitor with some specific substrate and metabolite.

Let us now discuss concrete data on the effect of metabolic inhibitors on growth and related processes. Cereal coleoptile segments are the main object for studying cell extension growth. Growing coleoptile segments were frequently used to investigate the effect of antibiotics on their growth and protein synthesis. Two main criteria were selected for such a test; 1) incorporation of labelled amino acids into proteins, and 2) incorporation of nitrogen-containing bases into the nucleic acids of extending cells in coleoptile segments. Usually these characteristics were determined on segments of quickly growing coleoptiles (in the presence of IAA) and relatively slowly growing coleoptiles (without IAA). Comparison of such data as incorporation of labelled (radioactive) precursors into nucleic acids and proteins, and also the effect of antibiotics on cell extension in coleoptiles enabled the investigators to draw a conclusion on the role of protein synthesis in the elongation process. As is known, protein is not accumulated in the course of coleoptile segment elongation which may be due to the fact that their production and decomposition occur at an equal rate.

For estimating the growth of wheat coleoptile segments studied in our experiments, we first incubated these segments during 20 hours in a 2% sucrose solution with or without IAA in a concentration of 7 mg/l. The growth of coleoptile segments was measured every hour.

The indirect measure of participation of a metabolic system in cell enlargement was a growth response of the coleoptile segments to so-called metabolic inhibitors by which are meant compounds (antiobiotics, poisons, herbicides) which produce a selective inhibiting action on individual links of the metabolic chain: biosynthesis of nucleic acids (actinomycin D, 8-azaguanine, 5-bromouracil), protein biosynthesis (chloramphenicol), oxidative phosphorylation (2,4-dinitrophenol), and photosynthetic phosphorylation (simazine, diurone). The sensitivity of the coleoptile segments to such inhibitors was studied on native and auxin-induced growth plants.

For experiments to study the action of inhibitors on growth processes, two characteristics were selected: I) the minimum duration of pre-incubation of coleoptile segments in inhibitor solution required to inhibit the subsequent growth stages; and 2) the minimum (critical) inhibitor concentration which does not yet affect endogenous growth, but already depresses IAA-induced growth.

The establishment of the minimum duration of pre incubation of coleoptile segments and cuttings in inhibitor solutions was necessary for making further experiments on the interaction of IAA with metabolic inhibitors. The scheme of these experiments comprised two stages: pre incubation in inhibitor solution and growing in water (control) or in IAA solution (experiment). When selecting pre incubation time it was necessary to take into account that this time had to be sufficiently long for the inhibitor to exert its effect and at the same time sufficiently short for IAA to act after the inhibitor.

The stimulating effect of IAA on coleoptile segment growth became obvious as early as the second hour of incubation, but was strongest after the first five hours of growth. Thus, after 3 hours of incubation IAA induced growth by 40 per cent, and after 5 hours it was already 170 per cent as against the control. In the subsequent period, the stimulating effect of IAA decreased and did not reach the activity displayed after the first 5 hours of growth till the end of the 10-hour experiment.

It may be asked how important are the processes which occur during the first five hours of incubation with IAA, for the growth of elongating cells in the subsequent period. To answer this question, IAA was not introduced into the medium immediately at the start of the experiment, but 1, 2, 3, 5, 6, 7, 10 and 12 hours after it has been started. It was found that IAA could display its stimulating effect on the growth of an elongating segment only if it was added into the solution no later than 5 hours after the start of the experiment (Figure 10.2).

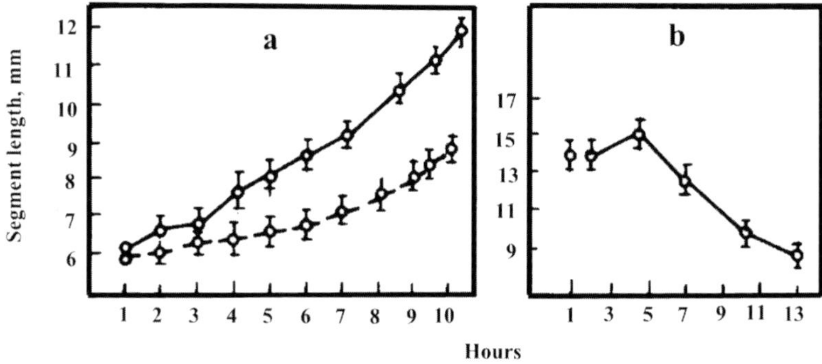

Figure 10.2 Growth of wheat coleoptile segments. a – effect of IAA on segment growth; 1 – in 2% sucrose without IAA; 2 – same with IAA (7 mg/l); b – growth of wheat coleoptile segments in 2% sucrose solution with IAA (7 mg/l) added at different time after start of experiment.

So, experiments with coleoptile segments were conducted in accordance with the following scheme: 5 hour incubation in inhibitor solution, three subsequent. It was washed out, and then additionally growing for 15 hours on 2% sucrose or in IAA solution.

As has been pointed out already, when adding inhibitors into a growing tissue it must be remembered that their action is not always highly specific, i.e. it by no means affects one particular progress involved in metabolism. In high concentrations, inhibitors produce a number of side effects, thus acting as structure-non-specific factors rather than highly specific blocking agents.

Following this recommendation, we have tested the action of metabolic inhibitors in different concentrations (in fractions of the saturated solution concentration). In this way, we have selected doses C_{50} so which depressed the growth of wheat coleoptile segments by 50 per cent, and found critical concentrations C_0 which no longer displayed significant inhibition of segment growth.

All inhibitors tested depressed the growth of coleoptile segments. None of the inhibitors was an electrolyte and, hence, its action depended on the Fergusson principle which states that the inhibitors which act specifically on some particular metabolic process must be effective in high dilutions of the saturated solution; on the contrary, if the action of a substance is non-specific (similar to alcohols and ethers), this substance inhibits growth only in high concentrations. The above data show that 2,4-DNP affects coleoptile segment growth in high dilutions of the saturated solution, whereas other inhibitors do it in low dilutions. Hence, it may be assumed that the segments of coleoptiles display the maximum sensitivity to 2,4-DNP, the uncoupling agent of oxidative phosphorylation (Table 10.1).

Table 10.1 Effect of inhibitors on coleoptile sections elongation.

System affected and inhibitor	*Concentration C_{50}*	*Concentration C_0*
General narcotic action	*Dilution from concentrated solution*	
heptanol	1/ 8	1/ 16
Oxidative phosphorylation 2,4-dinitrophenol (2,2-DNP)	1/ 128	1/ 256
Photosynthetic phosphorylation and photosynthesis		
diurone	1/ 2	1/ 4
simazine	1/ 2	1/ 8
Nucleic acid-protein metabolism		
actynomycin D	1/ 8	1/ 6
8-azaguanine	1/ 2	1/ 8
5-bromuracil	1/ 4	1/ 16
chloramphenicol	1/ 8	1/ 16

C_{50} concentration which decrease elongation by 50%;
C_0 concentration which does not effect elongation;

It was interesting to see whether the sensitivity of coleoptile segments to other inhibiting substances would increase if IAA was added to the incubation medium. In order to eliminate possible interaction of IAA and inhibitors in the solution, coleoptile segments were incubated with inhibitors during the first 5 hours of the experiment, then the inhibitors were washed off and the segments placed into the solutions of 2% sucrose (control) or IAA. The period of 5 hours was selected because this time is the limit of the sensitivity of coleoptile segments to IAA (see Figure 10.3) For inhibiting segment growth in IAA solution we have taken critical doses which did not cause significant growth inhibition.

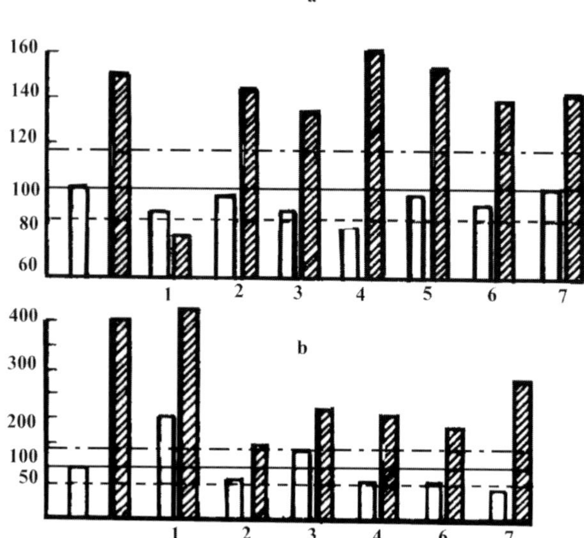

Figure 10.3 Response of biological tests to metabolic inhibitors (Turetskaya, Kefeli, Kof, 1968). a – growth of wheat coleoptile segments (% of control) during 15-hours incubation on 2% sucrose without IAA (white bars) aand in presence of 7 mg/l of IAA (hatched bars) depending on 5-hour pretreatment with different inhibitors; b – rooting of Phaseolus cutting (% of control) during 7 days in water (white bars) and 60 mg/l of IAA (hatched bars) depending on treatment with different inhibitors during first day. Inhibitors in doses not affecting growth of tests; 1 – 2,4 –DNP; 2 – chloramphenicol; 3 – actinomycin D,3' – actinomycin C,; 4 – 8-azaguanine; 5 – 5-Br-uracil; 6 – diurone; 7 – simazine.

It was found that none of the inhibitors, except for 2,4-DNP, could retard IAA-induced growth. This ability of 2,4-DNP cannot be attributed to its higher solubility in water, because, for instance, the solubility of chloramphenicol was also sufficiently high, but this substance did not inhibit IAA-induced growth. If inhibitor concentrations causing 50% inhibition of endogenous growth are used in experiments, the differences in the action of the inhibitors become considerably smaller (Kefeli, 1978).

The experiments with coleoptile segments permit a conclusion that elongating cells are most sensitive to the oxydative phosphorylation uncoupler. The testing of a number of phenolic growth inhibitors: p-coumaric acid, coumarin, naringenin, salicin, salicylic acid and others for coleoptile segment growth showed that, unlike 2,4-DNP, they inhibit growth gradually and an increase in the amount of natural inhibitors in the test solution is paralleled by a slow retardation in coleoptile segment growth.

This specific retarding action of natural growth inhibitors was noted by us earlier for natural growth inhibitors isolated from willow leaves and is illustrated in Figure 10.4. In contrast to auxins, none of the tested concentrations of natural growth inhibitors could strongly stimulate the growth of coleoptile segments. This gradual inhibitory action of natural growth inhibitors may be meaningful physiologically and manifested in a slow retardation of growth processes and in gradual transition of plants to dormancy.

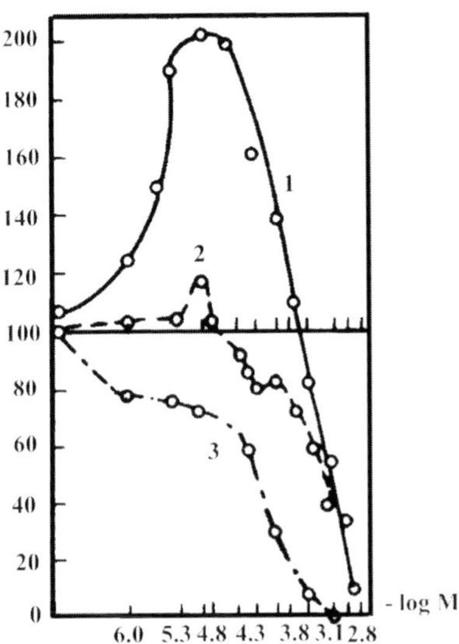

Figure 10.4 Effect of IAA (1), natural inhibitor from willow leaves (2),. 2,4-DNP (3) on growth of wheat coleoptile segments.

Stem segments (epicotyl and hypocotyl stems) are also frequently used for studying the effect of inhibitors. It has been showed that chloramphenicol, in a concentration of 3.7×10^{-3} M causes 50% inhibition of pea stem segment growth induced by auxins (IAA, 2,4-D or NAA). It should be pointed out that growth without auxin was depressed by this inhibitor in a weaker degree. Interestingly, 50% inhibition of segment elongation required a very high dose of chloramphenicol lying in the range of narcotic concentrations.

The reason for such a low response of elongating stem segments may be the fact that the synthesis of most proteins is not directly related to elongation growth. It has recently been demonstrated that if the segments of the subapical zone of 10-day pea seedlings are incubated for 30 minutes with actinomycin D (10 mg/ml) and then treated with IAA, no changes are usually observed in the biosynthesis of nuclear proteins. Separation of proteins (labelled with ^{14}C and ^{3}H) on a column with DEAE-cellulose showed that IAA affects the synthesis of only some of the proteins, while the synthesis of most proteins gives no response to the action of IAA. The inadequacy of data obtained with the aid of only one metabolic inhibitor becomes evident in the investigations of soybean hypocotyl segment growth. The authors showed that whereas 5-fluorouracil (2.5 mM) did not affect the growth of the segments induced with 2,4-D, cyclohexymide and actinomycin D in the concentrations of 1 and 10 mg/l, respectively, strongly inhibited this process. For actinomycin D it was even shown that RNA synthesis and segment elongation proceeded simultaneously. It has been stated above that the rate of the growth process determines its response to a metabolic growth inhibitor: thus, if growth is phytohormone-induced, its response to the action of a metabolic inhibitor will be stronger.

In this way, research into the inhibition of stem segment growth by antimetabolites and metabolic inhibitors makes the following two facts obvious: 1) stem segment growth and protein synthesis are depressed by growth inhibitors in a different degree, i.e. there is no direct proportional relationship between growth and protein synthesis; 2) at the same time in a number of cases it was proved that inhibition of some RNA forms and the synthesis of some proteins retards growth.

Rhizogenesis, root formation on cuttings, is the production of a new organ from female plant cells. Rhizogenesis is essentially a convenient model for studying such morphogenetic processes as formation of new tissues and whole organs. The formation of radicles from stem tissues may be regarded as a change of development programs which are defined by substances inducing successive transformations which had been present in the programs. It would seem that under the effect of endogenous phytohormones stem tissue cells alter the course of their metabolism and reveal biosynthesis of new products leading to initiation of radicles. Differentiation of few organs is not caused by the action of phytohormones alone, it is also closely associated with the synthesis of proteins specific for different stages of ontogenesis.

The rhizogenesis process is extremely sensitive to protein and nucleic metabolic inhibitors. Total inhibition of rhizogenesis processes in etiolated pea epicotyles was achieved with 40 mg/l of actinomycin D, 200 mg/l of 5-Br-uracil, 200 mg/l of azaguanine and 50 mg/l of chloramphenicol (Fellenberg, 1967). In Fellenberg's opinion, IAA lowers the melting point of native and denatured DNA 48 hours after rooting has started which maybe due to the breaking of the DNA-histone bond and DNA-DNA double-strength helix bond. This breakage is followed by the synthesis of new RNA molecules. Thus, by blocking the synthesis of new proteins, the metabolic inhibitors actually depress IAA inducing functions.

In our experiments the treatment of cuttings with all the inhibitors in various concentrations (expressed in fractions of saturated solution) caused strong inhibition of root formation in *Phaseolus* cuttings.

Table 10.2 Effect of inhibitors on rooting of bean cuttings

System affected and inhibitor	*Concentration C_{50}*	*Concentration C_0*
General narcotic action	*Dilution from concentrated solution*	
Heptanol	1/ 8	1/ 16
Oxidative phosphorylation 2,4-dinitroophenol (2,2-DNP)	1/ 64	1/ 128
Photosynthetic phosphorylation and photosynthesis		
Diurone	1/ 1024	1/ 2048
Simazine	1/ 2024	1/ 2048
Nucleic acid-protein metabolism		
actynomycin D	1/ 64	1/ 128
8-azaguanine	1/ 2048	1/ 4069
Chloramphenicol	1/ 1024	1/ 2048

C_{50} concentration which decrease elongation by 50%;
C_0 concentration which does not effect elongation;

The inhibitors of photosynthetic phosphorylation and nucleic acid-protein metabolism exhibited the strongest inhibitory effect, except for actinomycin D which depressed rooting in the same thermodynamic concentrations as 2,4-DNP. The effect of other inhibitors was 20 times as strong as that of 2,4-DNP. In this case it would appear that the inhibitors changed places: with respect to root formation they acted as specific agents which were effective in very high dilutions (Table 10.2)

If *Phaseolus* cuttings are treated with IAA solution, root formation is abruptly induced. It has already been pointed out that auxin treatment may be given not immediately after the cuttings are prepared, but on the next day after the beginning of the experiment, the stimulating action of IAA remaining unchanged. Taking this fact into consideration, on the first day of the experiment we added into the medium weak critical inhibitor concentrations which did not depress endogenous root initiation in the cuttings, and on the next day IAA was introduced in a concentration of 60 mg/l. The pretreatment with inhibitors appreciably attenuated the stimulating effect of IAA en the root formation process, with the exception of 2,4-DNP pretreatment. So, whereas in elongating cell growth the oxidative phosphorylation system exhibits the strongest response to inhibitors, the systems most sensitive to inhibitors in root formation are photosynthetic phosphorylation and nucleic protein metabolism.

Comparative experiments evidence that when green cuttings root in the dark, the sensitivity of photosynthetic phosphorylation to inhibitors in concentrations causing a 50% depression of growth (in fractions of saturated solution) decreases in a very substantial degree.

Table 10.3 Effect of some inhibitors on riiting of bean cuttings in the light and darkness.

Inhibitor	*Sensitivity under light*	*Sensitivity in the dark*
2,4-DPN	1/ 64	1/ 128
Diurone	1/ 1024	1/ 8
Simazine	1/ 1024	1/ 2

Consequently, photosynthetic phosphorylation is that specific source of energy which affects growth and organogenesis under light only the photosynthetic phosphorylation system does not function in the etiolated tissues of coleoptile segments and in the green tissues of cuttings placed in the dark (Table 10.3) As to oxidative phosphorylation, it is equally intensive in the dark and under light. Thus, in the elongating cells of coleoptile segments the oxidative phosphorylation processes are dominant, while the synthesis of protein and nucleic acids is weak, and this is the reason for a slow response of coleoptile segments to the inhibitors of nucleic acid-protein metabolism.

Root-forming cuttings in which cell division occurs are 500-1000 times more sensitive to the inhibitors of nucleic acid-protein metabolism and photophosphorylation than the extending cells of coleoptile segments. It is also significant that when coleoptile segment growth is auxin-induced, oxidative phosphorylation still functions as an inhibiting system and its depression with 2,4-dinitrophenol neutralizes the IAA induction effect. At the same time, substances causing inhibition of nucleic acid-protein synthesis and photophosphorylation are inert with respect to elongating coleoptile segments.

The reverse relationship is true for cuttings that form roots under light. Here, the inhibition of photosynthetic phosphorylation or nucleic acid-protein synthesis depresses IAA-induced growth, whereas 2,4-DNP does not affect this growth.

The experiments with inhibitors on systems of retarded (endogenous) and induced growth confirm the general rule that the more actively a system func-tions, the easier it lends itself for inhibition.

A growing coleoptile segment may be viewed as a system with highly inhibited nucleic acid-protein metabolism, and a rooting *Phaseolus* cutting as a system with active nucleic acid-protein metabolism. Then it becomes clear why in the former case nucleic-protein metabolism inhibitors produce an effect 1000 times weaker than in the latter.

The above facts suggest that the production of RNA and protein in growing coleoptile cells is extremely weak and does not limit the elongation process. Even induction of coleoptile segment growth in the presence of IAA fails to appreciably increase their sensitivity to the inhibitors of nucleic protein metabolism. This, however, in no way rules out the formation of pew protein RNA molecules, which accompanies elongation processes. The sensitivity of the object to nucleic protein inhibitors increases many times if cell division occurs in it.

Objects as different as coleoptile segments and rooting Phaseolus cuttings may be compared only if certain reservations are made, namely, that one object (coleopties) is a net growth system and the other, (Phaseolus cuittings) an induced root formation system; and that coleoptiles are a chlorophyll-free system, while the root-formation process is strongly affected by light. This is why, in view of the difficulties arising in comparative investigations of these two objects, Smirnov and his co-workers, 1970 found it desirable to use the third model an isolated lucerne root which was frequently used in the experiments. Unlike a *Phaseolus* cutting, this object is not affected by light since it has no chlorophyll-containing tissues. At the same time isolated roots are a system with dividing and enlarging cells, i.e. they can be contrasted to the enlarging cells of coleoptile segments. The experiments showed that isolated roots are extremely sensitive to metabolic inhibitors: their growth was 50% depressed by 8-azaguanine (1/400), 5-Br-uracil (1/4000) and chloramphenicol (1/1000). Let us now give a comparative description of the response of root-forming cuttings and isolated roots to metabolic inhibitors.

Formation of meristemic sites is most vigorous in rooting cuttings. However, there are three circumstances which interfere with the response of this object to metabolic inhibitors; 1) it has leaves which are able to affect indirectly the sensitivity of the roots being formed to metabolic inhibitors; 2) a transpiration flow in the cutting can rapidly transport the inhibitor throughout the cutting; 3) the stem cell differentiation process complicates the pattern of radicle initiation. Despite these complicating circumstances and the fact that the inhibitor treatment of the cuttings was different from the treatment of other objects, the rooting cuttings responded to the inhibitors applied in approximately the same way as lucern roots, i.e. they exhibited a high sensitivity to all the metabolic inhibitors, except 2,4-DNP. So, the conclusion that follows from the analysis of these data is that the sensitivity of growing objects to metabolic inhibitors increases if cell division processes occur in these objects.

The validity of this conclusion was verified on the following two models: growing lucern roots and Phaseolus cuttings, which had different rates of growth processes. The growth of the main root of lucern was found to follow a certain rhythmic pattern (Figure 10.5). In particular, the maximum growth of the root is observed on the 4th, 8th and 12th day. These peaks are associated with the intensification of cell division processes. In *Phaseolus* cuttings, cell division is activated on the second or third day which corresponds to the start of radicle formation. Taking into account that these critical periods in the growth of the above objects are accompanied by the acceleration of cell division, it becomes obvious that it is necessary to evaluate the sensitivity of isolated roots and cuttings to growth inhibiting substances in the periods of vigorous growth of these objects.

For this purpose, we added inhibitors on different days which was the technique we used previously. The experiments showed that if chloramphenicol and 8-azaguanine are not introduced into the cuttings and isolated root cultures right at the beginning of the experiment, but on the 2nd, 3rd, 4th, 5th, etc. day, the inhibition of the root formation processes in *Phaseolus* cuttings will be most pronounced if the inhibitor is applied during root initiation (on the 3rd or 4th day); inhibition in isolated roots will be optimum on the 3rd-4th day, i.e. in the period of the first maximum increment of the main root (Figure 10.6). These data are another argument in favor of the relationship existing between acceleration of cell division and the increase of the sensitivity of a growing system to metabolic growth inhibitors.

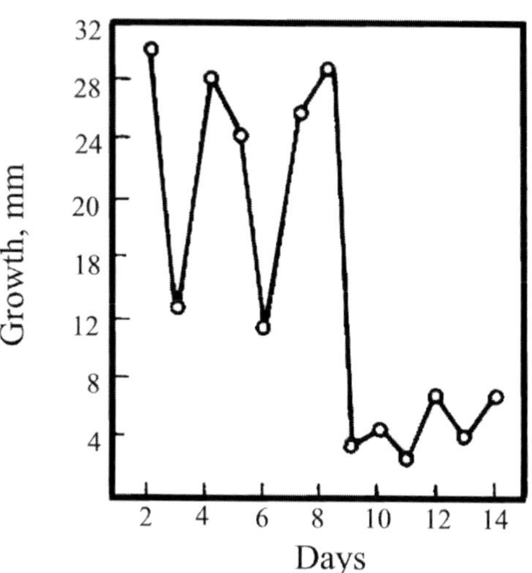

Figure 10.5 Growth rhythm of isolated lucern root (Kefeli et al., 1972)

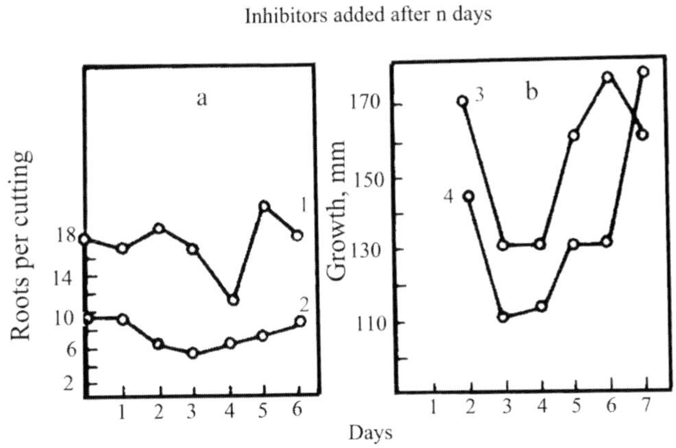

Inhibitors added after n days

Figure 10.6 Addition of metabolic inhibitors during formation and growth of roots (Kefeli et al., 1972). a – root formation in Phaseolus cuttings; b – growth of main root of lucern (isolated clone roots); 1 – chloramphenicol (C, 1/ 2048); 2 – 8-azaguanine (C, 1/ 2048); 3 – chloramphenicol (C, 1/ 30,000); 4 – 8-azaguanine (C, 1/400).

So, the analysis of the response of various objects to inhibitors showed that if active cell division occurs in a growing organ, it displays the highest response to the treatment with metabolic inhibitors (lucern roots, radicles on *Phaseolus* cuttings). The more active the cell-division processes, the more pronounced is the response of the object to the metabolic inhibitor. But if growth is mainly due to cell extension (coleoptile segments, stem segments) in this case the response to metabolic inhibitors significantly reduces.

After discussing the specific features of the growth of bio-assays and their response to metabolic inhibitors, we shall now proceed to the description of the effect of natural inhibitors on the growth of these objects.

10.2 Comparison of the Effects of Natural Growth Inhibitors, Antibiotics and Some Solvents on Growth of Coleoptile Segments and Root Formation in Phaseolus Cuttings

The intrinsic property of phenolic inhibitors is to exert inhibiting action only when they accumulate in tissues in large quantities. Due to this property, it was hypothesized that in this case they acquire the functions of narcotics. It seemed interesting td find out whether natural growth inhibiting substances (chalcones, coumarins, simple phenols) which do not possess electrolytic properties may be classed with non-specific compounds (narcotics) or with such specific agents as antibiotics; antimetabolites, etc. The magnitude of the effect of narcotic agents is known to depend not on the functional groups or the structure of their molecule, but on certain physical and chemical characteristics of these substances which favor their accumulation in some vital part of a cell which interferes with the sequence of respiratory and other metabolic processes. Usually in order to determine whether a substance is a non-specific agent, i.e. a narcotic, it is analyzed according to the Fergusson principle which states that if a substance acts non-specifically (as a narcotic), the relative saturation of its toxic concentration will be similar to that for some well known structurally non-specific agent. The main unit of the Fergusson scale is a thermodynamic concentration which is expressed as a ratio of a given concentration to that of the saturated solution. Using this technique, Ivanov, 1966, managed to separate the specific and non-specific effects of chloramphenicol on root growth processes.

The metabolic specificity of some natural phenolic inhibitors which are not electrolytes can be determined by analyzing the inhibitory effect of these compounds on the growth of bio-assays with the aid of the Fergusson scale.

For analysis of the toxicity of substances in accordance with the Fergusson scale, saturated solutions (C) are usually prepared and then a sequence of concentration diluted in geometric progression (1/2 C, 1/4 C, 1/8 C, etc.) is made. The substances we have selected as known narcotics were chloroform, carbon tetra-chloride, and primary normal alcohols: pentanol (C_5), heptanol (C_7), octanol (C_8), nonyl alcohol (C_9), lauryl alcohol (C_{13}) and cetyl alcohol (C_{16}). For comparison with natural growth inhibitors we have taken the following specific metabolic poisons and antibiotics: 2,4-dinitrophenol (DNP) which uncouples oxidative phosphorylation from respiration, actinomycins C and D which depress the synthesis of chromosome-ribosome RNA and inhibit transmission of information to ribosomes and also the protein synthesis inhibitor chloramphenicol. Besides, we have tested the substances inhibiting photosynthesis and photosynthetic phosphorylation - simazine and diurone. The compounds selected as natural phenolic inhibitors were phloridzin, an inhibitor readily soluble in water and inhibiting the growth of wheat coleoptile segments, and isosalipurposide which is a willow chalcone with a similar structure and properties. The saturated solutions of these 'substances were diluted according to the Fergusson principle and compared with similar dilutions of the narcotics. The bio-assays were wheat coleoptile segments (straight growth elongation test) and *Phaseolus* cuttings (cell division and elongation test) It was found that phloridzin and isosalipurposide affected coleoptile segment elongation in the same way as typical narcotics; alcohol, chloroform, carbon tetrachloride, etc.; in other words, in high concentrations they inhibited growth, but their retarding action ceased already in dilutions of 1/ 8 or 1/6 of the saturated solution.

Thus, despite considerable differences in the chemical structure, molecular weight, solubility, etc., all the substances tested behaved as structurally non-specific agents which lost their toxic effect in similarly high concentrations. In order to obtain 50% inhibition of coleoptile segment growth the thermodynamic concentration for C_7 alcohol was 1/ 16, for C_8 - 1/8, arid for C_{16} - 1/2 of the saturated solution (Figure 10.7).

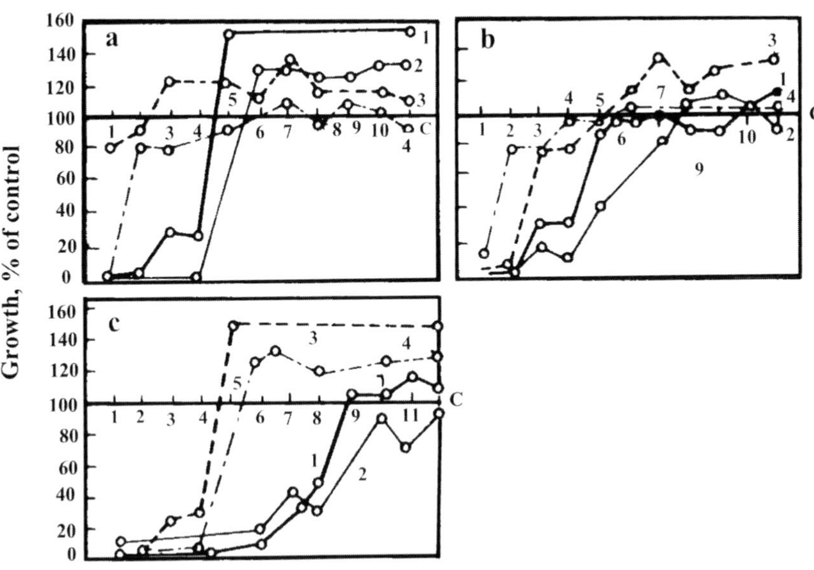

Figure10.7 Effect of solvents, endogenous inhibitors and antibiotics on growth of wheat coleoptile segments. C – concentration in dilutions of saturated solutions; a – effect of narcotics and endogenous inhibitors; 1 – phloridzin, 2 – isosalipurposide, 3 – $CHCl_3$, 4 – CCl_4; b – effect of polyatomic normal alcohols; 1 – C5; 2 – C7; 3 – C8; 4 – C16; c – effect of antibiotics, poisons and natural inhibitors: 1 – 2,4-dinitrophenoil, 2 – aurantine, 3 – phloridzin, 4 – isosalipurposide.

It was concluded, therefore, that natural willow and apple-tree inhibitors affect growth similar to structurally non-specific agents (narcotics). A question may be asked about the behavior of structurally specific agents (antibiotics and metabolic inhibitors). 2,4-DNP and the antibiotic aurantine (actinomycin C) were found to inhibit coleoptile segment growth even in concentrations 1/ 128 to 1/ 256, whereas natural growth inhibitors lose their retarding effect already in a concentration one order lower. Evidently, this is where one of differences between the two types of antibiotics lies; specific (antibiotics) and non-specific (narcotics and natural growth inhibitors). The point is that in low concentrations specific inhibitors affect one of the processes stronger than others. This property inherent in a specific inhibitor is lacking in a non-specific agent.

It is sufficient for a specific antibiotic to block one component of metabolism in order to inhibit the entire metabolic process; for this reason its concentration effective for this process may be considerably lower than the concentration of a non-specific agent (Ivanov, 1966).

It should be pointed out that the specificity of an inhibitor is not a constant so that an inhibitor may be either specific or non-specific depending on the nature of the system it acts upon. We have shown, for example, that chloramphenicol is not a specific agent with respect to elongating cells. It depresses their growth in rather high concentrations (C_{50} = 1/8). At the same time its inhibitory effect becomes rather specific for root-forming *Phaseolus* cuttings (C_{50} = 1/1024). The same refers to simazine and diurone. The photosynthetic phosphorylation system does not function in the etiolated tissues of coleoptile segments or in the green tissues of cuttings placed in the dark, and metabolic inhibitors lose their specificity in this case.

On the contrary, if biosynthetic processes depressed by inhibitors are active, the effect of metabolic inhibitors may become specific (see Figure 10.8). Green *Phaseolus* cuttings forming roots under light in which the cell-division process occurs are 500 to 1000 times more sensitive to substances inhibiting nucleic-protein metabolism and photophosphorylation than elongating coleoptile segments.

With respect to *Phaseolus* cuttings which are a model system reverse to that of coleoptile segments, the natural inhibitor phloridzin behaves as a non-specific agent, similarly to heptanol, and inhibits the root-formation process in very high concentrations (Figure 10.8).

Hence, unlike the metabolic inhibitors described above, natural phenolic growth inhibitors behave as non-specific agents with respect to two opposite systems of extending (coleoptiles) and dividing (rooting cuttings) cells inhibiting growth processes in very high concentrations. Such concentrations can upset the functioning of the most general metabolic systems, as for example, respiration and phosphorylation. It is interesting to note that phloridzin doses which cause the inhibition of the phosphorylation process and coleoptile segment growth are very similar ($5x10^{-4}$ M) which suggests that the inhibitors display their retarding action by interfering with the phosphorylation system. The oxidative phosphorylation system connected with ATP synthesis is extremely non-specific. If the ATP amount in a cell is limited, this inhibits all the synthetic processes; not all of these processes seem to be inhibited in the same degree, but this inhibition is not so specific as to provide selective control of the synthesis of various proteins.

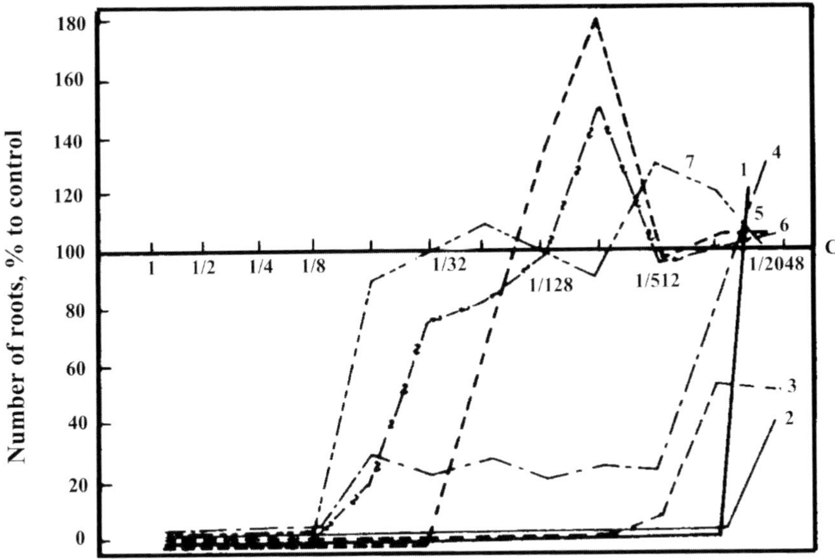

Figure 10.8 Effect of metabolic inhibitors and phloridzin on root formation in Phaseolus cuttings.

C – concentration in dilutions of saturated solution, 1 – chloramphenicol; 2 – 8-azaguanine; 3 – diurone; 4 – simazine; 5 – 2,4-DNP; 6 – phloridzin; 7 – heptanol.

The concept of the non-specific action of phenolic inhibitors on growth processes in plants needs further experimental corroboration; thus, it is not yet quite clear whether phenols can contact in cells with mitochondria, and what is in general their intracellular level in meristemic tissues. On the whole, it can be conjectured that some part of phenols is c6ntained not only in vacuole sap, but also in other parts of the cell, i.e. in the walls and chloroplasts. A known property of phenols is their ability of secretion from roots which indicates that they can translocate in the cells and tissues of roots. These fragmentary data, however, cannot give us a clear idea of phenol localization in cells. Therefore, at present we can go no further than assume that phenols inhibitors accumulating in tissues interfere with a sequence of metabolic processes, in the first place, by blocking energy transfer in a cell which obviously leads to non-specific disorganization of a number of biosynthetic processes.

Estimation of the specificity of inhibitor action in accordance with the Fergusson scale is applicable only for substances other than electrolytes. Therefore, abscisic acid and its analogs cannot be tested for specificity in this way describing the specificity of its effect on various metabolic processes.

Our experiments have demonstrated that abscisic acid is also able to inhibit a number of other processes which seem to be involved, directly or indirectly, in protein synthesis. Among such processes are inhibition of water supply into germinating apple seeds, retardation of chlorophyll synthesis and inhibition of the enlargement of cotyledons.

10.3 Comparison of the Effects of Natural and Synthetic Growth Inhibitors on Various Growth Forms

As is known, natural inhibitors occur in plants in the form of phenolic and terpenoid compounds. The characteristics of their influence on growth processes are usually studied on bio-assays from the simplest (elongation of coleoptile and stem segments) to the most complex ones (organogenesis, seed germination). Roughly speaking, all experiments on testing inhibitor action on plant growth fit into the following scheme: 1) direct action of natural growth inhibitors on the growth of bio-assays; 2) their influence on the growth-inducing activity of phytohormones.

The strongest inhibiting action on the growth of coleoptile segments is produced by abscisic acid (C_{50} = 0.08 mg/l). Abscisic acid has been found to inhibit the promoting effect of IAA on coleoptile segment growth and the inducing effect of gibberellic acid on the stem growth processes (Thomas et al., 1965). Besides, abscisic acid is an active inhibitor for more complex process, such as seed germination, bud opening, etc. At the same time, certain forms of growth, for example, curvature of oat coleoptiles, were insensitive to the effect of abscisic acid, and others were even stimulated by it, for example, rooting of stem cuttings.

Phenolic and aromatic compounds inhibited the growth of coleoptile segments in higher concentration than abscisic acid. Thus, resorcin, hydroquinone, quinic acid, ferulic acid, p-coumatic acid, anthranilic acid, morin and hesperidine inhi-bited the growth of segments in all concentrations (from 10^{-2} to 10^{-3} M). Other phenols (pyrocatechin, caffeic and gallic acids, phloridzin, vanillin and phloroglucinol) inhibited growth in high concentrations and weakly stimulated it in low concentrations. In another article the same authors reported that the IAA-induced growth of coleoptile segments could be inhibited by p-hydroxybenzoic and p-coumaric acid, and also by rutin and coumarin tested in concentrations from 10^{-2} to 10^{-3} M. The next group of compounds exhibited only synergism with respect to IAA (protocatechuic and gallic acids, van illin), while the effect of some phenols was reversed depending on the concentration used. Comparing the inhibitory effect of equimolar phenol concentrations on the growth of oat coleoptile segments, Fries (1968) showed that they may be arranged in the following sequence in the order of an increasing effect on endogenous and IAA-induced growth: tyrosine, p-hydroxyphenyllactic acid, p-hydroxyphenylpyruvic acid and p-coumaric acid. Wada & Nagao (1961) made an attempt at establishing a relationship between the inhibitory action of guaiacol, pyrocatechin and hydroquinone and their ability to regulate auxin-oxidase activity. According to Henderson & Nitsch (1962), some phenols, for example, chlorogenic acid in a concentration of 10^{-4} M can act as IAA synergists because they strongly inhibit decomposition of IAA.

The action of natural inhibitors on various growth processes has been studied far less profoundly than the effect of synthetic inhibitors, antibiotics and poisons. No adequate data are available on the response of bio-assays to substances inhibiting the synthesis of protein, nucleic acids, photo- and oxidative phosphorylation. No direct evidence is provided on the characteristics of the interaction of natural growth inhibitors with phytohormones and on the features of the influence of natural and synthetic inhibitors on the growth of bio-assays.

Since synthetic growth inhibitors have been studied much more profoundly than natural growth inhibiting substances, we decided to compare naturally occurring growth inhibitors (coumarin, phloridzin, abscisic acid) with synthetic compounds (MH, TIBA, CCC and morphactin). Our plan was to analyze the effect of these substances on various forms of the growth process, from simple to most complex. For this purpose the following growth forms were selected: 1) cell extension growth (wheat coleoptile segments); 2) stem growth (etiolated pea seedlings); 3) root formation (*Phaseolus* cuttings); 4) shoot formation (opening willow buds); 5) formation of an entire plant (seed germination and seedling growth). Apart from studying the effect of inhibitors on native growth we tested the ability of these compounds to inhibit growth induced by phytohormones - auxins and gibberellins. Natural growth inhibitors were selected so that one of them was a terpenoid compound (abscisic acid), and the two others aromatic (coumarin) and phenolic (phloridzin) substances. All these substances are sufficiently well soluble in water and are known as compounds inhibiting coleoptile segment growth, abscisic acid inhibiting this process in concentrations 2 or 3 orders lower than phloridzin and coumarin. Each of the selected natural inhibitors posesses a number of peculiar features: abscisic acid, for example, can cause leaf abscission and inhibit RNA synthesis, coumarin affects cell-division, phloridzin can inhibit phosphorylation processes.

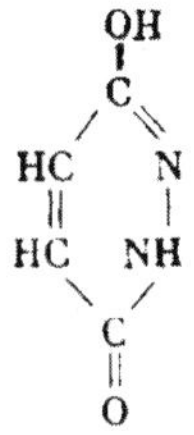

Maleic hydrazide

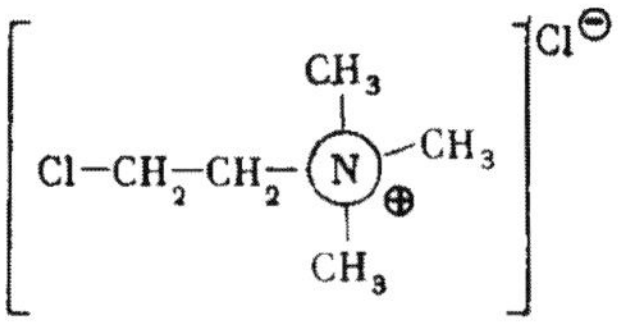

Chlorocholine chloride

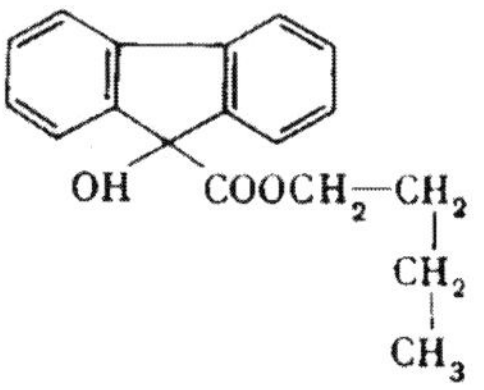

Morphactin

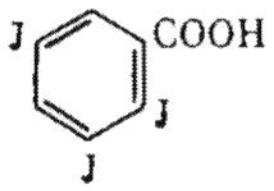

2,3,5-Triiodbenzoic acid

* MH – Maleic acid;
CCC – Chlorocholine chloride;
TIBA – 2,3,5-Triiodbenzoic acid.

The synthetic growth inhibitors we used in our experiments were maleic hydraride (MH), which is a compound hardly soluble in water; triiodbenzoic acid (TIBA) which depresses auxin transport in a plant; chlorocholine chloride (CCC), and morphactin (chlorfluorenol) which were chosen as stem growth inhibitors with an effect opposite to that of gibberellins. Indolylacetic acid (IAA) in coleoptile experiments was used in a dose of 7 mg/l, and in *Phaseolus* cutting experiments in concentrations of 60 mg/l. Gibberellic acid (GA) in pea experiments was used in the doses of 0.1 and 0.2 mg/l. The effect of the inhibitors on auxin induced growth was estimated with the aid of a pretreatment technique developed specially for the purpose and described in detail earlier. Natural and synthetic inhibitors have been tested on five different forms of growth. The criteria for estimating the retarding action of the inhibitors was the dose which caused 50% inhibition of growth (Figure 10.9). Abscisic acid exerted the strongest inhibiting action on the growth of coleoptile segments (50% inhibition was achieved with a dose of 0.08 mg/l) and pea stem growth (1.25 mg/l). Its effect was the weakest on seed germination.

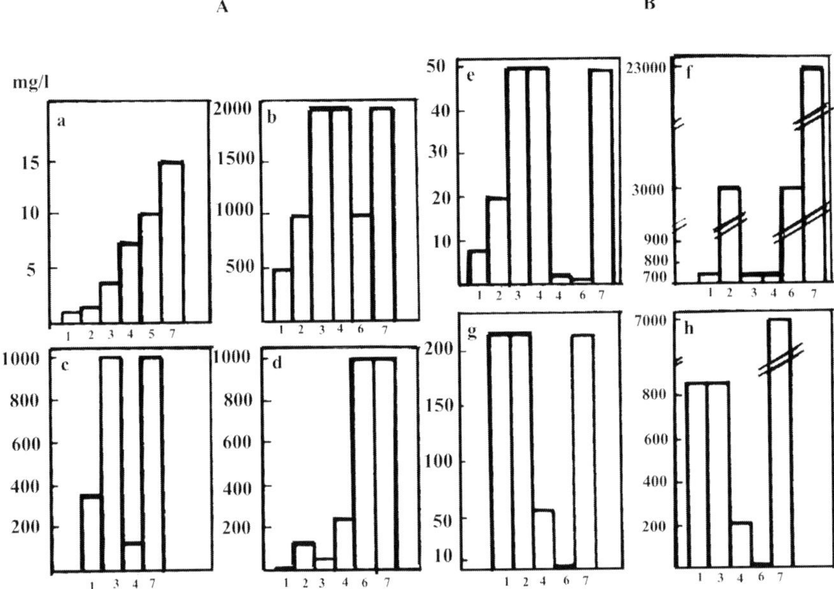

Figure 10.9 Effect of natural (A) and synthetic (B) inhibitors on various forms of growth. 1 – elongation; 2 – pea stem growth; 3 – wheat stem growth; 4 – wheat root growth; 5 – bud openings; 6 –root formation; 7 – seed germination; a – abscisic acid; b – phloridzin; c – p-coumaric acid; d – coumarin; e – chlorofluorenol; f – CCC; g – TIBA; h – MH.

Coumarin depressed easily the growth of coleoptile segments and the growth of a wheat seedling (50% inhibition was achieved with 10 and 62.5 mg/l, respectively), exerted a weaker effect on root growth (250 mg/l) and practically did not affect the root formation process and wheat seed germination at all. Phloridzin inhibited root formation processes (50% inhibition was achieved with 250 mg/l) and wheat coleoptile segment growth (500 mg/l) stronger than the growth of a wheat seedling (about 2000 mg/l) or pea stem growth (1000 mg/l).

On the whole, it can be stated that the stronger inhibitor of growth processes among natural growth inhibiting substances is abscisic acid, the inhibiting effect of coumarin is weaker, and that of phloridzin is still weaker. The comparison of the effect of natural inhibitors on coleoptile segment growth with its effect on the growth of a coleoptile itself permits a conclusion that the former process is easier to inhibit than the latter.

Synthetic inhibitors affected growth processes in a different way than natural inhibitors. Morphactin produced the strongest effect on root formation processes (50% inhibition was achieved with a dose of 0.47 mg/l) and bud opening (50% inhibition with 3 mg/l) and affected much weaker the growth of a seeding and a root as well as wheat seed germination. Typical of morphactin (chlorofluorenol) was to cause active changes in the functions of the growing points. The apical meristems of willow roots and buds seemed to stop their activity, and the development of branch roots was observed. Even low concentrations of morphactin (0.2 mg/l) that practically failed completely to affect the growth of pea seedlings interfered with their georeaction which was not observed with the control and gibberellin-treated seedlings.

Chlorocholine chloride (CCC) produced the strongest inhibitory action on wheat coleoptile segment growth (50% inhibition was achieved with 730 mg/l), and affected weaker pea stem growth (3000 mg/l) and still weaker seed germination.

Triiodbenzoic acid (TIBA) strongly inhibited root formation (50% inhibition was achieved with 3.5 mg/l), and somewhat weaker wheat root growth (56 mg/l), and did not practically affect other processes. The same properties with respect to this bio-assay were displayed by MH.

Discussing the features of the action of synthetic growth inhibitors it should be mentioned that the strongest inhibitor is morphactin. TIBA and MH are effective in doses one order higher and finally, CCC exerts inhibitory action in concentrations 2 or 3 orders higher. Unlike natural growth inhibitors, synthetic compounds are weaker in their effect on a more simple form of growth (coleoptile segment growth) and are usually most active in depressing more complex growth processes (morphactin, TIBA, MH, CCC).

The following features are common for all the tested inhibitors: I) none of them can depress all five studied growth forms in the same degree; 2) none of the tested inhibitors in low concentrations exhibited a hormonal stimulating effect on any growth form. Thus, whereas IAA promoted 219% growth of coleoptile segments, MH in a dose of 110 mg/l caused stimulation exceeding the control by 16%, coumarin in a dose of 1 mg/I by 21%, and morphactin in a dose of 5.5 mg/l by 29%.

Whereas the criterion for estimating the retarding effect of growth inhibitors was a concentration causing 50% inhibition of native growth, this characteristic proved hardly suitable for describing induced growth depression.

To study the retarding effect of the inhibitors on IAA- and GA-induced growth processes, we have selected critical inhibitor concentrations which did not inhibit native growth, but were able to affect induced growth.

The strength of their antagonistic effect can be classified as follows; total antagonism with respect to a phytohormone (abbreviated TA), i.e.total canceling out of the hormonal effect; partial antagonism with respect to a phytohormone (abbreviated PA), i.e. partial neutralizing of the hormonal effect and, finally, no antagonism (abbreviated O), when the inhibitor in critical concentrations does not affect hormone-induced growth.

An example of total antagonism is the action of TIBA on IAA-induced root formation in *Phaseolus* cuttings; control (water) 100, TIBA (0.4 mg/l) 103, IAA (60 mg/l) - 362, TIBA + IAA 65%. An example of partial antagonism is the effect of coumarin on the IAA-induced growth of wheat coleoptile segments: control (water) - 100, coumarin (8 mg/l) - 10, IAA (7 mg/l) - 303, coumarin + IAA - 218%. An example of lack of antagonism is the effect of CCC on the growth of coleoptile segments: control (water) - 100, CCC (180 mg/l) - 97; IAA - (7 mg/l) - 319, and CCC + IAA - 330%. The criterion selected for estimating the antihormonal effect of the growth inhibitors was used on the three tests: the growth of coleoptile segments and *Phaseolus* cuttings treated, with IAA solution, and also on the growth of the stem of a GA-treated pea plant (Table 10.4).

Table 10.4 Interaction of growth inhibitors with phytohormones and their effects on various growth forms

Growth form (object)	*Phyto-hormones, mg/l*	*Natural inhibitors*			*Synthetic inhibitors*			
		ABA	*Coumarin*	*Phloridzin*	*MH*	*TIBA*	*CCC*	*Morphactin*
Cell extension (wheat coleoptile)	IAA	PA	PA	PA	PA	O	O	PA
	7	(0.04)	(0.8)	(125.0)	(430)	(14.0)	(180)	(7.0)
Stem growth (pea)	GA	PA	O	O	PA	-	O	PA
	0.1	(0.04)	(1)	(20.0)	(12.0)	-	(750)	(0.2)
Root formation Phaseolus cuttings	IAA	-	PA	O	TA	TA	PA	TA
	60	-	(12.5)	(12.5)	(3.5)	(0.4)	(460)	(0.06)

Note: TA – total antagonism, PA – partial antagonism, O – no antagonism. Bracketed figures are critical inhibitor concentrations (in mg/l) which do not affect native growth. For inhibiting coleoptile segment growth, morphactin was used in the form of a solution in 0.1% ethanol.

Below the main characteristics are listed of the groups of natural and synthetic inhibitors, without structural features for individual representatives of each group.

1. Natural growth inhibitors tested with respect to IAA and GA do not exhibit total antagonism, in other words their limiting concentrations cannot completely cancel out the effect of these hormones on growth.

2. Each of the synthetic inhibitors depresses antagonistically one or two tested growth processes.
3. The same inhibitor can display the same form of antagonism with respect to two quite different processes, one induced with gibb4rellin and the other with auxin.
4. The same inhibitor (for instance, TIBA) can exhibit total antagonism with respect to IAA when the latter acts as a rooting inducer, or show no antagonism at all to the same IAA if it affects cell extension. The same applies to the other inhibitors: MH, CCC and morphactin.

On the whole it would appear that there are no specific antihormones (anti-auxins or antigibberellins) among the tested inhibitors, but there are substances which act as inhibitors on some forms of growth processes. The same substance (for example, TIBA) can antagonistically depress a phytohormone or show no effect on it whatsoever, depending on the form of growth induced by this phytohormone. Thus, analysis of data on the action of natural and synthetic inhibitors on growth processes makes it possible to define natural inhibitors (abscisic acid, coumarin and phloridzin) as compounds depressing predominantly the growth processes of extending cells, MH and TIBA as inhibitors of the processes of organogenesis (for example, root formation), morphactin as a substance inhibiting the normal function of the apical meristem, and CCC as a stem growth inhibitor. This conclusion by no way means that any of the above compounds may not posses a number of other properties, such as the ability to accelerate leaf abscission (abscisic acid), inhibit auxin transport (TIBA) or retard the geotropic processes (morphactin).

There are considerable differences between phenol derivatives and abscisic acid even inside the group of natural growth inhibitors. Phenol derivatives inhibit growth processes in high concentrations and retard root formation in *Phaseolus* cuttings; abscisic acid does not exert inhibitory action on root formation and depresses other processes in lower concentrations than phenols. Despite specific differences between phenolic inhibitors and abscisic acid, both subgroups of endogenous inhibitors have one property in common, namely: there is no narrow hormonal specificity in their action on growth, i.e. these substances are not anti hormones in a strict meaning of this term.

Whatever form of growth is induced by a phytohormone (stem growth by gibberellin, root growth by auxin, seed germination by kinin), natural growth inhibitors can act antagonistically by inhibiting this induction. Here lies to a certain extent the non-specificity of their effect on growth processes. At the same time the above scheme does not rule out possible specific action of the growth inhibitors on general metabolic pathways, hormone synthesis and on hormone decomposition pathways. These problems will be discussed in the next section.

10.4 Effect of Natural Inhibitors on Phytohormone Activity and on Some Metabolic Processes in Plants

Almost all the numerous works devoted to metabolic aspects of the action of natural inhibitors on plant growth fall into the following categories: the effect of natural growth inhibitors on respiration and energy exchange, the effect of growth inhibitors on nucleic and protein metabolism, and their action on the activity of phytohormones.

It remains totally vague whether natural growth inhibitors can act as uncoupling agents in a plant itself. Another problem which has not been properly studied yet is the possibility of interaction of natural growth inhibitors with chloroplasts and mitochondria in plant cells.

Another aspect of the action of natural inhibitors on growth is connected with their effect on enzyme synthesis. While phenols act mainly on the activity of enzymes, abscisic acid affects predominantly their synthesis. Thus, Jacobsen & Varner (1967) reported that abscisic acid in concentrations of 1 and 10^{-6} mM inhibited the synthesis of an amylase and protease and also cancelled out the effect of gibberellins on the synthesis of a-amylase. Such an effect could be attributed to the inhibition of synthesis of all the nucleic acid fractions. If abscisic acid (10^{-6}M) was added to chromatin with RNA-polymerase activity, this caused 22-38% inhibition of the polymerase activity in case the inhibitor was introduced into a homogenized seedling system. On the other hand, the addition of abscisic acid directly to pure chromatin did not affect its activity.

Phenolic compounds, among them natural phenolic inhibitors, regulate the activity of the enzyme auxin-oxidase which decomposes indolylacetic acid, i.e. these substances are involved in the control of auxin catabolism.

It has been proved that each process of auxin metabolism is regulated by a particular group of phenolic compounds. Below some phenolic compounds are given which exhibit a specific regulatory effect on the IAA content.

Participation in auxin catabolism	*Compounds*
IAA synergists or auxin-oxidase inhibitors	Phenolcarbonic acids; protocatechuic, caffeic, syringic, gallic, chlorogenic; flavonoids: quecitrin, luteolin, astragalin, myricetin
IAA antagonists, or auxin-oxidase cofactors	Phenolcarbonic acids; salicylic, p-hydroxybenzoic, vanillic, p-coumaric; flavonoids: apigenin, naringenin, naringin, flavonoids with a para-position of the OH group in B ring

The location of the OH-group in a para-position imparts IAA antagonist properties to phenol. This antagonism is accomplished through the intensification of the activity of the IAA-breaking enzyme (auxin-oxidase). If the OH-group is placed in the meta- or ortho-position, the activity of phenol is either attenuated or ceases completely.

The introduction of the second or third hydroxy group into a phenol molecule imparts the opposite properties; phenol becomes an IAA synergist. Affecting IAA through the auxin-oxidase mechanism, dixydroxyphenol retards decomposition of IAA, i.e. maintains preserves it in an active state. Example, a flavonol belonging to the group of IAA

Synergists may become an antagonist in case if its molecule is acylated. The monophenol-dixydroxyphenol system not only controls elongation of cells, but also regulates geotropic curvatures of the stem and tissue culture growth.

To compare the inhibitory ability of polyphenols and their quinones two substances were selected: para-benzoquinone and hydroquinone. It is known, however, that most water-soluble polyphenols are ortho-forms. This seems to testify against para-isomers as the model pair of natural inhibitors. The circumstances described below explain to a certain extent why we have selected ortho-quinone and para-quinone as test compounds. The argument against high specificity of natural inhibitors is a great variety of substances in this group from the viewpoint of their chemical nature. It is sufficient to say that the substances included in the group of natural growth inhibitors frequently belong even to different classes of chemical compounds. Despite certain differences, ortho- and para-benzoquinones as well as ortho- and para-diphenols are rather close to each other so far as a whole number of their chemical and physico-chemical properties are concerned. On the other hand, our preliminary experiments have shown that plant tissues used as bio-assays oxidized hydroquinone much slower than pyrocatechin. This peculiarity of para-forms helps discriminate in time between the effect of polyphenol on physiological processes and the effect of its phenol on physiological processes and the effect of its quinone. Another advantage of para-forms is higher stability of para-benzoquinone as compared with ortho-benzoquinone which makes it easier to distinguish between the inhibitory effect of quinone itself and the effect of its oxidative condensation products.

Judging the inhibitory properties of the tested compounds by a concentration causing 50% growth inhibition, para-benzoquinone proved to be more active than its original phenol-hydroquinone (Table 10.4) in experiments with wheat seedling segments and *Phaseolus* cuttings (in bio-assays used when testing inhibitors and indole auxins). Thus, for example, para-benzoquinone depressed coleoptile segment growth and root formation in *Phaseolus* cuttings, respectively, 15 times and 2 times stronger than hydroquinone.

As can be seen from Table 10.4 the ability to inhibit elongation of wheat coleoptile segments or root formation in *Phaseolus* cuttings is very similar for resorcin and hydrdquinone. On the other hand, the ortho-form of pyrocatechin is several times more active than the para- and meta-forms: hydroquinone and resorcin. Taking into account the fact that, as has been found earlier, the tissues of cereal coleoptiles can oxidize at a high rate ortho-, but not para- or meta-forms of dioxybenzene, it may be concluded that strong inhibiting properties of pyrocatechin solutions do not contradict the assumption of the leading role of the oxidation products of ortho-phenols (quinones) in the display of inhibitory properties by the latter; on the contrary, these properties are rather an argument in favor of this assumption.

However, in order to be able to discuss a possible physiological role of pyrocatechin oxidation products, in particular ortho-benzoquinone, it is necessary to obtain reliable

Table 10.5 Polyphenol doses (mg/l) causing 50% inhibition of wheat coleoptile growth and root formation in Phaseolus cuttings

Substance	*Coleoptile segment growth*	*Root formation per cutting*
Para-benzoquinone	120	500
Pyrocatechin	470	125
Resorcin	1725	1000
Hydroquinone	1825	1000

data for proving their formation when polyphenols act upon coleoptile segments. As ortho-benzoquinone is an extremely unstable compound, its formation as the result of contact of pyrocatechin with bio-assay tissues was proved by converting quinone into its stable derivatives - ortho-dihydroxy. diphenylsulfone with the aid of benzenesulfinic acid.

The use of chromatographic analysis and color reactions made it possible to isolate and identify ortho-dihydroxy-diphenilsulfon from a mixture of pyrocatechin solutions (2.5 x 10^{-3} M) and benzenesulfinic acid (2.5 x 10^{-3} M) after segments of wheat seedling coleoptiles had been incubated on this mixture (Table 10.5). Sulfone could be produced only if ortho-benzoquinone was first formed, so that the formation of ortho-dibydroxy-diphenylsulfone is a proof of the oxidation of pyrocatechin to its quinone in experiments with bio-assays.

In preliminary experiments it was shown that benzenesulfinic acid does not affect the rate of oxidation of polyphenols both by polyphenol-oxidase enzyme compounds and by whole plants. The oxygen uptake was estimated polarographically.

When studying the inhibiting properties of benzenesulfinic acid it was found that its retarding effect is weaker than that of original phenolic compounds, and still weaker than the effect of their respective quinones. Besides, sulfones, products of interaction between benzenesulfinic acid and quinones, are much less toxic than phenolic compounds. The production of ortho-benzoquinone in the contact of pyrocatechin solutions with the tissues of the growth bio-assay necessitated investigation of the inhibiting properties of

Table 10.6 Action of some inhibiting products on growth of Albidum wheat coleoptile segments (% to control)

Substance	*Concentration mg/l*				
	2000	1000	5000	250	125
α- Naphthoquinone	0	0	0	0	5
Para-Quinione-dioxime	0	0	7	13	24
β-Naphthoquinone	0	0	0	0	0
Pyrogallol	0	15	38	65	71
Hydroquinone	50	62	65	77	88
Pyrocatechin	0	17	57	85	96

ortho-benzoquinone. By virtue of their low sensitivity and a slow response such common bio-assays as coleoptile segment growth, seed germination and rooting of *Phaseolus* cuttings are inapplicable for estimating the inhibiting activity of ortho-quinones which have a very short life (half-life of ortho-benzoquinone is 20 min. at pH 1.9). Therefore, the bio-assay used in this series of experiments was the elongation of garden pepper-grass roots which, due to their small size and a high surface-to-volume ratio, can respond with changes in their growth even to a 2 min treatment with quinone solutions.

The decomposition of ortho-benzoquinone in water solutions is accompanied by the production of free radicals - semiquinones, various polyphenols and polymerization products. But experiments showed that as ortho-benzoquinones decay in the solutions at pH from 4 to 7 (initial concentration 5×10^{-3}M) they lose their inhibitory properties, i.e. it is evidently ortho-benzoquinone which is mainly responsible for the inhibitory properties of the pyrocatechin oxidation products. In these experiments two-minute treatment of the roots of garden peppergrass seedlings with pyrocatechin solutions did not affect their subsequent growth.

It must be pointed out that naphthoquinones, similar to benzoquinones, produce a strong inhibitory effect which is much stronger than that of polyphenols (Table 10.6). In this way we have obtained the following data: the production of phenolic quinones when incubating plant tissues on solutions of these phenolic compounds; a high inhibitory activity of quinones which exceeds the inhibiting properties of polyphenols many times; stronger inhibition of growth on the solutions of those dioxybenzene isomers which are oxidized in a higher degree than others by coleoptile tissues. Reduction in the inhibiting activity when passing from ortho-benzoquinone to its further conversion products testifies to an important role of polyphenol oxidation products and, in the first place, quinones, in growth inhibition by exogenous ortho-phenols.

It is appropriate here to point to a number of questions which need experimental study. We do not know, for example, what the inhibition mechanism of non-phenolic inhibitors is, i.e. those which cannot form quinones as aggressive intermediate agents, or what a further pathway of growth inhibition is in quinones themselves. Other questions to be answered are what centers in a cell are first of all affected by these compounds and how easily they penetrate through vacuolar and other intracellular membranes.

However, the problems of the specificity of mechanisms underlying such interaction at each stage have not been given due attention. At the same time, the 'adjustment', or 'fitting' of each phytohormone to its natural inhibitor counterpart is tremendously important. The study of the biosynthesis of auxins and phenolic inhibitors, on the one hand, and abscisic acid, on the other, makes it obvious that at this early level of their formation each group of phytohormones is associated with a specific group of inhibitors, though such an association no longer exists at the functioning stage.

At the stage of biosynthesis a natural growth inhibitor could specifically block the formation of a phytohormone at the precursor conversion level. In our experiments, for example, the natural inhibitor isolated from pea tissues, quercetin-glycosyl-coumarate, inhibited 88% of the synthesis of IAA-derivatives and retarded conversion of L-tryptophan into indole auxins by 65%.

p-Coumaric acid isolated from maize inhibited synthesis of IAA derivatives by 40% and conversion of L-tryptophan by 34%. There might be no inhibition of L-tryptophan conversion if it was totally incorporated into a molecule of a more complex products, for instance, glucobrassicin, as was observed in experiments with cabbage. However, even in this case the synthesis of IAA derivatives was strongly inhibited. It is also known that abscisic acid inhibited production of gibberellins in the seedlings of a number of plant species.

In other words, at the stage of the synthesis of phytohormones and inhibitors each phytohormone seems to correspond to a specific inhibitor. However, the existence of this rule needs further experimental corroboration.

At the next stage when synthesized compounds start functioning, the 'one phytohormone - one inhibitor' specificity no longer exists. Although various forms of interaction between phytohormones and inhibitors become apparent at this stage, none of these forms may be called specific. For this reason we have to admit that in the functioning stage the functions of all the phytohormones are suppressed non-specifically through the inhibition of general metabolic processes which are of vital importance for growth.

The above data corroborate this statement experimentally. Natural inhibitors can reduce the rate of growth by partially inactivating phytohormones or by directly retarding individual metabolic processes. In any case the growth process will be inhibited, and the inhibition will be the stronger, the higher the concentration of the inhibitor and the lower the concentration of a phytohormone in the tissue.

At the third stage of inactivation of phytohormones and natural inhibitors, specific and non-specific processes seem to co-exist. On the one hand, the processes of forming complexes and complication of the molecules of natural growth regulators may be regarded as non-specific: thus, the production of low activity glycosides has been observed for all types of phytohormones and inhibitors. On the other hand, certain forms of inactivation are displayed by separate groups of phytohormones and inhibitors. In the previous chapters we pointed to the existence of mechanisms of IAA complexes with phenols and emphasized the regulatory role of phenols in IAA oxidation by means of the auxin-oxidase system.

On the whole, it must be admitted that natural growth inhibitors and phytohormones involved in various stages of synthesis, functioning and decomposition, interact through specific and general (non-specific) mechanisms.

Natural phenolic inhibitors accumulating in plant tissues in large amounts retard the growth of elongating (coleoptile segments) and dividing cells (rhizogenesis in cuttings) in relatively high concentrations. Substances inhibiting the synthesis of protein and nucleic acids also retard the former growth process in high concentrations, and the latter form in low concentrations. Natural phenolic inhibitors (phloridzin, naringenin, salicylic acid, etc.) are able to act in experiments in vitro as agents uncoupling oxidative and photosynthetic phosphorylation.

Unlike phenolic inhibitors, abscisic acid inhibits extension of coleoptile segment cells in low concentrations, and produces no inhibiting action at all on rhizogenesis. Abscisic acid can inhibit the growth of cotyledons and water supply in germinating apple seeds, as well as retard chlorophyll synthesis in such cotyledons.

The tests of natural and synthetic inhibitors with respect to various forms of growth processes - from simple to most complex, have shown that the strongest inhibitor of growth processes is abscisic acid; coumarin is weaker and phloridzin is weakest.

Among the tested synthetic inhibitors the strongest inhibitory action was produced by morphactin. TIBA and MH are effective in doses one order higher, and, finally, CCC exerts inhibiting activity in doses 2 or 3 orders higher.

Natural growth inhibitors are most effective towards coleoptile segment growth, whereas synthetic inhibitors produce the strongest effect on more complex forms of growth - root formation (morphactin, TIBA, MH) or stem growth (CCC).

The study of the interaction of IAA and GA has demonstrated that natural inhibitors in critical concentrations which do not affect endogenous growth can not completely neutralize the effect of these phytohormones on the growth process. Natural phenolic and terpenoid inhibitors are able to depress both auxin- and gibberellin-induced growth processes, i.e. they do not possess anti hormonal specificity.

Contrary to natural inhibitors, some synthetic growth inhibitors display total antagonism relative to phytohormones.

Phytohormones and natural inhibitors may interact at the stages of synthesis, functioning and decomposition of these compounds. During biosynthesis this interaction is predominantly specific due to the fact that the pairs: auxins-phenolic inhibitors and gibberellins-abscisic acid are produced, respectively, from common metabolic precursors - shikimic and mevalonic acids. At the second stage the interaction of phytohormones and inhibitors is no longer specific, i.e. inhibitors retard any phytohormone-induced form of growth. At the third stage (decomposition) this interaction again becomes partially specific and is manifested, for example, in that phenols are involved in the auxin oxidation process. However, apart from specific interaction, there are also general pathways of inactivation of phytohormones and natural growth inhibitors (glycosiding, formation of complexes, etc.).

10.5 Bio-assay methods

The bio-assay or biological test method has been playing a leading role in study of the functions of phytohormones. By a biotest is meant a procedure applied to such an object (a plant or its separate parts) which possesses high sensitivity to phytohormones or growth inhibitors supplied from the outside.

10.5.1 Cell expansion tests

One of the first biotests used for estimating auxins was a curvature test for etiolated oat coleoptiles elaborated by Went (Went, 1926, 1963; Went, Thimann, 1937). The test procedure is as follows; an isolated organ is placed on an agar block in a moist chamber and after some time the agar is transferred onto an etiolated coleoptile. The curvature of the coleoptile is a measure of the auxin content in the coleoptile.

Although developed in the early 1920's, this test remains the most specific and sensitive test for auxins. With this test, it is possible to detect indolyl-acetic acid in a concentration down to 10^{-7} M.

The limitations of this test are that it is labor intense, also the necessity for the researcher to stay for a long time in a dark room and in a highly humid atmosphere and, most important, high sensitivity of the coleoptiles to changes in environmental conditions. On account of these drawbacks, the oat coleoptile curvature test is little used in modern laboratory work.

Though less sensitive to auxins as compared with the previous test, this test is convenient in that coleoptiles are prepared in scattered light, their segments are grown in water solutions, and the laborious technology of agar blocks is thus made redundant. Coleoptile segments are incubated in test solutions for 17 to 20 hours after which measurements are made. In order to avoid the action of microorganisms, Balin (1967) suggested that the period of growing segments can be reduced to 6 or 8 hours. His argument is that apart from the risk of microbial contamination, prolonged exposure masks differences between various test versions. Control sections of coleoptiles grow linearly during 20 hours, whereas their growth in IAA solution is linear during the first 8 hours and then remains constant. Thus, prolonged incubation obscures differences between the control and test solutions. Baum believes that 6 hours of incubation is sufficient for segment growth; besides, this preserves specificity to auxins which diminishes in the subsequent period under the effect of other factors, for example, the action of inhibiting exudates of the coleoptiles.

The test for the growth of wheat coleoptiles is applicable both for estimating auxins and natural growth-inhibiting substances. This test is practically insensitive to gibberellins and kinins. The drawbacks of the method are time consuming and labor intense procedure of preparing coleoptile segments, a tiring procedure of pushing the initial leaf out of the coleptile cylinder, and a long process of selecting a wheat variety suitable for the test. The selection of a wheat variety for this biotest is usually guided by the following considerations: 1) the germination of the seeds of the variety selected must be high; 2) the seedlings must reach 18-20 mm in length within 48 hours of growth; 3) during the incubation period the coleoptile segments must increase in length so that their measured increments fit a single-peak Gaussian curve which allows estimation of the degree of standardization of the material, and 4) the segments of coleoptiles must possess good sensitivity for IAA, i.e. an increase of a coleoptile segment length in IAA solution must be 250-300% that of the control segment.

The segments of coleoptiles are measured on graph paper which permits measurement with an accuracy to within 0.5 mm. The measurement accuracy can be appreciably improved through using a magnifying glass or a microscope.

Apart from the two tests described above the method of elongation of the oat first internode had become recently used. After incubation in IAA or gibberellic acid solutions, 4 mm segments of the first internode of an etiolated oat seedling exhibit a considerably larger rate of growth as compared with their control counterparts grown in water. With this method, it is possible to detect down to 1 µg/l of the phytohormone. The main limitation of this method is the lack pf hormone specificity, since an oat internode segment is sensitive both to auxins and gibberellins.

Another biotest used for phytohormone detection is the method of young wheat coleoptiles. After the first 24 hours of growth, young wheat coleoptiles isolated from the caryopsis show a high growth response to gibberellin and kinetin. When wheat coleoptiles are incubated in the solutions of these substances, their growth can be induced even with the dose of 10^{-8}M. The disadvantage of the method, similar to the previous test, is that it is not specific, since coleoptiles respond readily to both phytohormones.

Other methods which are less common are auxin tests: curvature of *Phaseolus* hypocotyls, elongation of pea stem segments, curvature of cotyledon petioles in radish and dissected pea stems, and also a gibberellin test - elongation of the first internode of *Phaseolus*. Thus, the most common of the above cell-expansion tests is that for growth of wheat coleoptiles which gives easily reproducible results provided all conditions are observed. This method is usually used for estimating the activity of growth-inducing substances (auxins) and the activity of growth inhibitors.

10.5.2 Cell division tests

With a view to detecting cytokinins, auxins, gibberellins and growth inhibitors, a number of methods have been suggested that are based on the investigation of cell division.

Isolated cells of tobacco, tomato, bean, carrot, lettuce and other plants were placed in a microchamber with a drop of the medium in which a cell was incubated. The method is used to identify cytokinins which affect the cell division process.

The method of cultivating individual isolated cells makes it possible to study the action of a number of compounds on the rate of the cell-division process, formation of cell wall elements, changes in the number of mitochondria, etc. This method is time consuming and has a rather limited application for a large scale investigations.

Much more frequent tests for determining cell division factors (cytokinins) are soybean and apple calluses grown in sterile conditions. To produce a material suitable for use as a biotest, calluses grown on a medium with kinetin and without kinetin for several times. This procedure induces a growth response in calluses even when 10^{-8} M of the kinetin solution is introduced.

In addition to the above methods, the action of growth substances on mitosis can be estimated using the endosperm of some immature seeds, for instance, the endosperm of the African blood lily *Haemathus katheriniae* (Baker). This method is generally used for studying the properties of colchicine-type substances which affect normal mitosis.

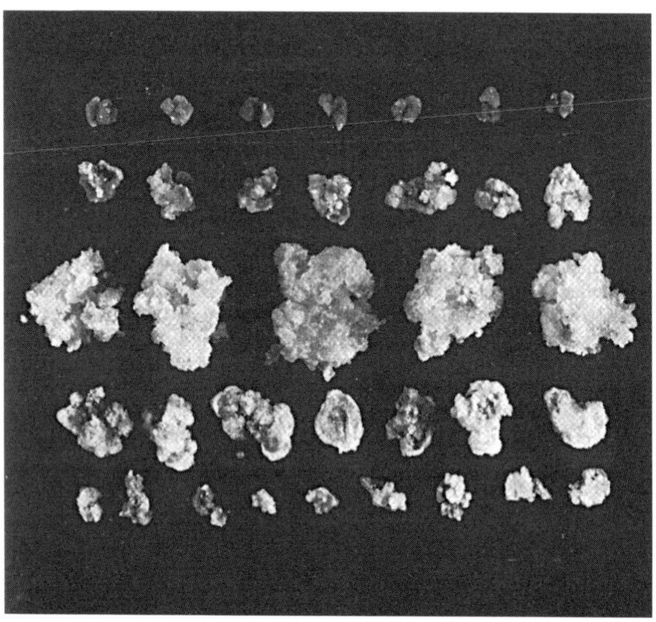

Figure 10.10 Effect of various doses of kinetin on growth of pith parenchyma callus in Wisconsin 38 tobacco (F. Skoog's method). Experiments by Prof. V.Kefeli. Top line – control; second – kinetin (0.01 mg/l); thirdline – 0.1 mg/l; fourth line – 1 mg/l; bottom line – 10 mg/l.

The callus formation ability of sterile pieces excised from the pith of the tobacco stalk is used for identification of cytokinins (Figure 10.10). This method was used for testing a broad variety of cytokinin-type compounds (Skoog et al., 1967). *Phaseolus* cuttings which are ten-day seedlings with a removed root system usually serve for identifying auxins. With this method,

IAA can be identified in the concentration of 50 mg/l and above. Coleus cuttings are sometimes used for the purpose.

As we can see, the cell and tissue culture is most often used to identify kinetin and kinin-like substances, where as coleoptile and etiolated stem segments and rooting cuttings serve for detecting auxins. Intact plants show a weak response to these phytohormones.

10.5.3 Complex tests

Above, consideration was given to cell-expansion and cell-division tests. There are also a number of complex tests such as for example, seed germination tests. Most often, a seed germination test is used for detecting inhibitor compounds such as phenolic growth retardants and ABA. The test is simple and sufficiently fast. Thus, for example, mustard seeds are incubated during 15 to 17 hours. It is recommended to count germinated seeds after 50 per cent of control seeds have been germinated, i.e. after germination of half the seeds in Petri dishes with control solutions or eluates. Estimation of germinated seeds at a later period may diminish the differences between versions, particularly, at low concentrations of the inhibitor in a test solution.

The number of seeds germinating in a control, i.e. on water, is taken to be 100%. The content of ABA or other inhibitors will be given by the ratio of the number of germinated seeds in test versions to that of the control. All data are expressed in % and are presented in the form of charts or histograms. Unlike auxin and kinetin, gibberellin is capable of accelerating the growth of a whole plant. Therefore, for identifying gibberellin, use is made of various plants, their seeds, and sometimes the above-ground parts of plants, for example, pea and maize, in which the growing point is preserved.

Bioassays have a varying sensitivity to different gibberellins which is another peculiar feature of gibberellin biotests. Thus, one of the bioassay systems, for example, cucumber hypocotyl, is responsive to gibberellin A_4, the lettuce hypocotyl exhibits response to A_7, while the dwarf d-3 maize mutant is most sensitive to gibberellin A_5.

Apart from growth-based bio-assays, other methods are available which depend on the action of phytohormones on certain biochemical reactions. For example, the method of inducing α-amylase in barley endosperm was used to identify gibberellins (Paleg, 1961); the chlorophyll decomposition delay methods were used for detecting kinins (Osborne et al., 1961; Kende, 1964).

Abscisic acid and its related substances are now being identified in a bioassay for abscission of petioles of cotton explants. This bioassay was developed by Addicott with co-workers. Estimation of cytokinin activity in natural and synthetic compounds depended on the stimulation of betacyanin formation by seedling of *Amaranthus caudatus L.*

A proportional relationship between the contents of exogenous substances with cytokinin activity and the biosynthesis of betacyanins enables one to estimate quantitatively the amount of these substances in plant tissue extracts.

So, it should be admitted that bioassays are convenient techniques which reveal the activity of phytohormones or growth inhibiting substances. They prove to be particularly effective in the initial stage of investigation for primary isolation of a pure product out of various complex organic substrates. However, researchers also resort to the methods of biological control of phytohormones at various stages of chemical purification , the more so that the available chemical methods for identifying phytohormones are either insufficiently sensitive or possess low specificity.

10.6 Analysis of interaction of growth regulators

The final stage in the analysis of plant growth substances is estimation of their growth activity by means of bio-assays. The test for growth of wheat or oat coleoptile segments is known to be one of the most common bio-assays. With this test, stimulation and inhibition zones (spots} are identified on chromatograms. Let us consider certain advantages of the coleoptile growth test.

The most important advantage of this test is its fast response. During 20 hours of incubation, a wheat coleoptile segment (Moskovka variety wheat) with an initial length of 4 mm increases up to 8 mm in 2% sucrose solution (control). In IAA solution (1 mg/l), the segment reaches the length of 16-18 mm, in other words, its increase in length is 300 per cent and above as against the control. In the solution of the growth inhibitor p-coumaric acid (175 mg/l), the growth of the segment is 6 mm, i.e. 50 per cent as compared with the control. Thus, the physiological activity of the compound under test is manifested already within a short time period. The other advantage of the coleoptile growth test is its specificity for auxins. It has already been pointed out that this test is almost insensitive to gibberellin and shows no response whatsoever to kinetin. At the same time, this test possesses certain limitations, one of which is inadequate growth information due to the fact that the test describes growth only in terms of elongation. One more disadvantage is lack of specificity with respect to growth substances isolated from tissues of various plants. Indeed, coleoptiles may contain, and probably contain, a quite different set of enzymes than the tissues of donor plants from which the growth substances have been isolated. This difference in the set of enzymes which destroy and synthesize growth regulators as well as a difference in the residual amounts of plant growth substances may, in the long run, determine the degree and the nature of the sensitivity of wheat coleoptiles and, say, segments of maize and pea seedlings with respect to the maize inhibitor p-coumaric acid.

In Figure 10.11 one can see that segments of maize and pea stems from which tissues p-coumaric acid is isolated are much less sensitive than wheat coleoptile segments. The above mentioned properties of donor plant tissues (namely, differences in the enzymes present and in the residual amounts of growth substances) can even induce a reverse negative growth response in wheat coleoptile segments as compared with the test prepared from the tissues of donor plants. Thus, cabbage hypocotyl segments respond negatively to phenolcarboxylic acid isolated from cabbage leaves, i.e. their growth is almost 50% inhibited as compared with that of the control, whereas the growth of wheat coleoptile segments is induced 40-50 per cent by the same amount of phenolcarbonic acid.

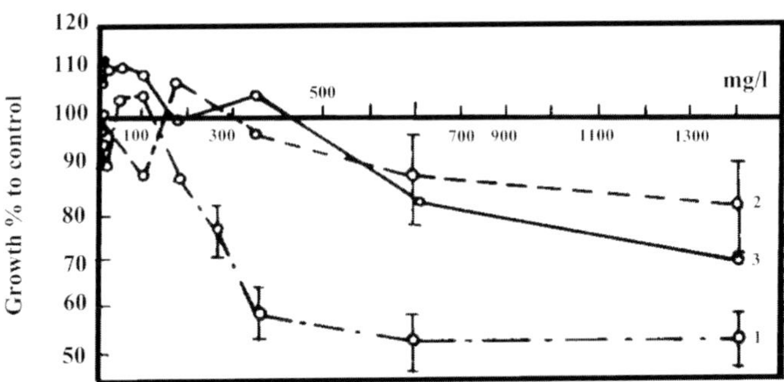

Figure 10.11 Effect of various concentrations of p-coumaric acid on growth of wheat coleoptile segments of Moskovka variety of wheat (1), Voronezhskaya 76 maize (2) and Ranniy zelenyi pea (3).

If a researcher is interested in the growth functions of the isolated compound with respect to the donor plant, it is common practice to use the tissues of this plant as a bio-assay. The bio-assays prepared from the tissues of donor plants are of some advantage because, first, they are specific towards the isolated plant growth substances, second, because they are simple to prepare (the seeds are wetted, stems are cut into segments, and leaves are excised) as compared with coleoptile cylinder section, and, third, due to a much more abundant information they provide in the effect on the growth process (growth is estimated both in terms of division and in terms of expansion). The disadvantage of the bio-assays is a wide variability of the plant material which obviously compels the experimenter to make several times more biological replications than the replications of the coleoptile segment test if one wants to achieve a 5% -10% significance level. Another disadvantage is slow growth in the tests. Whereas the coleoptile segment growth test lasts 20 hours, the seed germination test takes at least three or five days and sometimes even longer. The growth of stem segments (hypocotyls and epicotyls) lasts from 48 to 72 hours.

It should be taken into account that a prolonged (more than 20 hours) contact with plant tissues results in almost total decomposition of inhibitors and auxins. Therefore, incubation during several days occurs practically in the absence of the substances being tested. In the case of short-time use of the coleoptile growth test (i.e. under 20 hours), this phenomenon is pronounced in a much smaller degree. To compensate for this loss, in some cases the researchers add the same concentration of the plant growth substance after decomposition of the first portion This action, however, is physiologically senseless, because introduction of additional portions makes it impossible to estimate quantitatively the amount of the growth substance.

Another limiting factor for some tests from donor plant tissues is correlation between the organs of a cutting or a seedling observed during their growth. Thus, a question arises whether, for instance, we may consider salicylic acid a stimulator if it induces to some extent the growth of the epicotyl of *Phaseolus* cuttings while suppressing rooting of the same cuttings, or regard IAA as a growth inhibitor because it retards the growth of the epicotyl as the result of lavish root formation.

Of course, this question cannot be answered on the basis of only one test, no matter how specific it is. In order to determine the function of an isolated product (auxin or a growth inhibitor), a number of tests should be made.

When analyzing the results obtained in several bio-assays it often becomes obvious that a substance being tested (for example, IAA) may induce some growth" processes in *Phaseolus* cuttings and inhibit other processes (seed germination in some crops, the growth of the epicolyl in cuttings, etc.), while there are also processes which remain unaffected by this substance (for example, an increase of the hypocotyl in length, growth of the leaf area, etc.). Similar differences in the action on various growth tests can also be observed when analyzing natural growth inhibitors, for example, inhibitors isolated from a willow plant. These inhibitors depress bud opening and rooting processes in spring willow cuttings, but do not inhibit whatsoever the growth of buds in autumn. Thus plant growth substances are not universal in their functions. Auxins stimulate certain forms of growth processes and do not affect other processes which are induced by other hormones, for example, gibberellin. Therefore, it would be wrong to conclude that a compound does not perform any regulatory function only if the data obtained during the analysis of its properties evidence that it does not affect the growth of some bio-assays or is not involved in some processes.

What are the characteristics that make it possible to define a compound as a plant growth regulator and to differentiate its stimulating effects from the effects caused by other nonhormonal compounds? One of the basic characteristics is the stimulating effect of the substance being tested on a growth process. For establishing this effect the concentration curve of the tested substance should be compared with the concentration curve of a known growth regulator, for example, IAA or gibberellin. Thus, if the compound being tested induces, similar to IAA 200-300% growth of coleoptile segments (at the 5-10% significance level) it can be considered as a growth regulator. Weak stimulating effects caused by low amounts of alcohols, toxins and other compounds will remain below such a high level of activation.

Sometimes the stimulating effect seems to be sufficiently strong, but the accuracy of the data is much higher than 5 per cent (for example +15%). Then, after subtracting a three-fold deviation from the mean value, 50% stimulation will be as follows: 150% - 3 x 15% = 150% - 45% = 105%, that is, actual stimulation, or percentage in excess of the control, appears to be only 5%. Such cases may often be observed with a test for rooting of *Phaseolus* cuttings. Control cuttings which have taken roots in water have usually four roots each (100%), those grown in IAA solution (60 mg/l) - 60 roots (1500%), and cuttings which were grown in 2,4-dinitrophenol solution form 8 roots (200%). It would first seem that 2,4 DNP causes a hormonal effect. However, taking into account that the accuracy of the experiment is 15% and permissible deviations from the control are ±45%, the stimulating effect of DNP is reduced to 155%, and that of IAA to 1455%. Comparison of these figures shows that DNP does not exert a hormonal effect. So, the magnitude of the stimulating effect is the first property inherent in a growth regulator.

Another characteristic is a stimulating concentration range. If the high stimulating effect of the substance being studied is manifested for a sufficiently large number of concentrations as is the case with IAA (10^{-6} and $5x10^{-4}$M) this substance can be considered a plant growth regulator. The range of such compounds includes many indoles, dioxyphenylcarbonic acids (ferulic acid, caffeic acid, etc.) and also some phenoxy acids. Similar criteria should evidently be applied to investigation of the properties of an inhibiting compound. One of such criteria in estimating the growth retarding effect of inhibitors is C_{50} i.e. concentration of the inhibitors which depresses the growth of a bio-assay by 50% as compared with the control. For determining the value of C_{50} was suggested that the doses of substances should be taken in the ratios of 1:2:4:8:16. The plotted inhibition curves serve as a basis for estimating the value of C_{50}. This value can be used for estimating the growth retarding effect of not only natural, but also metabolic inhibitors. Another criterion which defines the action of an inhibitor is the type of its interaction with a phytohormone.

Generally, inhibitors taken in doses close to C_{50} failed to show specific differences in their action on IAA- or gibberellin-induced growth.

A difference between growth inhibitors when acting upon phytohormone-induced growth became obvious only in those critical doses which no longer inhibited native growth. In other words, the native growth of the bio-assays treated with these inhibitors was the same as that of the control or close to the control (Figure 10.12). In this case, the bio-assays started to exhibit retardation in their growth after they were treated with phytohormones. It is convenient to take phytohormones in suboptimal doses. The data obtained made it possible to classify all the growth inhibitors studied into four types depending on their antihormonal action. The first-type inhibitors totally neutralized the action of the phytohormone (P) when applied in a dose which did not retard growth (I_o). This type of interaction between an inhibitor and a phytohormone may be called competi-tive or total antagonism (TA). Such an effect may be written in the following

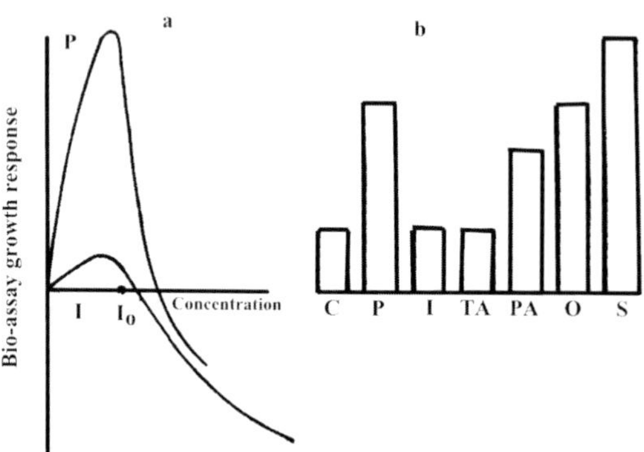

Figure 10.12 Schematic illustration of interaction of critical concentrations of inhibitor (I_o) with phytohormone (P); a – concentration curves of phytohormones and inhibitors; b – forms of interaction; C – control, TA – total antagonism, PA – partial antagonism, O – no antagonism, S – synergism.

form: P + I_0 x C, i.e. a phytohormone when acting with inhibitors, induces growth equal to, or even smaller than, the growth of the control (C).

Other inhibitors, used in the same critical dose (I_0) ,eliminated the effect of the phytohormone (P) but partially, i.e. some stimulating action was still evident and the growth they induced exceeded that of the control. This type of interaction can be called partial antagonism (PA) and written as follows: $C < P + I_0 < P$. The third type of inhibitors are those which produce no effect whatsoever on growth induced by a phytohormone (P), i.e. there was no antagonism at all (O) in the dose I_0: $P + I_0 = P$. The fourth type are inhibitors which stimulated the effect of a phytohormone (S). This latter type of interaction can be described as; $C < P + I_0 > P$. Schematically, these four types TA, PA, 0 and S are illustrated, see Figure 10.12.

In order to find the critical concentration (I_0) it is necessary to test at least 10-12 doses of the growth inhibitors. When using this critical concentration I_0 in phytohormone interaction experiments, it is convenient to take two more adjacent concentrations, one higher and the other lower than the concentration under study.

The use of the above scheme for classifying the effect of natural inhibitors on the growth of bio-assays appears more advantageous than the plotting and analysis of enzyme-inhibitor inhibition curves which are generally employed for experiments.

Gibberellins-hormones of the stem elongation. Therefore pea test developed by Brian is a model with total concentration on the stem enlargement. One of first procedures with pea is the cutting of the root in order to break root-stem relations and activate only stem elongation processes (Figure 10.13) show that only gibberellin has a specific effect on the stem elongation

These pictures possess the common structure-determination the standard level of each bio-test, its sensitivity to the phytohormone or inhibitor, evaluation of the specifity-reaction of the biotest on main growth substances. Thus pea test is sensitive only to gibberellins, which stimulate the test growth on 350%. ABA inhibits this test much more intensive than phenolic inhibitors (Figure 10.13).

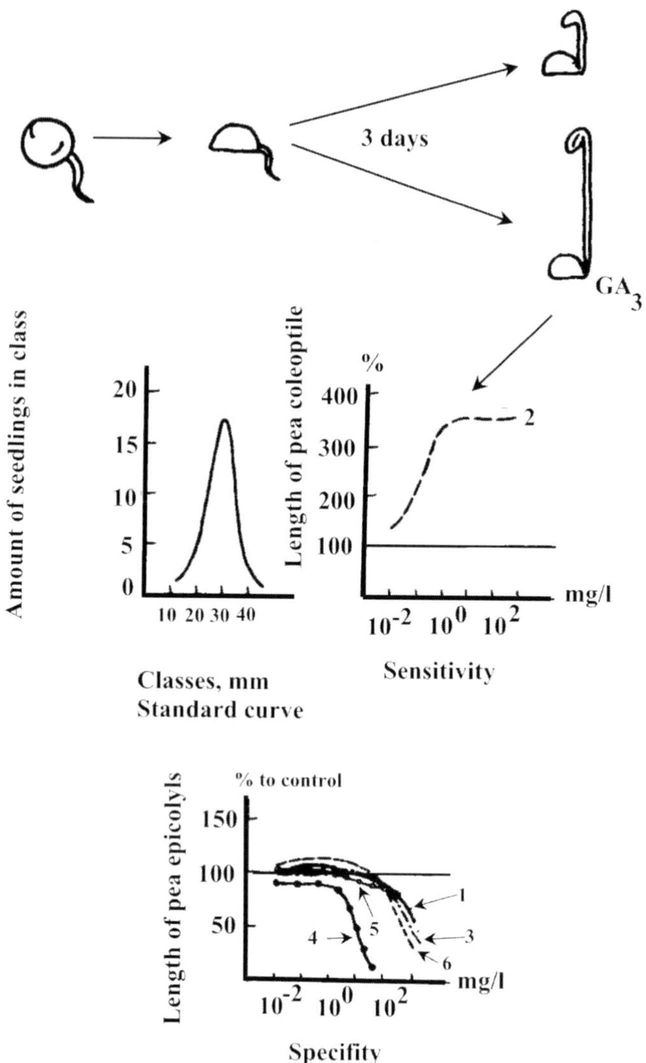

Figure 10.13 Pea test for gibberellins identification.
1 – IAA; 2 – GA_3; 3 – kinetin; 4 – ABA; 5 – Coumarin; 6 – p-coumaric acid.

More simple and less specific is lettuce test which is also reactive to gibberellin (GA_3). However this test is mostly used for the investigation of Gibberellin A-4. None of the other tested substances are able to stimulate the stem elongation (Figure 10.14). The property of gibberellin is to elongate the cells in the intact stem system.

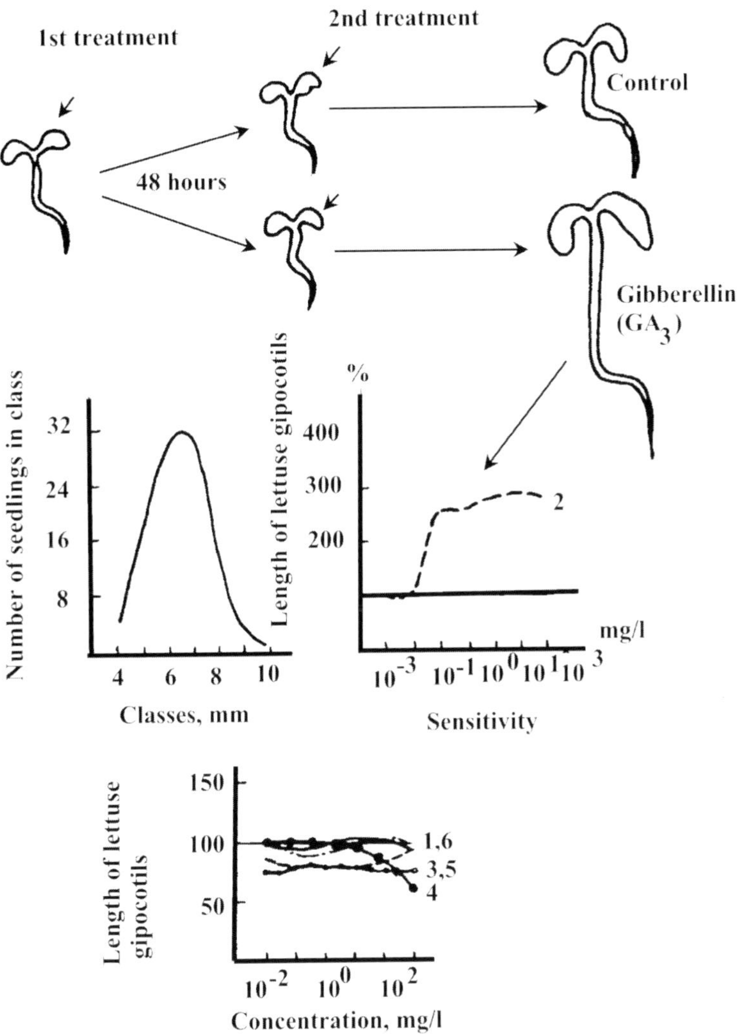

Figure 10.14 Lettuce hypocotyle growth test for gibberellin identification.
1 – IAA; 2 – GA_3; 3 – kinetin; 4 – ABA; 5 – Coumarin; 6 – p-coumaric acid.

Just opposite property possess auxin, like indolyl acetic acid IAA). This class of hormones is able to activate the stem elongation mostly in the complex of isolated cells. As an example, wheat or oat coleoptile tests-test of isolated sections of coleoptile with the isolated elongating cells. Natural growth inhibitor -ABA-abscisic acid in this test acts as negative factor and retards the cells elongation (Figure 10.15).

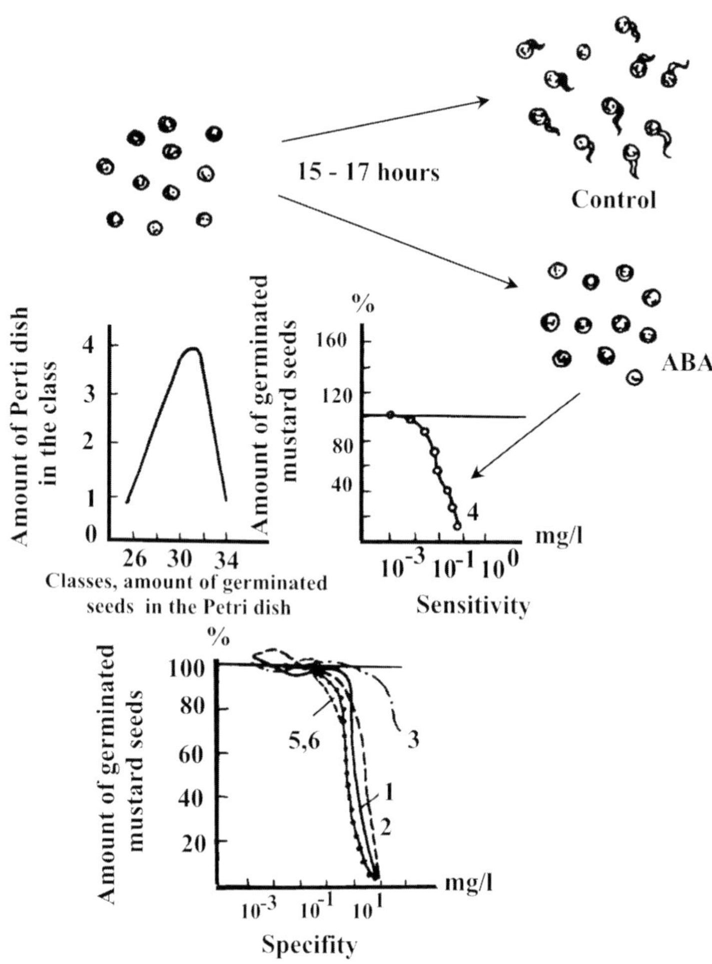

Figure 10.15 Effect of ABA on seed germination.
1 – IAA; 2 – GA_3; 3 – kinetin; 4 – ABA; 5 – Coumarin; 6 – p-coumaric acid.

ABA is not only the inhibitor of the cell elongation but also the retarding factor of seed germination. (Figure 10.16) Previous experiments showed that ABA is able to imitate the dormancy of seeds, preventing shoots and roots development. Not only ABA but also the other substances are able to depress the seeds germination -there are phenolic inhibitors which are located in seeds coats, however they are not such strong inhibitors as ABA.

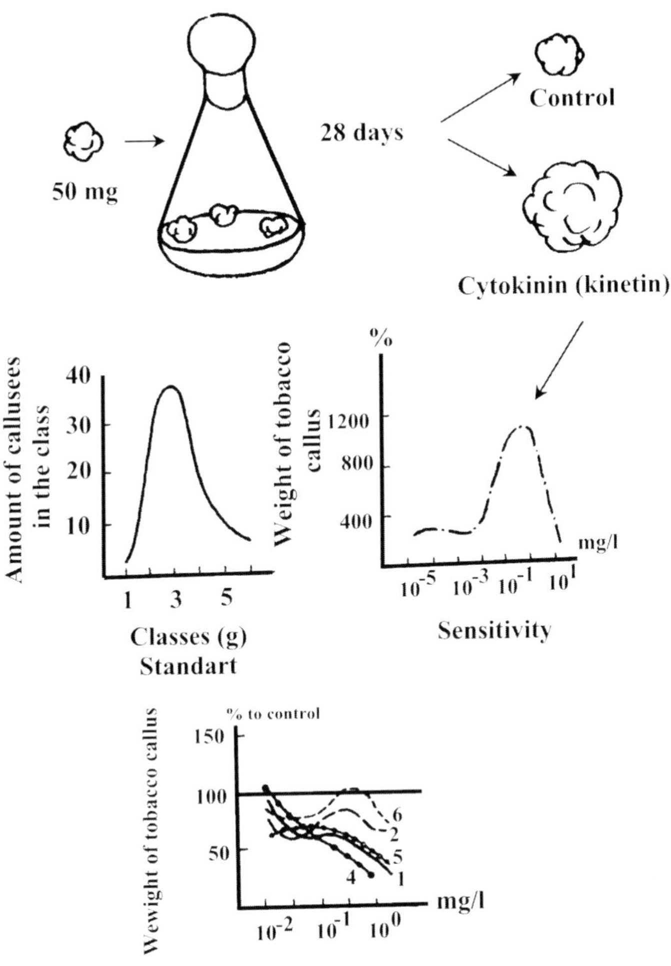

Figure 10.16 Callus test for cytokinin identification.
1 – IAA; 2 – GA_3; 3 – kinetin; 4 – ABA; 5 – Coumarin; 6 – p-coumaric acid.

Previous tests deals mostly with cell elongation and the tests on cytokinins are connected ether with cell division (Figure 10.17). It is interesting that this test also deals with the isolated cells which could long time divide but not differentiate. Tobacco test could also used as a model of organ differentiation. This last process is based on the balance of auxins.

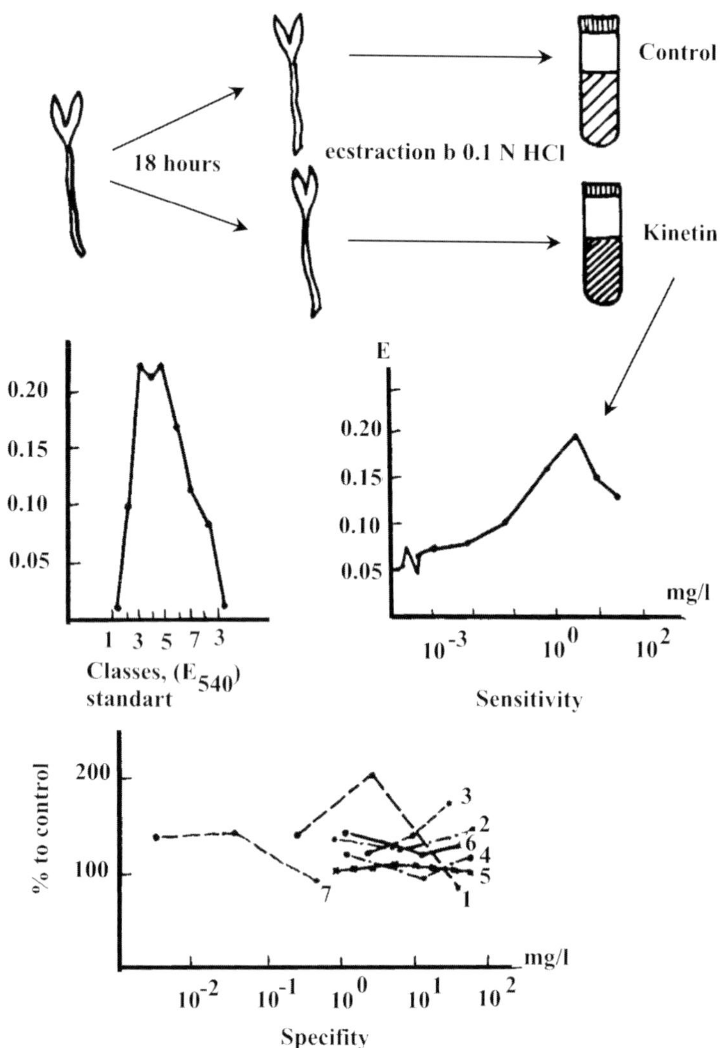

Figure 10.17 Amaranthus test for cytokinin identification
1 – adenin; 2 – adenosin; 3 – benzimidasol; 4 – GA; 5 – IAA; 6 – coumarin; 7 – ABA.

It is important to mention that some cytokinin tests do not need any sterility and are based on pigment reaction but not on the callus growth (Figure 10.18). Amarathus test is sensitive to cytokinins. In its cotyledons the pigment amarathin-beta-cyanine is formed. The process of its formation is just activated by cytokinins.

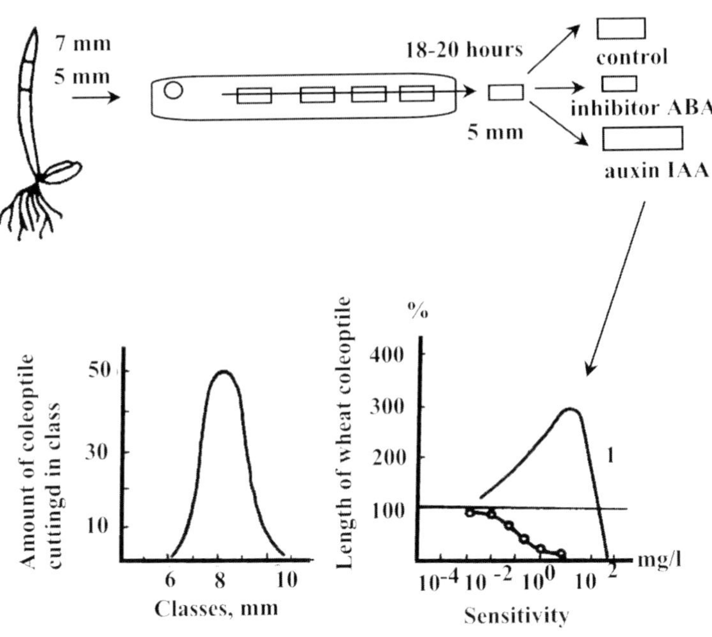

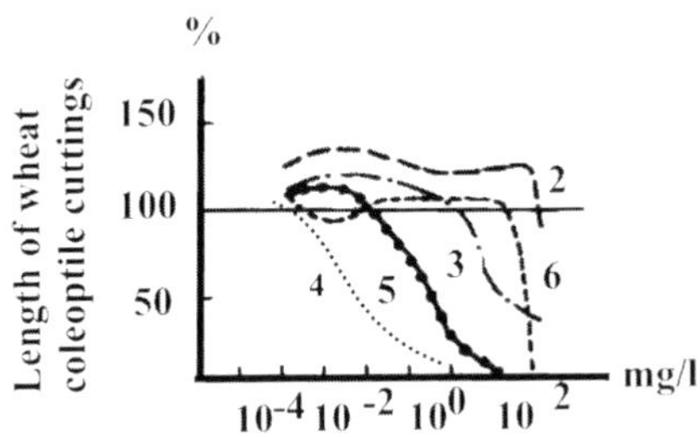

Figure 10.18 Wheat coleoptile cuttings test for auxin identification.
1 – IAA; 2 – GA_3; 3 – kinetin; 4 – ABA; 5 – Coumarin; 6 – p-coumaric acid.

This system of biotests shows the specifity of each class of phytohormones and used for the research by R.Turetskaya (1961), M. Chailakhjan, (1964) and by authors of this book In the previous chapters we describe more about tests for the allelopathic factors such as botanical herbicides and others.

Thus, our procedure for investigation of natural inhibitors and phytohormones comprises of the following stages:

1. Description of the growth process; estimation of its intensity and rate: elimination of organs (leaves buds, roots) saturated with phytohormones and growth inhibitors and control over changes in the nature of growth of the plant or its parts.
2. Fixation and extraction of plant material containing phytohormones and growth inhibitors, purification of the extract, its chromatographic separation and development of the chromatograms by means of color reactions and bio-assays.
3. Isolation of pure phytohormones and growth inhibitors by preparative chromatography, their detailed description and accumulation in sufficiently large quantities (from 20 to 200 mg).
4. Staging of model experiments aimed at studying the regulatory functions of the isolated phytohormones and growth inhibitors:

a) the study of the effect of natural growth inhibitors on formation of phytohormones from radioactive and non-radioactive precursors;

b) investigation of the features of interaction between phytohormones and natural inhibitors on various bio-assays;

c) introduction of natural growth inhibitors into growing and dormant tissues of donor plants;

d) the study of the decomposition of phytohormones and growth inhibitors in vivo and in vitro;

e) comparative studies of the features of the inhibitory effect of natural and synthetic growth inhibitors.

General observations and experimental proofs of hormonal biological tests permits to conclude that they could be also applied for the investigation of natural growth inhibitors. At the same time it is important to mention that the natural growth inhibitors never possess hormonal properties.

CONCLUSION

A well-known fact is that plant growth, development and senescence are under the regulation of the system of natural growth regulators: natural inhibitors and phytohormones. These substances could be transported in the cell or even between the plant tissues and organs. These processes are part of the so-called “biological clock” and rhythms that take place in plants.

The fate of these regulating substances, their role in biosynthesis and other transformations in the plant, in the soil and in the allelopathic ecosystems is the main topic of this book.

The leaf as an organ of photosynthesis is a center of formation of the primary products, their metabolites, also it is involved in subsequent evacuation of hormones into reserve organs, and later aging. The role of photosynthesis in the growth as a determining factor of productivity is obvious.

In addition to this, the main primary mechanisms of ontogenesis have been examined which, in parallel with the growth, are involved into the yield of crop formation.

The primary mechanisms of ontogenesis include formation of morphological structures, functioning of meristems, formation of leaf primordial, and correlations in plant growth and development. All these processes are controlled by genome and by the system of phytohormones within the space and time.

Each organ, including the leaf, experiences together with a whole plant the main stages of development such as embryo development, youth, maturity and aging. The final stage, dying off, is followed by abscission of the plant organ.

Undoubtedly, the leaf activity depends on the well being of the whole plant including its age. This interconnection serves as a basis for development of the leaf itself that is formed in the system of an integral plant organism.

One of the most important functions of the leaf is the accumulation of photosynthetic assimilates and the ability to transport these assimilates into the vital organ of the plant. The leaf capability to evacuate metabolites sideways the growth point of stem or root is of no less importance. In this case a peculiar cascade of "transfer stations" is formed. These stations can be considered as a system of independent pairs, that is the donors and acceptors. The pair leaf parenchyma-phloem ends is localized at the beginning of this system. Then the pair leaf cutting-stem phloem is positioned, and so on. In each of the given pairs the hormone gradients play a great role in determining the factors channeling metabolites from the leaf into the vital organ.

The auxins and cytokinins are the most active hormonal metabolites participating in the formation of sink (acceptor) center. The role of abscisic acid (ABA) in the evacuation of metabolites is not yet definitely clear.

The process of leaf fall (abscission) is closely connected with crop and yield formation. In reality, the leaf is the source of assimilates, that are components of vital product exports the main metabolites to the final stage of ontogenesis.

The regulatory mechanism of any reaction or process implies operation not only of stimulating, but also have inhibiting (suppressing) factors. The higher the complexity of a phenomenon or a process, the greater the number of activators and repressors involved in this phenomenon. This principle is quite applicable to such a complex process as growth.

The use of additional doses of exogenous phytohormones in experiments with growth systems had made it possible to show that natural growth regulators (gibberellins and cytokinins) can affect the nucleus DNA functions and induces such processes as the "de novo" synthesis of a number of enzymes. Another phytohormone, auxin, which is also an activator of biosynthesis can, in contrast to the above mentioned two hormones, induce mainly synthetic processes at the cytoplasm level by controlling RNA functions.

Synthesis in cytoplasm can be regulated not only through promoting or restricting RNA functions, but also at the level of enzymes involved in producing constitution substances. Such feedback regulation results usually in the accumulation of physiologically inhibitory products (including phenols), which inhibit the functions of individual enzymes or enzyme systems involved in the synthesis of these products, the activity of other enzymes and even diminishes the activity of some cell organelles. Accumulated inhibiting products (natural inhibitors) can retard both normal growth processes and their induced forms. Specific pairs of endogenous regulators, of which one is an activator and the other is a growth inhibitor, control successive stages of the growth process.

The class of natural growth inhibitors is extremely large and includes not only terpenoid compounds (abscisic acid and its analogs), but also phenol derivatives (some phenolcarbonic acids, chalcones, flavanones), coumarins and furocoumarins.

The basic property of phenolic inhibitors is their ability to inhibit the growth of coleoptile segments, suppress seed germination and retard rhizogenesis processes. Some phenolic inhibitors are known to be active uncoupling agents of phosphorylation and respiration, photophosphorylation inhibiting agents and regulators of formation of nucleic acids.

It should be mentioned that the content of phenolic growth inhibitors depends very strongly on the external effects which may be schematically represented as

$$\textbf{\textit{external effects}} \xrightarrow[\textbf{\textit{phytohormones}}]{\textbf{\textit{natural inhibitors}}} \textbf{\textit{growth.}}$$

In a plant natural growth inhibitors interact closely with phytohormones. One of the first stages of this interaction is mutual influence at the level of formation of these substances. In our earlier work we showed that some phenolic growth inhibitors were able to retard the synthesis of indole auxins. Recently, first data have been obtained on the reverse effects, i.e. on the ability of phytohormones to regulate the production of phenolic inhibitors.

In the review Kefeli (1997) we considered the actual role of hormones and inhibitors in the plant ontogenesis. H. Fuku et al., (1996) described the role of genomes in the xylem differentiation and lignin synthesis.

Lignin is one of the most important components of secondary walls, and it had been more intensively studied that any other macromolecules which synthesis is closely associated with secondary wall formation. However, the synthesis of lignin is also induced by wounding or infection. Lignification that occurs in association with xylem differentiation and phenolic precursors can act as growth regulators during first stage of cell elongation and differentiation.

During lignin biodegradation (Lewis and Yamamoto, 1990) low molecular weight phenolics could play a role of inhibitors and could be inactivated by such enzymes as laccasses and peroxidases.

Thus phenolic inhibitors could act as anti-growth substances on the early stages of cell elongation and differentiation and during the lignin biodegradation.

Raskin (1992) emphasized that salicylic acid produced in the rhizosphere of some plants functions as an allelopathic chemical that inhibits growth of the surrounding vegetation.

So, natural inhibitors are involved in various growth regulation processes. The functions of some of these processes are well known to the researchers, others still remain vague. Our understanding of the properties and regulatory activity of natural growth inhibitors will obviously become more profound if we disclose the primary mechanism of phytohormone activity. The knowledge of the role of this important component in a system regulating growth of plants will help our better understanding of the peculiar features in the activity of natural inhibitors.

Of the great importance for interpreting the mechanism of action of natural inhibitors will be information on the penetration of these compounds through vacuole, chloroplasts and mitochondria membranes. This information will make it possible to find out when and how enzymes and natural inhibitors 'meet' in a cell and what is the mechanism of blocking of the main metabolic processes.

The aspect of genetic regulation of phytohormone and inhibitor synthesis is totally unclear so far. This problem is difficult to solve separately from the growth process itself. Promising trends in this area might be experiments on tall and dwarf plants whose genome differs in one gene (allele). Usually such genetic mutants produce other than normal amounts of phytohormones and natural inhibitors.

The development of plants with an intentionally altered hormonal and anti-hormonal apparatus will permit the establishment of a closer relationship between growth stimulators, inhibitors and plant growth, and will enable regulation of this relationship as desired by the scientists.

Finally, after interpreting the mechanism of action and the structure of the main growth inhibitors, the scientists could try to create artificial analogs of inhibitors, which would differ from the known synthetic toxic substances, and herbicides in a subtle effect. Their quick and specific incorporation into processes involved in plant growth and ability to be readily inactivated without serious toxic side effects also could be of future interest.

Like any new problem, the mechanism of action of natural growth inhibitors needs profound investigation bearing in mind that some problems can be solved at the level of a whole plant, while others require simpler and particular systems such as genome and DNA manipulation.

Within the period of its development the leaf was a consuming organ, an acceptor, at the beginning of ontogenesis, and then it was converted into donor, the metabolites from which were transported into the other parts of a plant. Subsequent to the export of metabolites the process of leaf aging occurs, the final stage of ontogenesis, followed by defoliation.

Meanwhile, the principle of "export integrity" of metabolites is essential. The export in some low-productive varieties can happen with a low rate in acceptor or cannot occur at all but can direct into the localized vegetative organs. All of the above shows the variety of specificity in function, which to a greater extend, complicates defoliation. That is why chemical regulators and, in particular, the defoliants "correct the mistakes of breeding" and facilitate the leaf abscission.

A special emphasis is placed in the book on phytohormones and inhibitors as integrators of functioning shoots and roots of the plant. While presenting the observations on physiological processes, an attempt has been made to correlate functions of plant and soil interactions on the level of agro-ecosystems, as well as to describe the role of these properties in plant ontogenesis.

The book describes approaches to the modeling of plants favorably developed under conditions of modern agroecosytem. The authors of the book pay special attention to the interrelations of such fundamental processes such as photosynthesis and plant growth.

The problem of photoregulation correlates with the ecological conditions of phytocenosis. The book also deals with the effects of red, blue and UV--light on such physiological processes as differentiation of plastids and biosynthesis of the secondary by-products and metabolites. The peculiarities of light effect of different quality on the level of phytohormones was as well described in the book. The approaches to creating the optimal plant habitats have been defined.

For the past few decades research was done on accumulation of some natural growth inhibitors and phytohormones by microorganisms. The role and function of these substances in the heterotrophic cells of bacteria and fungi are not yet well understood. But it is now obvious that they play an important role in plant growth and development. The biosynthetic processes in the soil are also based on the presence of plant organic matter, the transformation of which proceeds under the microbiological activity. Therefore the interaction of soil and air nutrition of plants can be viewed only by cooperative investigations of the group of specialists

Hence, normal regulation of plant growth may be regarded as a balance between phytohormones and their antagonists whose relative amounts in tissues are precisely coordinated. Hormones which regulate growth processes in a plant are supercellular regulation mechanisms. Formed in one site of a plant organism, they are transported inside the plant and function in some other part. The complex of endogenous inhibitors control the biosynthesis and level of endogenous phytohormones at each stage of their existence, starting from their synthesis and up to the point of their use.

Modern studies of the mechanism of action of natural growth inhibitors will be most effective if a comprehensive approach is used including handling of the main problems at the cell, tissue, organ levels and the level of a whole plant and their environment.

Therefore the investigation of natural growth regulators is one of the tools for the regulation of the biological processes in individual plants and in the ecosystems. The investigation of natural growth inhibitors and phytohormones is still in progress. The new research projects which we propose to students and scientists will help to approach the problem of integrative regulatory aspects of natural growth inhibitors and phytohormones in the native ecosystems and agrocenosis.

REFERENCES

Abdre Xin Xu., van Lammeren A.M., Vermeer E., and Vreugdenhil D. 1998. The Role of Gibberellin, Abscisic Acid, and Sucrose in the Regulation of Potato Tuber Formation in Vitro. Plant Physiol. 117:575-584

Abzalov M.F., Nadzhimov U.K. Masaev D.A., and Fahtullaeva G.N. 1985. Genetic Aspects of Growth Regulation. Genetic acpects of cotton growth. In Plant Growth and Its Regulation. (Genetic and Physilogical Aspects) (In Russian), Kishinew "Schiintza",pp:5-10.

Addicott F.T. 1965. Physiology of abscission. Handbuch der pfianzenphysiologie, Bd. XV/2. Springer Verlag, p. 1094

Addicott FT., Cams H.R., Lyon J.L., Smith O.E., and McMeans J.L. 1964. On the physiology of abscisins. Regulateurs naturels de la croissance vegetale. Colloq. Internat. CNRS, 123. Paris, p.687.

Albrecht von Arnim and Xing-Wang Deng. 1996. Lifgt Control of Seedling Deveopment. Ann. Rev. Plant Physiol. Mol. Biol. 47:215-243

Arapetyan E. 1985. Influence of growth regulators and desintegration on the anthocyan appearance in maize seedlings. In: Plant Growth and its Regulation. Kishinev, Stinitza, pp.88-91

Arshavski I.A., Kalevitch A.E., and Kefeli V.I. 1992. To the role of the cytosceletyon in processes of plant growth and development.(in Russian). Biophysics. 37(5):983-994

Averina N.G., Shegai I.D., Kefeli V.I., and Kof E.M. 1992. Chlorophyll Biolsynthesis in Cotyledon Leaves of a Chlorophyll_Deficient Cotton Mutant. Russian Plant Physiology. 39-1:57-61

Ballin G. 1967. Kritische Bemerkungen zur biologiscben Auxinbestimmung. Konference ve Starem Smokovci, Abstracts, Praha, 13.

Bardinskaya M.S. 1964. Rastitelnye kletochnye stenki i ikh obrazovaniye (Plant cell walls and their formation).Nauka, Moscow (in Russian).

Bartel B. 1997. Auxin Biosynthesis. Annu. Rev. Plant Physiol. Plant Biol. 48:51-66

Bassman J.H., Robberecht R., and Edwards G. 2001. Effects of Enhanced UB-B Radiation on Growth and Gas Exchyange in Populus deltoides Bartr. Ex Marsh. Int. J. Plant Sci. 162-1:103-110.

Bassman J.H., Robberecht r., and Erwards G.E. 2001. Effects of Enhqanced UV-B Radiation on Growth and Gas Exchange in Populus deltoides Bartr. Ex Marsh. Int. J. Plant Physiol. 162(1):103-110.

Battaglia P.R and Brennan T.M. 2000. Differential Effect of Short_term Exposure to Ultraviolet-B Radiation Uponn Photosynthesis in Cotyledon of a Resistant and a Susceptable Species. Int. J. Plant Sci. 161-5:771-778

Bilyeu K. D., Cole J. L., Laskey J. G., Riekhof W. R., Esparza T. J., Kramer M. D., and Morris R. O. 2001. Molecular and Biochemical Characterization of a Cytokinin Oxidase from Maize. Plant Physiolgy, 125:378-386

Bonotto S., Kefeli V.I., and Puiseux-Dao S. (eds.) 1978. Developmental biology of Acetabularia: proceedings of the Round Table on Acetabularia. K.A. Timiriazev Institute of Plant Physiolgy.Moscow.

Borisova T and Bonavanture N. 1992. Phytohormones and the Regulation of the Water Relations in Wheat Seedlings. Root Ecology and its Practical Application. In: Kutschera L., Hubl E., Lichtenegger E., Persson H., and Sobotik M. (eds.). 3 ISRR Symp. Wien, Univ. Bodenkultur. pp. 331-334

Briggs W.R. and Olney M.A. 2001. Photoreceptors in Plant Photomorphogenesis to Date. Five Phytochromes, Two Cryptochromes, One Phototropin, and One Superchrome. Plant Physiology, 125:87-91

Bronwyn B. J. and Pantoja O. 1996. Physiology of Ion Transport Across the Tonoplast of Higher Plants. Ann. Rev. Plant Physiol. Mol. Biol. 47:159-184

Burden R.S., Firr R.D., Hiron R.W., Taylor H., and Wright S.T. 1971. Induction of plant growth inhibitor xanthoxin. Nature, New biol. 234:95.

Butenko R.G., Kutachek M., Guskov A.K., Makarova R.V., Karanova S.L., and Kefeli V.I. 1979. Metabolism of indolic compounds in cell lines of Dioscorea. Sov. Plant Physiol. 26(2):318-322.

Chailakhyan M.K. 1957. Vliyaniye gibberellinov na rost I tsveteniye rasteniy (The effect of gibberellins on the growth and flowering of plants). Proc. Acad. Sd. USSR 117:1077 (in Russian).

Chailakhyan M.Kh. 1937. Gormonal'naya teoriya razvitiya restenii (Hormonal Theory of Plant Development). Akad. Nauk SSSR. Moscow. P.342

Chappel J. 1995. Biochemistry and Molecular Biology of the Isoprenoid Biosynthetic Pathway in Plants.Annu. Rev. Plant Physiol. Plant Mol. Biol. 46:521-547

Chilton M. 2001. Agrobacterium. A Memoir. Plant Physiology, 125:7-13

Chkhaidze N.M., Mikaberidze V.E., Prusakova L.D., and Kefeli V.I. 1993. The Effect of Kamposan on Ethylene Evolution and Abscisic Acid Content in Shoots of Citrus Plants during Autumn and Winter. Russian Journal of Plant Physiology, 40-6:897-900

Cholodny N.G. 1928. Novyie dannye k obosnovaniyu gormonalnoy teorii tropismov (New data for substantiation of hormonal tropism theory). Journal of Russian Botanical Society 13:191 (in Russian).

Coruzzi G., and Bush D.R. 2001. Nitrogen and Carbon Nutrient and Metaolite Signaling in Plant. Plant Physiology, 125:63-67

Cosgrove Daniel J. 1998. Cell Wall Loosening by Expansins. Plant Physiology, 118: 333-339

Dmitrieva G.A. and Kefeli V.I. 1991. Plant Physiology (Manual). Text-Books for Universities. Moscow. P.75

Doorenbos J. 1953. Review of the literature on dormancy in buds of woody plants. Med. Landbouwhogeschool, Wageningen 53:1

Dorftling K. 1963. Uber das Wuchsstoff-Hemmstoffsystem von Acer pseudop Itanus L. Planta 60:390.

Drakina T.I. and Kefeli, V.I 1967. K probleme immuniteta. Vzaimodeystviye Indoliluksusnoy kisloty I fenolnykh soedinenjy pri indutsirovanii patologicheskogo rosta Taphrina (On immunity problems. Interaction of indolylacetic and phenolic compounds in inducing pathological growth of Taphrina) Gen Biol. J.XXVIII:93 (in Russian)

Engelsma C. and Mayer C. 1965. The influence of light of different spectral regions on the synthesis of phenolic compounds, I. Biosynthesis of phenolic compounds. Acta hot. neerl. 14:54.

Ensikat H.J., Neinhuis C., and Barthlott W. 2000. Direct Access to Plant Epicuticular Wax Crystals by a New Mechanical Isolation Method. Int. J. Plant Sci. 161-1:143-148

Estelle M. 1998. Polar Auxin Transport: New Support for an Old Model. The Plant Cell. 10:1775-1778

Fellenberg G. 1967. Beemfiussung der Auxinwirkung dureb Glucose bei Wurzelbildung. Wiss Z. Univ. Rostock 16:535.

Filonova V. 1985. Influence of kinetin on phenol metabolism in cytokinin-sensitive models. In: Plant Growth and its Regulation. Kishinev, Stinitza, pp.87-88

Fry S.C. 1983. Feruloylated pectins from the primary cell wall: Their Structure and possible functions. Planta.157:111-123.

Fry S.C. 1986. Cross-linking of matrix polymers in the growing cell walls of angiosperms. Annu. Rev. Plant Physiol. 37:165-186.

Fukuda Hiroo 1996. Xylogenesis: Initiation, Progression, and Cell Death. Ann. Rev. Plant Physiol. Mol. Biol. 47:299-325

Furuya M. and Galston A. 1965. Flavonoid complexes in Pisum sativum L J. Phytochem. 4:285.

Galston A.W. 1961. The life of the green plant. Englewood Cliffs. Prentice-Hall, Inc.

Galston, A.W., Jackson L., Kaur-Sawhney R.N.P.. and Meudt W.J. 1964. Interactions of auxins with macromolecular constituents of pea seedlings. Regulateurs naturels de la croissance vegetale. Colloq. internat. CRNS, 123, p.251

Grotewold E., Chamberlin M., Snook M., Siame B., Butler L., Swenson J., Maddock S., Clair G., and Bowen B. 1998. Engineering Secondary Metabolism in Maize Cells by Ectopic Expression of Transcription Factors. The Plant Cell, 10:721-740

Guifoyle T., Hagen G., Ulmasov T., and Murfett J. 1998. How Does Auxin Turn On Genes? Plant Physiology, 118: 341-347

Hahlbrock K. and Scheel D. 1989. Physiology and Molecular Biology of Phenylpropanoid Metabolism. Annu. Rev. Plant Physiol. Plant Mol. Biol. 40:347-369

Harper D., Douglas A., and Smith H. 1970. The photocontrol of precursor incorporation into the Pisum sativum flavonoids. Phytochemistry 9:497.

Hemberg T. 1961. Biogenous inhibitors. Handbuch der Pflanzen physiologie, Bd. XlV, Berlin, Springer Verlag, p.1162.

Herrmann K. and Weaver L. M. 1999. The shikinate pathway. Annu. Rev. Plant Physiol. Plant Mol. Biol. 50:473-503

Hillman J.R. and Hocking T.J. 1974. Abscisic acid and regalation of dormancy. In: Plant growth substances. Hirokawa Publishing Co. Tokyo, p.882.

Horne J.E., Kalevitch A.E., and Filimonova M.V. 1994. Germination of Five Wheat Varieties Under Various Soil Conditions; Implicationsfor Sustainable Agriculture. KCSA. Oklahoma, USA. Reaserch and Education Papers.100:11

Horne J.E., Kalevitch A.E., and Filimonova M.V. 1995. Acidity Stress and Wheat Growth and Development. Proceedings. Annual Meeting of ASPP. Abstract 236. J. Plant Physiol. 108:56

Ikegawa T., Mayama S., Nakayashiki H., and Kato H. 1996. Accumulatio of diferulic acid during the hypersensitive response of oat leaves to Puccinia coronata f. sp. Avenae and its role in resistance of oat leaves to cell wall degrading enzymes. Physiol. Mol. Plant Pathol. 48:245-255

Ivanov V.B. 1966. O spetseficheskom i ne spetseficheskom deistvii khoramfenikola na rost kornya (On specific and non-spesific effect of chloramphenicol on root growth) Z. obshch. Biol. (J. General Biol.). 3:229 (in Russian).

Jackson MB. and Osborne D.J. 1967. Ethylene, the natural regulator of leaf abscission. Nature 225:1019.

Jacobsen J. and Varner J.E. 1967. Gibberellin-acid-induced synthesis of protease by isolated aleurone layers of barley. Plant Physiol. 42:1596.

Jalilova F.K., Rakitina T.Y., Vlasov P.V., and Kefeli V.I. 1993. Growth and Ethylene Evolution in Three Genetic Lines of Arabidopsis thaliana as Affected by Unltaviolet Radiation (UV-B). Russian Journal of Plant Physiology, 40-5:764-769

Jarret J.M. and Williams A. 1967. The flavonoid glucosides of Salix purpurea. Phytochemistry 6:1585.

Jensen P.J., Hangarter R.P., and Estelle M. 1998. Auxin Transport Is

Required for Hypocotyl Elongation in Light-Growth but Not Dark-Grown Arabidopsis. Plant Physiol. 116:455-462

Jung H.G., Buxton D.R., Hatfield R.D., and Ralph J. 1993. Forage Cell Wall Structure and Digestibility. (Madison, WI: Americac Society of Agronomy Inc.)

Kalevitch A.E. and Filimonova M.V. 1995. Monochromatic UV-Light and K^+ Transport. Proceedings. Annual Meeting of ASPP. Abstract 257. J. Plant Physiol.108:61

Kamachaarya N. and Kefeli V. 1998. Effect of Low Temperatures and UV-C-Light on Wheat Seedlings. Proceedings of Annual Meeting of CPUB. Abstract 9.

Karanova S.L. 1999. The principles of the obtaining of practicaly valuable strains of cultivated plant cells. Moscow, Institutre Of Plant Physiology. Ph.D. Thesis. pp. 1-42 (in Russian).

Karanova S.L., Makarova R.V., and Kefeli V.I. 1979. Physiological properties of auxin independent cell line of Dioscorea.

Kefeli V. and Lozhnikova V. 1995. Light as Factor of Stress: Plant Growth and Development. The 11th Congress of FESPP. Bulgarian Journal of Plant Physiology. Special Issue. Abstarct 337

Kefeli V.I., Dashek W.V. 1984. Non hormonal stimulators and inhibitors. Biol. Rew. Cambridge. 59:273-288.

Kefeli V., Borsari B., Steglich C., and Welton S. 201. Botanical herbicides and conception of selectivity. 65th Ann. Meeting of NE Section of ASPP. Abstract 44

Kefeli V., Liguori A., and Filimonova M. 1999. Growth Regulating Properties of Two Types of Leaves of Hackberry - Growing on the Tree and Abscised. Annual Meeting. Nothern Section of ASPP. Abstarct 69.

Kefeli V., Shotwell M., and Banko A. 2000. Effect of Ultraviolet (UV) Light on the Growth of Some Plants. Proceedings. 4th Annual Meeting, Nothern Section of ASPP. University of Connecticut. pp.37

Kefeli V., Shotwell M., Banko A., Vlasov P., and Rakitina T. 1998. UV-Light, Hormones and Plant Growth. The 11th Congress of FESPP. Bulgarian Journal of Plant Physiology. Special Issue. Abstarct 328.

Kefeli V.I. 1968. Native Wachstuminhibitoren, ihre physiologische Rolle und ihr Wirkungsmechanismus. Wiss. 7. Univ. Rostock 17:383.

Kefeli V.I. 1978. Natural Plant Growth Inhibitors and Phytohormones. Dr W. Junk b.v. Publishers. The Hague/Boston. P 277

Kefeli V.I. 1992. Phytohormones, Genome and Prpperties. Genetics and Breeding. 25:213-226

Kefeli V.I. 1997. Natural growth inhibitors. Russian Journal of Plant Physiology. Vol:44, No:3, pp. 471-480

Kefeli V.I. 2002. Fabricated Soils for Landscape Restoration: An Example for Scientific Contribution b a Public-Private Partnership Effert. SME Annual Merting. Phoenix, AZ pp1-3

Kefeli V.I. and Kudejarov V.N.(eds.) 1991. Experimental Ecology. Nauka, Moscow, p 248

Kefeli V.I. and Sidorenko O.D. 1991. Plant Physiology with Basics of Microbiology. Agropromisdat. Moscow. P.255 (in Russian)

Kefeli V.I. and Turetskaya R.K. 1965. Uchastiye fenolnykh soedineniy v ingibirovanli aktivnosti auksinov i v podavlenii rosta pobegov ivy (Participation of phenolic compounds in inhibiting activity of auxins and in depressing growth of wollov shoots.) Fiziol. Rast 12:638 (in Russian).

Kefeli V.I. and Turetskaya R.K. 1966. Lokalizatsiaya prirodnyich fenolnykh ingibitorov V kletkakh listyev ivy (Localization of natural phenolic inhibitors in cells of willow leaves). Doklady Acad Nauk. SSSR 170:472 (in Russian).

Kefeli V.I. and Turetskaya R.K. 1967. Sravnitelnoye deystviye prirodnykh ingibitorov rosta, narkotikov i antibiotikoy na rost rastenjy (Comparative effect of natural growth inhibitors, narcotics and antibiotics on plant growth.) Fiziol. Rast 14:796 (in Russian).

Kefeli V.I., Borsari B.,and Welton S. 2001. The isolation of inhibiting compounds from the leaves of the red maple (Acer rubrum L.) for the germination and growth of lettuce seeds (Lactuca sativa L.). Annual Meeting of ASPP.J. Plant Physiol. Abstarct 433.

Kefeli V.I., Kalevitch A.E.and Protasova H.H. 1992. Growth, Photosynthesis and Mineral Nutrients in . Phisio9logy abd Biochemistry of Cultivated Plants. 1: 64-82 (in Russian)

Kefeli V.I., Kof E.M., Vlasov P.V., and Kislin E.N. 1989. Prirodhyi inhibitor rosta - abstsizovaya kislota (Abscisic acid: A Natuiral Growth Inhibitor). Nauka, Moscow. P.233

Kefeli V.I., Komizerko E.I and Kutachek M. 1969. Obrazovaniye auksinov v kulture tkaney kapusty (Auxin formation in cabbage tissue culture). Proc. of Growth Regulation Conf., Vilnyits, Mintis (in Russian).

Kefeli V.I., Komizerko E.I., Turetskaya, R.K., Kof E.M. and Kutachek M. 1972. Evolutsionnyi aspekt formirovaniya sistemy gormonalnoy regulyatsii (Evolutionary aspect of formation of hormonal regulation system). In: Immunitet I pokoi rasteniy. Nauka, Moscow. P. 200 (in Russian)

Kefeli V.I., Morrow S.M., and Snow R. 2000. Plant Roots as Organs for Secretion of Ions and Allelopathic Compounds. Proceedings of Annual Meeting of CPUB.Clarion University. pp.27

Kefeli V.I., Rakitina T.Y., Vlasov P.V., Jalilova F.H., and Kalevitch A.E. 1993. Various Conditions of Illumination and Ethylene Evolution. In: Cellular and Molecular Aspects of the Plant Hormone Ethylene. Pech et al. (eds.). Kluwer Academic Publishers. pp. 347-352

Kefeli V.I.,Turetskaya R.K., Smirnov, A.M., Zakharova, A.A., Kof E.M. and Kuzovkina I.N. 1972. Chuvstvitelnost nekotorykh izolirovannykh organov k ingibitoram metabolizma (Response of some isolated organs to metabolism inhibitors.) USSR Acad. Sci. Bulletin, Biol. Ser. 182 (in Russian).

Keller C.P. and Van Volkenburgh E. 1998. Evidence That Auxin-Induced Growth of Tobacco Leaf Tissue Does Not Involve Cell Wall Acidification. Plant Physiol. 118:557-564

Kende H. 1964. Preservation of chlorophyll in leaf sections. Science 145:1066

Kende H. 2001. Hormone Responcse Mutants. A plethora of Surprises. Plant Physiology, 125:83-87.

Klee H. 1991. Molecular Genetic Approach to Plant Hormone Biology. Annu. Rev. Plant Physiol. Plant Mol. Biol. 42:529-551

Klee H. and Extelle M. 1991. Molecular Genetic Approach to Plant Hormone Biology.Annu. Rev. Plant Physiol. Plant Mol. Biol. 42: 529-551

Kof E.M. and Kefeli V.I. 1994. Growth and Morphogenesis of Acetabularia. Nauka, Moscow. P. 125

Kof E.M., Chuvasheva E.S., Kefeli V.I., and Zelenov A.N. 1993. The Influence of IncreasingLight Intensities on the Growth of Pea Plants with a Abnormal Leaves. Russian Joournal of Plant Physiology. 40-5:734-741

Kof E.M., Gostimski S.A., and Kefeli V.I. 1994. Phytohormones in a

chlorophyll-deficientpea mutant of the "xantha" - type. Pisum Genetics. 261-10
Kof E.M., Kefeli V.I., Kutachek M., Eder J., Chermak V., and Gostimskii S.A. 1994. Effect of Light and Chlorophyll-Deficient Mutation on Phytohormones in Pea Plants. Fiziol. Rast.(Russian J. Plant Physiol.). 41(5):675-681
Kof E.M., Opatrny Z., Vackova K., Keefli V.I., Kutachek M., and Vol;fova A. 1983. Effect of p-Coumaric Acid on the Growth of Tobacco Cell Suspension Culture. Dok. Acad. Nauk SSSR. 270(3): 764-768
Kutacek M. and Kefeli V.I. 1968. The present knowledge of indole compounds in plants of Brassicaceae family. In: Biochemistry and physiology of plant growth substances. Ottawa, Range-Press,. p.127.
Kutacek M. and Kefeli V.I. 1970. Metabolism of tryptophan in plants. I. Biogencsis of indole compounds from D- and L-tryptophan. Biol. Plantarum 12:145.
Kutachek M., Opatrny Z., Vackova K., Kefeli V.I., Butenko R.G., Karanova S.L., and Makarova R.V. 1981. Comparison of Anthranilate Synthase Activity and IAA Content in Normal and Auxin Habituated Dioscorea Tissue Cultures. Bioch. Physiol. Pflanzen. 176(3):244-250
Kutchan T.M. 2001. Ecological Arsenal and Develp[pmental Dispatcher. The Paradigm of Secondary Metabolism. Plant Physiology, 125:59-63.
Letham D. 1964. Isolation of kinin from plum fruits and other tissues. Regulateurs natureles de la croissance vegetale. Colloq. internat. CNRS, 123, p.109.
Leung J. and Giraudat J. 1998. Abscisic Acid Signal Transduction. Ann. Rev. Plant Physiol. Mol. Biol. 49:199-222
Lewis N.G. and Yamamoto E. 1990. Lignin: Occurrence, Biogenesis and Biodegradation. Annu. Rev. Plant Physiol. Plant Mol. Biol. 41:455-496
Libbert E. and Kunert R. 1966. Uber einen angeblichen histochesehen Nachwis des Auxins EIS. Flora 156:573.
Lichtenthaler H. K. 1998. The plants' 1-deoxy-D-xylulose-5-phosphate pathway for biosynthesis of isoprenoids. Fett/Lipid. 100(4-5_: 128-138
Long S.R. 2001. Genes and Signals in the Rhizobium - Legume Symbiosis. Plant Physiology, 125:71-75
Lozhnikova V.N., Komarova E.N., Duska N.D., and Vyskrebentseva E.I. 1995. Variation in the Activity of Cell Wall Lectines and Some Phytohormones in Tobacco Leaves During Photoperiodic Induction.Doklady Biochemistry. 343(1-6): 108-111. Translated from Doklady Akademii Nauk. 343(4):547-550
Macoskey V.M., Morrow S.M., Kefeli V.I., and Steglich C.S. 2000. Phenols in Water Extracts of Italian Parsley. Proceedings of Annual Meeting of CPUB.Clarion University. pp.23
Makarova R, Borisova T. Vlasov P., Machackova I., Ederr J., Cermak V., and Kefeli V. 1990. Physiology and Biochemistry of Cytokinins in Plants. Proceedings of the International Symposium on Physiology and Biochemistry of Cytokinins in Plants, Liblice, Czechoslovakia. Pp.115-117
Makarova R.V., Andrianov V.M., Borisova T.A., Piruzyan E.S., and Kefeli V.I. 1997. orphogenetic Manifestation of the Expression of the Bacterial ipt Gene in Regenerated Tobacco Plants in vitro. Russian Journal of PlantPhysiology. 44-1:6-13.
Makarova R.V., Borisova T.A., Vlasov P.V., Machackova I., Andrianov V.M., Piruzyan E.S., and Kefeli V.I. 1997. Phytohormone Production by Tobacco ipt-Regenerants in vitro. Russian Journal of PlantPhysiology. 44-5:662-667
Mayer A.M and Poljakoff-Mayber A. 1963. The germination of seeds. Oxford Pergamon Press.
McCourt P. 1999. Genetic analysis of hormone signaling. Annu. Rev. Plant Physiol. Plant Mol. Biol. 50: 219-243

McFadden G.I. 2001. Chloroplast Origin and Intergration. Plant Physiology, 125:51-55.
Meyerowitz E.M. 2001. Prehistory and History of Arabidopsis Reaserch. Plant Physiology, 124:13-19
Miller C.O. 1961. Kinetin and related compounds in plant growth. Annual Rev. Plant Physiol. 12: 395.
Mohr H. 1972. Lectures on photomorphogenesis; Berlin-Heidelberg, N-Y. Springer Verlag.
Mok D. and Mok M. 2001. Cytokinin metabolism in higher palnts. Annu. Rev. Plant Physiol. Plant Mol. Biol. Vol. 52,p.89-118
Monties B. 1969. Presence de composes polyphenoliques dans les chioroplastes d'angiospermes. Bull. Soc. Franc. physiol. veget. 15: 29.
Mothes K. 1964. The role of kinetin in plant regulation. Regulateurs naturels de la croissance vegetale. Colloq. internat. CNRS, 123, p.131
Muzafarov E.N. ad Zolatareva E.K. 1989. Uncoupling effect of hydrocinnamic acid derivatives in pea chloroplasts. Biochem Phisiol. Pflanzen, 184:363-369
Nadzhimov U.K. and Abzalov M.F. 1988. Cotton dwarfs and physiologically active substances. In: Growth Gegulators and Plant Development. Moscow, Nauka p.124
Nadzhimov U.K., Kefeli V.I., Polyakov A.S., and Chetverikov A.G. 1985. Possibility of Blocking Photosynthesis at the Early Stages of Growth and Morphogenesis in Cotton Seedlings. Izv. Acad. Nauk SSSR. Sr. Biol. 1:90-95
Naqvi S.M. and Engvild K.C 1974. Action of abscisic acid on auxin transport and its relation to phototropism. Physiol. plantarum 30: 283-287.
Neumann J. and Avron M. 1967. Oxidation of phloridzin by isolated chioroplasts. Plant and Cell Physiol. 8:241.
Osborne D. 1968. Defoliation and defoliants. Nature 5154:564.
Osborne D. and McCalla D. 1961. Rapid bioassay for kinetin and kinin using senescening leaf tissue. Plant Physiol. 36:219.
Ouellet F., Carpentier E., Cope M.J.T.V., Monroy A.F., and Sarhan F. 2001. Regulation of a Wheat Actin-Depolymerizing Factor during Cold Acclimation. Plant Physiol. 125:360-368
Paleg L. 1961. Physiological effects of gibberellic acid. III. Observation on its mode of action on barley endosperm. Plant Physiol. 36: 829.
Parr A.J., Waldron K.W., and Parker M.L. 1996. The wall-bound phenolics of chinese waterchestnut (Eleochans dulais). J.Sci. Food Agric. 71:501-507
Pieniazek J., Saniewski M.,and Jankiewiez L. 1970. The effect of growth regulators on cambial activity. Tagungsber Dtsch. Akad. Landwiss. Berlin 99: 61.
Polyakov A.S., Kof E., Gostimskii S.A.,Kvarzhava aand Kefeli V.I. 1985. Phytohormones and Pigments in Albino Mutants of Cotton and Pea. Biologia Plantarum 27(2-3): 139-144
Post-Beittenmiller D. 1996. biochemistry and Molecular Biology of Wax Productionin Plants.Ann. Rev. Plant Physiol. Mol. Biol. 47:405-430
Post-Beittenmiller D. 1996. Biochemistry and Molecular Biology of Wax Production in Plants. Annu. Rev. Plant Physiol. Plant Mol. Biol. 47:405-430
Protasova N.N. and Kefeli V.I. 1982. Photosynthesis and Growth of Higher Plants, Their Interrelation and Correlations. Fiziologiya Fotosinteza. Nauka, Moscow. P. 251
Protasova N.N., Lozhnikova V.N., Nichiporovich A.A., Kefeli V.I., and

Chailakhyan M.K. 1980. Growth, Activity of Phytohormones and Inhibitors, and Photosynthesis in Dwarf Mutants of Pea under Different Conditions of Illumonation. Izv. Akad. Nauk SSSR, Sr. Biol.1:94

Pryce R.J. 1971. Lunularic acid, a common endogenous growth inhibitor of liverworts. Planta 97: 354.

Putnam A.R. and Chung-Shih Tang. 1986. The Science of Allelopathy. A Wiley-Interscience Publication. John Wiley & Sons.

Qin X. and Zeevaart J.A.D. 1999. Inaugural Article: The 9-cis-epoxycarotenoid cleavage reaction is the key regulatory step of abscisic acid biosynthesis in water-stressed bean. Proc. Natl. Acad. Sci. U.S.A. 96:15354-15361

Rakitina T.Y., Vlasov P.V., Jalilova F.K., and Kefeli V.I. 1994. Abscisic Acid and Ethylene in Mutants of Arabidopsis thaliana Differing in Their Resistance to Ultraviolet (UV-B) Radiation Stress. Russian Journal of Plant Physiology. 41-5:599-603

Raschke K. 1974. Abscisic acid sensitizes stomata to C02 in leaves of xanthium strumarium L Plant growth subst. 8th Intern. Conf., Tokyo, Hirokawa Publishers.

Raskin I. 1992. Role of salicylic acid in plants. Annu. Rev. Plant Physiol. Plant Mol. Biol. Vol. 43,p.439-463

Rezonoca E., Fklury N., Meins Ir. F., and Beffa R. 1998. Transcriptional Down-Regulation bu Abscisic Acid of Pathogenesis-Related bete-1,3-Glucanase Genes in Tobacco Cell Cultures

Ribnicky D.M., Shulaev V., and Raskin I. 1998. Intermediates of Salicylic Acid Biosynthesis in Tobacco. Plant Physiol. 118:565-572

Richard D.E., King K.E., Ait-ali Tahar, and Harberrd N. P. 2001. How gibberellin regulates plant growth and development: A moleculal Genetic Analysis of Gibberellin Signaling. Annu. Rev. Plant Physiol. Plant Mol. Biol. Vol. 52,p.67-88

Rubin L.B. 1975. Nekotorye voprosy evolyutsli sistem fotoregulyatsii (Certain aspects of evolution of photoregulation systems). In: Fotoregulyatsiya metabolizma i morfogeneza rasteniy.Nauka, Moscow. P.82 (in Russin)

Russel D.W. and Gaiston A.W. 1969. Blockage by GA of phytochrome effects on growth. Plant Physiol. 44:1211.

Sabinin D.K. 1963. Fiziologiya razvitiya rastenly (Physiology of plant development). M., USSR Acad. Sci. Publishers (in Russian).

Sakamoto T., Kobayashi M., Itoh H.i, Agiri A., Kayano T., Tanaka H., Iwahori S., and Matsuoka M. 2001. Expession of a Gibberellin 2-Oxidase Gebe around the Shoot Apex Is Related to Phase Transition in Rice. Plany Physiology, 125: 1508-1516

Sarapuu L.P. 1970. Fizlologiya i biokhirniya floridzina (Physiology 'and biochemistry of phloridzin). Doct. thesis, Tartu (in Russian).

Sarapuu L.P. 1964. Floridrin v kachestve beta-ringibitora i sezonnaya dinamika produktov ego metabolizma v pobegah yabloni (Phloridzin as a beta-inhibitor and seasonal variations in its metabolic products in apple shoots). Fiziol. Rast. 2:607 (in Russian).

Sarapuu L.P. and Kefeli V.I. 1968. Fenotnye soedineniya i rost rasteniy (Pliendlic compounds and plant growth). In: Fenolnye soedineniya' i ikh biologicheskiye funktsii. Nauka, Moscow. P.129 (in Russian).

Shen-Miller L., Cooper P., and Gordon S.A. 1969. Phototropism and photoinhibition of basipolar of auxin in oat coleoptiles. Plant physiol. 44:491-496

Sinnot E.M. 1963. Morfogenez (Morphogenesis). Mir Publishers (in Russian).

Skoog F., Hamz, Q., Szweykowska L., Leonard N., Carraway K., Fuji T., Helgeson J. and Loeppky R. 1967. Cytokinins: Structure activity. Phytochemistry 6:1169.

Smith H. 1972. The photocontrol of flavonoids biosynthesis. In:

Phytochrome. K. Mitrakos and W. Shropshire (eds.) Acad. Press, New York. P.433
Somerville C.R. 2001. An Early Arabidopsis Demonstration. Resolving a Few Issues Cocerning Photorespiration. Plant Physiology, 125:19-25.
Steward F.C. and Caplin S.M. 1952. Investigation of the growth and metabolism of plant cells III. Ann. Bot. 16:477
Steward F.H. and Caplin S.M. 1952. Investigation on the growth and metabolism of plant sells. III. Ann. Bot. 16:477
Suge H. and Murakami Y. 1968. Occurrence of a rice mutant deficient in gibberellin-like substances. Plant and Cell Physiol. 9:11
Sze-Chung Clive Lo and Nicholson Ralph L. 1998. Reduction of Light-Induced Anthocyanin Accumulation in Inoculated Sorghum Mesocotyls. Implication for a Compensatory Role in the Defense Response. Plant Physiol. 116:979-989
Tamagnone L., Merida A., Stacey N., Plaskitt K., Parr A., Chang Chi-Feng., Lynn D., Dow J.M., Roberts K., and Martin C. 1998. Inhibition of Phenolic Acid Metabolism Results in Precocious Cell Death and Altered Cell Morphology in Leaves of Transgenic Tobacco Plants. The Plant Cell. 10:1801-1816.
Thieme H. 1964. Die Phenoiglucoside det Salicaceaen. 2. Miss. Pharmazie 19: 471.
Thimann K.V. 1960. Plant growth. Fundamental aspects of normal and malignant growth. Amsterdam. Elsevier Publ., p.748.
Thimann K.V. 1965. Toward an endocrinology of higher plants. In: Recent progress in hormone research. Acad. Press. 21:579
Thomashow M.F. 2001. So What's New in the Field of Plant Cold Acclimation? Lots! Plant Physiology, 125:91-97.
Titova O.V. and Kefeli V.I (eds.) 1991. Ecological Aspects of Regulation of Plant Growth and Plant Productivity.(in Russian). Iaroslavl State University. Iaroslavl. P. 111
Tomaszewski M. 1964. The mechanism of synergistic effects between auxin and some plants. Bull. Acad. polon. sci., cl. 8: 65.
Touchnine V., and Kefeli V. 1995. Corn root reaction to light. Proceedings of CPUB Annual Meeting. Slippery Rock University. pp.34
Tumanov I.I., Kuzina G.V. and Karnikova LD. 1970. Vliyaniye gibberellinov na period pokoya i morozostoikost rasteniy (Effect of gibberellins on dormancy and frost-resistance of plants). Fiziol. rast. 17: 885 (in Russian).
Turetskaya R., Kefeli V., Kutachek M., Vackova K., Tschumakovski N., and Krupnikova T. 1968. Isolation and Some Physiological Properties of Natural Plant Growth Inhibitors. Biol. Plant. 10(3):205-221
Turetskaya R.K. 1961. Fiziologiya korneobrazovaniya u cherenkov i stimulyatory rosta (Physiology of rooting in cuttings and growth stimulators). M., USSR Acad. Sci. Publishers (in Russian).
Turetskaya R.K., Kefeli V.I., and Kof .M. 1966. ole of Natural Growth Regulators in Onthogenesis in Cherry and Grape Cuttings. Sv. Plant Physiol. Eng. Translation. 13(1): 29-37
von Arnim Albrecht and Xing-Wang Deng 1996. Light Control of Seedling Development. Annu. Rev. Plant Physiol. Plant Mol. Biol. 47:215-243
Wain R.L. 1977. Root growth inhibitors. In: Plant growth regulation (P.E. Pilet, ed.). Sprin-ger-Verlag, Berlin/Heidelberg/New York, p.109.
Waldron K.W., Ng A., and Parker M.L. 1997. Ferulic acid dehydrodimers in the cell walls of Beta vulgaris and their posible role in texture. J.Sci.Foo Agric. 74:221-228.
Wallace G. and Fry S.C. 1994. Phenolic compounds of the plant cell wall. Int.

Rev. Cytolo. 151:229-267.
Welton S., Borsari B., Kefeli V., and Steglich C. 2001. Transformation of Phenolic Compounds in Red Maple. Proceedings of Annual Meeting of CPUB. Bloomsburg University of Pennsylvalia. Abstarct B-14
Went F.W. 1926. On growth-accelerating substances in the coleoptile of Avena. Proc. Koninki. Akad. wet. Amsterdam 30:10.
Went F.W. 1963. The plants. N, Y., Time Inc.
Went F.W. and Thimann K.V. 1937. Phytohormones. N.Y., MacMillan.
Wheeler A.W and King H.G. 1968. Conversion oftryptophan to auxin by phenolic estes in leaves of dwarf French bean. Phytophemistry. 7:1057
Williams P.M., Ross S.D. and Bradbeer J.W. 1973. Studies in seed dormancy. VII. The abscisic acid content of the seeds and fruits of Corylus avellana L. Planta 110: 303-310.
Xin Xu, van Lammeren A.A.M., Vermeer E. and Vreugdenhil D. 1998. The role of Gibberellin, Abscisic Acid, and Sucrose in the Regulation of Potato Tuber Formation in Vitro. Plant Physiol. 117:575-584
Yamamoto T., Yokotani-Tomita K., Kosemura S., Yamada K., and Hasegawa K. 1999. Allelopathic Substances Exuded from a Serious Weed, Germinating Barnyard Grass (Echinochloa crus-galli L.), Roots. JU. Plant Growth Regul. 18:65-67
Yxley J.R., Ross J.J., Sherriff L.J., and Reid J.B 2001. Gibberellin Biolsynthesis Mutation and Root Development in Pear. Plant Physiology, 125: 627-633
Zaprometov M.N. 1964. Biokhimiya katekhinov (Biochemistry of catechins). Nauka, Moscow. (in Russian).
Zaprometov M.N. 1968. Dostizheniya i perspektivy biokhimii fenolnykli soedinenly (Progress and penpectives in the biochemistry of phenols). In: Fenolnye soedineniya i ikh biologieheskiye funktsii. Nauka, Moscow. (in Russian).
Zaprometov M.N. and Kolonkova S.V. 1968. Khloroplasty kak mesto synteza vodorastvorimykh fenolnykh soedinenly v rastitelnoy kletke (Chloroplasts as a site of synthesis of water-soluble phenol compounds in a plant cell). In: Fenolnye soedineniya I ikh biologieheskiye funktsii. Nauka, Moscow. (in Russian).
Zaprpmetov M.N. and Kolonkova S.N. 1967. On the Biosynthesis of Phenolic Compounds in Chloroplasts. Dokl. Akad. Nauk SSSR. 176(2): 470-473
Zeevaart J.A.D and Creelman R.A. 1988. Metabolism and Physiology of Abscisic Acid. Ann. Rev. Plant Physiol. 39:439-473
Zhohg R., Morrison III H.W., Negrel J., and Zheng-Hua Ye. 1998. Dual Methylation Pathways in Lignin Biosynthesis. The Plant Cell. 10:2033-2045
Ziemienowicz A., Merkle A., Schoumacher T., Hohn F., and Rossi B. 2001. Import of Agrobacterium T-DNA into Plant Nuclei: Two Distinct Functions of VirD2 and VirE2 Protein. Plant Cell 13:369-384
Zucconi F. 1996. Decline del suolo e stanchezza del terreno. Edino de: SPAZIO VERDE. Riviera dei Ponti Romani 22, Prodova. P.323

GLOSSARY

Abscisic acid.
A plant hormone having growth-inhibitory action. It promotes abscission and associated with the onset and maintenance of dormancy.

Abscission zone.
Zone at base of leaf flower, fruit, or other plant part that contains an abscission (or separation) layer and a protective layer, both involved in tile abscission of the plant part.

Abscission.
The shedding of leaves, flowers, fruits, or other plant parts, usually after formation an abscission zone.

Absorption.
Sucking up or imbibing, as a sponge absorbs water. The absorption of water by roots, brought about by osmotic and other sources developed in the roots.

Adaptation.
An ecological or evolutionary change of structure or function that produces better adjustment of an organism to its environment and hence enhances its ability to survive and reproduce.

Adaxial meristem.
Meristematic tissue on the adaxial side of a young leaf that continues to increase in thickness of the petiole and midrib.

Adenosine triphosphate, called ATP.
The major source of usable chemical energy in metabolism On hydrolysis ATP loses one phosphate and becomes adenosine diphosphate (ADP) pH usable energy, and on further hydrolysis it tis changed to adenosine monophosphate (AMP) with further release of energy.

Adventitious.
Refers to structures arising not at their usual sites, as roots originating On stems 0 leaves instead of on other roots, buds developing on leaves or roots instead of in leaf axils on shoots.

Aerobic.
Growing or proceeding only in the presence of oxygen.

After-ripening.
Term applied to the metabolic changes that must occur in some dormant seeds before germination can occur.

Agar.
A gelatinous substance obtained from red algae. Used as a solidifying agent in preparation of nutrient media for growing microorganisms and for other purposes.

Aggregate fruit.
Fruit developed from a flower having a number of pistils, all of which ripen together and are more or less coherent at maturity.

Agrobacterium tumefaciens.
This organism causes tumors, called crown galls, on susceptible plants. Tumor induction results from the transfer of a small piece of DNA, called T-DNA, from the bacterium to the plant cell during infection.

Alkaloids.
Organic compounds with alkaline properties, produced by plants. These substances which constitute the active principles of many drugs and poisons of plant origin, have a bitter taste and sometimes are poisonous. An alkaloids contain carbon, hydrogen, and nitrogen; most contain oxygen also. Examples: nicotine, morphine, quinine, and caffeine.

Aleurone layer.
Outermost layer of endosperm in cereals and many other taxa that contains protein bodies and enzymes concerned with endosperm digestion.

Aleurone.
Granules of protein (aleurone grains) present in seeds, usually outermost layer, the aleurone layer of the endosperm (protein bodies is the perfect term for aleurone grains)

Alternation of generations.
The alternation of gametophytic and sporophytic generations in the life cycle. The sporophyte develops from the zygote and produces spores. The gametophyte develops from the spore and produces gametes. The cells of the sporophyte generation contain twice as many chromosomes as those of the gametophyte.

Amino acids.
Nitrogen-containing organic acids, the "building stones or units from which protein molecules are built.

Ammonification.
The formation of ammonia by decay organisms of the soil acting upon proteinaceous compounds.

Amylase.
An enzyme that hydrolyzes starch.

Amyloplast.
A colorless plastid containing one or more starch grains.

Anabolism.
Constructive, or synthetic metabolism, such as photosynthesis, assimilation, and the synthesis of proteins.

Anaerobic respiration.
A type of respiration, found only in bacteria, in which the hydrogen released in glycolysis is combined with the bound oxygen of inorganic compounds, such as carbon dioxide.

Anaerobic.
Proceeding in the absence of or not requiring oxygen.

Anaphase.
A stage in mitosis in which the chromatids of each chromosome separate and move to opposite poles.

Annual .
(L. annuus, lasting a year) living for a single growing season.

Annual ring.
In secondary xylem, growth ring formed during one season. The term is deprecated because more than one growth ring may be formed during a single year.

Annular cell wall thickening.
In tracheary elements of the xylem; secondary wall deposited in the form of rings.

Anthesis.
The time of full expansion of flower; from development of receptive stigma to fertilization.

Anthocyanin.
A water-soluble blue, purple, or red flavonoid pigment occurring in the cell sap.

Antibiotics.
Substances of biological origin that interfere with metabolism of microorganisms.

Apex .
(L., apex, summit)-the upper or distal end of an organ; plural, apices.

Apical dominance.
Influence exerted by a terminal bud in suppressing the growth of lateral buds.

Apical meristem.
A group of meristematic cells at the apex of root or shoot that by cell division produce the precursors of the primary tissues of root or shoot; may be vegetative, initiating vegetative tissues and organs, or reproductive, initiating reproductive tissues and organs.

Asexual reproduction.
(1) any reproductive process that does not involve the union of gametes;(2) in lower form, reproduction by spores not associated with the sexual process.

Autotroph.

Any organism that can synthesize all its organic 'substances from inorganic nutrients, using light or certain inorganic chemicals as a source of energy. Green plants are the principal autotrophs.

Auxin.

Plant hormone that regulates various functions, including cell elongation, root formation, and bud growing.

Axillary meristem.

Meristem located in the axil of a leaf and giving rise to an axillary bud.

Axillary.

Term applied to buds or branches occurring in the axil of the leaf.

Axis.

(L.,axis, an axle) the central stem of an inflorescence.

Bacillus.

A rod-shaped bacterium

Back-crossing.

The crossing of a hybrid with one of its parents or with a genetically equivalent organism

Bacteriophage.

A virus that attacks bacteria.

Bark.

An inclusive term for all issues outside the cambium. A nontechnical term applied to all tissues outside the vascular cambium or the xylem; in older trees may be divided into dead outer bark and living inner bark, which consists of secondary phloem.

Bifurcation.

Refers to branching of the metabolic pathway from common metabolic precursor.

Biosphere.

The overall ecosystem. It is the sum total of all the biomes and smaller ecosystems, which ultimately are all Interconnected and Interdependent through global processes such as water and atmospheric cycle.

Biosynthesis.

The formation of the organic substances in the living organism

Biotest.

Intact plant or plant parts, which are sensitive to natural growth substances.

Bisexual .

(L.,bi+sexualis, twice+pertaining to sex)- a flower possessing both an androecium and gynoecium

Blade .
(ME blad, a leaf) The flattened expanded portion of a leaf or petal,

Branch Root.
A root arising from another, older root; also called secondary root if the older root is the primary root, or taproot.

Brassinolide.
Steroid growth substances isolated primary from Brassica pollen grains. Natural Brassinolide and its synthetic analogues imitate some effects of auxins and cytokinins

Broad-spectrum pesticides.
Chemical pesticides that kill a wide range of pests. They also jail a wide range of non-pest and beneficial species; therefore, they may lead to environmental upsets and resurgences. The opposite of narrow spectrum pesticides and biorational pesticides.

Bud.
(1) an embryonic shoot; (2) vegetative outgrowth from a yeast cell.

Budding.
(1) the method of vegetative reproduction in yeast; (2) a form of grafting.

Bulb.
(L., bulbus, bulb)- a subterranean plant structure consisting of a series of overlapping leaf bases insened on a much-reduced stem axis, as in the onion.

Bundle cap.
Sclerenchyma or coflenchymatous parenchyma appearing like a cap on the xylem and/or phloem side of a vascular bundle.

Bundle sheath.
Layer or layers of cells enclosing a vascular bundle in a leaf may consist of parenchyma or sclerenchyma.

Button.
An immature mushroom before the expansion of the cap.

Calcicole.
A plant that thrives on calcareous soil.

Calcifuge.
A plant that thrives on an acid soil.

Callus.
A tissue composed of large thin-walled cells developing as a result of injury, as in wound healing or grafting, and in tissue culture.

Calyx.
The first (beginning from below) of the series of floral pans, composed of sepals. The calyx is usually green and somewhat leaf like, but may be colored like the petals.

Cambial initials.

Cells so localized in the vascular cambium or phellogen that their periclinal divisions can contribute cells either to the outside or to the inside of the axis; in vascular cambium, classified into fisiform initials (source of axial cells of xylem and phloem) and ray initials (source of the ray cells).

Cambial zone.

A region of thin-walled cells between the xylem and phloem. Composed of cambium and its recent derivatives that have not yet differentiated into mature cells.

Cambium.

A meristem with products of periclidal divisions commonly contributed in two directions and arranged in radial files. Term preferably applied only to the two lateral meristems. The vascular cambium and the cork cambium, or phellogen.

Common wall.

Constituent in the sieve areas of sieve elements; also develops rapidly in reaction to injury in sieve elements and parenchyma cells.

Carbohydrates.

A large group of organic compounds composed of carbon, hydrogen, and oxygen, the ratio of hydrogen to oxygen being 2:1, as in water.

Carotene.

A reddish pigment found in chloroplasts and certain kind of chromoplasts; unsaturated hydrocarbons, such as B-carotene ($C_{4}0H5_{6}$).

Carotenoids.

Orange or yellow (occasionally red) pigments found in the plastids of plants.

Cell interactions and differentiation.

The behavior of the cell. in multi cellular organisms acquire different identities in an ordered special arrangement. It is supposed that nutrients and phytohormones are the factors of cell orientation, as well as shifting patterns of gene expression could accomplish cell differentiation and position dependent interactions.

Cell plate.

A partition appearing at telophase between the two nuclei formed during mitosis (and some meioses) and indicating the early stage of the division of a cell (cytokinesis) by means of a new cell wall; is formed in the phragmoplast.

Cell.

Structural and physiological unit of a living organisms The plant cell consists of protoplast and cell wall; in nonliving state, of cell wall only, or cell wall and some nonliving inclusions. The basic unit of life; the smallest unit that still maintains all the attributes of life, Many microscopic organisms consist of a single cell. Large organisms consist of trillions of specialized cells functioning together.

Cellulase.

An enzyme that hydrolyses cellulose.

Cellulose.

The organic macromolecule that is the prime constituent of plant cell walls and hence the major molecule in wood, wood products, and cotton. It is composed of glucose molecules, but because it cannot be digested by humans its dietary value is only as fiber. bulk, or roughage. Cellulose is a polysaccharide, B-l,4 glucan- the main component of cell walls in most plants; consists of long chainlike molecules the basic units of which are anhydrous glucose residues of the formula $C_6H_{10}O_5$

Central cylinder.

A term of convenience applied to the vascular tissues and associated ground tissue N in stem and root. Refers to the same part of stem and root that is designated stele.

Central mother cells.

Rather large vacuolated cells in subsurface position in apical meristem of shoot in gymnosperms.

Chlorophyll.

The green pigment in plants responsible for absorbing the light energy required for photosynthesis. The green pigment responsible for photosynthesis.

Chloroplast.

A chlorophyll-containing plastid with thylakoids organized into grana and frets, or strome thylakoids, and embedded in a stroma.

Chromoplast.

A coloured plastid, or solid cell inclusion.

Chromosome.

Structural units in the nucleus that preserve their individuality from one cell generation to another; the site of the genes. One of the small bodies in the nucleus of the cell on which the genes that determine the hereditary characteristics of the organism are carried.

Clone.

A group of genetically identical individual derived from the asexual propagation a single individual. A group of plants, often many thousand in number, that have had a common origin and that have been produced only by vegetative means such as grafting, cutting, or division rather than from seed. The members of a clone may be regarded as the extension of a single plant.

Coleoptile.
The sheath enclosing the epicotyl in the embryo of Poaceae; sometimes interpreted as the first leaf of the epicotyl.

Companion cell.
A parenchyma cell in the phloem of an angiosperm that has a common origin with a sieve tube.

Complete flower.
A flower having all types of floral parts: sepals, petals, stamens, and carpels, or tepals, stamens, and carpels.

Compound leaf.
A leaf divided into two or more parts, or leaflets.

Controlling gene.
A gene that controls the expression of a structural gene. Controlling genes are of two types; operator genes, which directly control structural genes, and regulator genes, which control operator genes.

Cork cell.
A phellem cell derived from the phellogen, nonliving at maturity, and having suberized walls; protective in function because the walls are highly impervious to water.

Cortex.
The primary ground tissue region between the vascular system and the epidermis in stem and root. Term also used with reference to peripheral region of a cell protoplast.

Cotyledonary trace.
The leaf trace of cotyledon located within the hypocotyl and connecting the vascular system of the root with that of the cotyledon.

Cotyledons.
The leaves (seed leaves) of the embryo, one or more in number.

Critical light period.
The dividing line between day length favorable to vegetative growth and that which induces flowering.

Cytokinesis.
The process of division of a cell as distinguished from the division of the nucleus, or karyokinesis.

Cytokinins.
Hormones associated primarily with cell division.

Cytology.
The science dealing with the cell.

Cytoplasm.
In a strict definition, the visibly least differentiated part of the protoplasm of a cell that constitutes the ground mass enclosing all other components of the protoplast. Also called hyaloplasm.

Day-neutral plant.
plant who's flowering is insensitive to length of day.

Deciduous.
(1) falling of parts at the end of the growing period, such as leaves in autumn, or fruits, or flower parts at maturity; (2) broad-leaved trees or shrubs that drop their leaves at the end of each growing season, as contrasted with plants that retain their leaves for more than one year.

Decomposers.
Organisms whose feeding action results in decay or rotting of organic materials. The primary decomposers are fungi and bacteria.

Dedifferentiation.
A reversal in differentiation of a cell or tissue which is assumed to occur when a more or less completely differentiated cell resumes meristematic activity..

Denitrification.
The process by which nitrogen is released from the soil by the action of denitrifying bacteria.

Deoxyribonucleic acid (DNA).
One of the two main kinds of nucleic acids; the genetic material of organisms and some viruses.

Derivative.
A cell produced by division of a meristematic cell in such a way that it enters the path of differentiation into a body cell; its sister cell may remain in the meristem.

Desmogen.
Merirtematic strand destined to differentiate into a vascular bundle. May be primary, that is, derived from a cambium in plants in which the secondary body consists of vascular bundles embedded in secondary parenchyma tissue.

Desmotubule.
Tubule (often appearing as a solid rod) connecting the two-endoplasmic reticulum cisternae located at the two opposite ends of a plasmodesma.

Determinate growth.
Formation of a restricted number of lateral organs by an apical meristem; characteristic of a floral meristem.

Determination.
The process in which a group of meristematic cells (such as an organ primordium) is fixed in a particular developmental pathway.

Development.
The change in form and complexity of an organism or part of an organism from its beginning to maturity; combined with growth.

Dicot.

A semi technical group name for those flowering plants which typically have two cotyledons (hence the name), net venation, and flower parts in 4's, 5's or multiples there of.

Dicotyledonous.

(Gk., two + a vessel)-having two cotyledons.

Differentiation.

A physiological and morphological change occurring in cell, a tissue, an organ, or a plant during development from a meristematic, or juvenile, stage to mature adult stage. Usually associated with an increase in specialization.

Diffusion pressure.

The activity of a specific kind of a molecule (particle) as a result of combined effects of concentration, temperature, and pressure.

Dioecious.

(Gk., two+ a house)-having staminate and pistillate flowers on separate pit species; the term is best applied to species, never to individual flowers.

Diploid.

Having two sets of chromosomes; the 2n number characteristic of the spc generation.

Division.

The largest category of classification of plants according to rules of nomenclature aggregation of classes, synonymous with phylum as used by zoologists.

Dominant.

A plant that dominates a plant community, owing to its tallness, or superior or both.

Dormancy.

A period of growth inactivity in bulbs, buds, seeds, and other plant organs.

Dwarf form.

Plants with the short stem, which could be more the gibberellins, than tall forms.

Ecology.

The science of the relationships between organisms and their environment.

Ecosystem.

An interacting System of one to many 'Mug organisms and their nonliving environment.

Embryo sac.
The female gametophyte of angiosperms, generally composed of seven cells: the egg cell, two synergids, and three (or more) antipodals (each with a single nucleus), and the central cell (with two nuclei).

Embryo.
The rudimentary plant formed in a seed or within the archegonium of lower plants.

Embryoid.
An embryo, often indistinguishable from a normal one, developing not from an egg but from a Somatic cell, often a tissue culture.

Endodermis.
The layer of ground tissue forming a sheath around the vascular region and having the casparian strip in its anticlinal walls; may have secondary walls later. It is the innermost layer of the cortex in roots and stems of seed plants.

Endoplasmic reticulum.
(Usually abbreviated to ER). A system of membranes forming cisternoid or tubular compartments that permeate the cytoplasm The cysternae appear like paired membranes in sectional profiles. The membranes may be coated with ribosomes (rough or granular ER) or be free of ribosomes (smooth ER).

Endosperm.
The nutritive tissue formed within the embryo sac of angiosperms from the central cell containing the primary endosperm nucleus.

Epicotyl.
The shoot part of the embryo or seedling above the cotyledon or cotyledons consisting of an axis and leaf primordia.

Epiphyte.
(Gk., upon + a plant)- a plant which grows upon another plant for position or support, but which doesn't parasitize it.

Essential elements.
Elements required by plants for normal growth and development.

Ethylene.
A gaseous hormone that has an inhibitory effect upon growth. It inhibits elongation and promotes ripening of fruits.

Etiolation.
(1)The condition characterizing plants grown in the dark or in light of very low intensity. (2)Stress condition of light deficiency. Plants developed in darkness called etiolated plants

Eukaryotic (also eucaryotic).

Refers to organism having membrane-bound nuclei, genetic material organized into chromosomes, and membrane-bound cytoplasmic organelles. Opposite of prokaryotic.

Evolution.

The history of the development of a race, species, or larger group of organisms that, following modifications in successive generations, has acquired characteristics that distinguish it from other groups.

Expansion.

Term of the increase of size of leaf, Cotyledon and the plant rozette form environment. The combination of all things and factors external to the individual or population of organisms in question. F1: the first filial generation following a cross. F2 and F3 are the second and the third generation.

Enzyme.

A proteinaceous catalytic agent that increases the rate of a particular transformation of materials in animals and plants.

False berry.

An accessory fruit in which both the ovary wall and the floral tube are fleshy at maturity, as in a cranberry.

Family.

A category of classification above a genus and below an order; composed of one or (usually) a number of genera. The suffix of the family name is usually -aceae.

Fats.

Organic compounds containing carbon, hydrogen, and oxygen, as in carbohydrates. The proportion of oxygen to carbon is considerably less in fats than it is in carbohydrates. One of the three kinds of plant foods. Fats in the liquid state are called oils.

Fermentation.

A respiratory process in which the hydrogen released in glycolysis is recombined with the pyruvic acid to form alcohol, lactic acid, or other products. No additional energy is made available during these steps. Fertilization. The union of two gametes to form a zygote. The fusion of two gametes, especially of their nuclei, resulting in the formation of a diploid (2n) zygote.

Fertilizers.

Materials added to the soil to provide elements essential to plant growth or to bring about a balance in the ratio of nutrients in the soil.

Fibers.

Greatly elongated, thick-walled, supporting cells, tapering at both ends. The two principal types are wood and sclerenchyma fibers.

Fiber.
An elongated strengthening cells with a thick wall, usually, but not necessarily, lignified.

Fibril.
Submicroscopic threads, composed of cellulose molecules, that constitute the form in which cellulose occurs in the cell wall.

Fibrous roots.
A root system in which the roots are finely divided. Filament. The stalk supporting the anther in a stamen.

Floral tube.
A tube or cup formed by the united bases of sepals, petals, and stamens, often in perigynous and epigynous flowers.

Florigen.
The postulated flowering hormone. A hypothetical hormone assumed to be concerned with the induction of flowering.

Flower.
(L.flos, a flower) the assemblage of reproductive structures and enveloping perianth; a complete flower consists of a calyx, corolla, androecium, and gynoecium.

Flower bud.
A bud that contains only one or more embryonic flowers.

Food.
An organic compound that can be respired to yield energy and that can be used in assimilation.

Fruit.
(L.,fructus, fruit or produce) a ripened ovary along with any adnate structures which mature along with it.

Fruit wall.
The outer part of the fruit derived either from the ovary wall (pericarp) or from the ovary wall plus accessory floral pans associated with ovary in fruit.

Gamete.
One of the two cells that Rise during sexual reproduction. A protoplasmic body capable of fusion with another gamete.

GAS (General Adaptation Syndrome).
Adaptation of the living organism to the complex of environmental stresses.

Gene interaction.
The condition in which the usual expression of one gene is modified by the presence of other genes.

Gene mutation.
A change in the nucleotide sequence of DNA as a result of errors in duplication.

Gene pool.
The sum total of all the genes that exist among all the individuals of a species.

Gene.
A segment of a DNA molecule, composed of several hundred nucleotides, that specifies the number, kind, and arrangement of the amino acids in the polypeptide chain unit of hereditary material attached to a chromosome.

Genes.
Segment of DNA that codes for one protein, which in turn determines a particular physical, physiological, or behavioral trait.

Genetic bank.
The Concept that natural ecosystems with all their species serve as a tremendous repository of genes.

Genetic code.
The sequence of basis in the DNA molecule which determines the arrangement and kinds of the amino acids in a polypeptide chain.

Genotype.
The genetic constitution of an organism; contrasted with phenotype.

Genetic control.
Selective breeding of the desire plant or animal to make it resistant to pest attack. Also attempting to introduce harmful genes – for example – to sterilize pest population.

Genetic engineering.
The artificial transfer of specific genes from one organism to another.

Genetic makeup.
Refers to all the genes that an individual possesses and that determine all of the individual's inherited characteristics.

Genetic variation.
An expression of the range of genetic (DNA) differences that occur among individuals of the same species.

Genetic.
The study of heredity and the processes by inherited characteristics are passed from one generation the next.

Geotropism.
Growth the direction of which is determined by gravity.

Germination.
The resumption of growth by the embryo in a seed; also beginning of growth of a spore, pollen grain, bud, or other structure.

Gibberellins.
A group of growth hormones that promote both cell division and cell elongation.

Gland.
(L., glans, an acorn) a secretory structure; used more broadly for any warty protuberance; also refers to the corpusculum in the milkweed flower; glandular, having glands.

Glucose.
Grape sugar, or dextrose, a 6-carbon sugar.

Granum.
A water-like body in a chloroplast, containing chlorophyll.

Growth regulator.
A synthetic chemical that affects growth.

Growth regulators.
Natural and synthetic substances with growth inhibiting or stimulation properties.

Growth substances.
Phytohormones and its analogues which stimulate or inhibit the main growth processes

Growth ring.
A growth layer of secondary xylem or secondary phloem as seen in transection of stern or root; may be an annual ring or a false annual ring.

Growth.
Irreversible increase in number and size of cells due to division and enlargement; usually accompanied by cell differentiation.

Guard cells.
A pair of cells flanking the stomatal pore and causing the opening and closing of the pore by changes in turgor.

Habit.
(L.,habitus, appearance). The general appearance of a plant.

Haploid.
The reduced, or n, chromosome number, characteristic of the gametophyte generation.

Hardwood.
(1) Wood produced by broad-leaved trees, such as maple, oak, ash, and elm; the xylem of woody dicotyledons; characterized by the presence of vessels. (2) Any tree having hardwood as defined under (1).

Heartwood.
Nonliving and commonly darker-colored wood surrounded by sapwood.

Hemicelluloses.
Polysaccharides more soluble and less ordered than the cellulose; common component of the cell wall matrix.

Herb.

(L., herba, plant)- an annual or perennial plant which dies back to the ground at the end of the growing season because it lacks the firmness resulting from secondary growth; herbaceous, having the features of an herb,

Herbaceous.

A term refers to any non-woody plant.

Herbicide.

A chemical used to kill or inhibit the growth of undesired plants.

Hermaphrodite.

A plant that has 'rude and female sex organs in the same flower.

Heterothallism.

The condition in which the two kinds of gametes that fuse, forming a zygote, are derived from different plants of the same species. Self-fertilization is not possible.

Heterotrophic.

Organisms that must obtain some or all of their food from external sources. In general, applied to animals and non- photosynthetic plants.

Hormone.

A chemical substance produced in one pan of an organism and transported to another part in which it has a specific effect.

Humidity.

Dampness. Relative humidity is the water vapor content of air as a percentage of the saturation content at the same temperature.

Humus.

A complex mixture of incompletely decomposed and synthesized newly organic materials in the soil.Humus is a dark brown or black, soft spongy residue of organic that remains after the bulk of dead leaves, wood, or other organic matter has decomposed. Humus does oxidize, but relatively slowly. It is extremely valuable in enhancing physical and chemical properties of sod.

Hybridization.

The process of crossing individuals of unlike genetic constitution.

Hydrolysis.

The disassembly of large molecules by the addition of water.

Hypocotyl.

Axial part of embryo or seedling located between the cotyledon or cotyledons and the radicle.

Hypocotyl-root axis.

Axial part of embryo or seedling comprising the hypocotyl and the root meristem, or the radicle if one is present.

Immunity.
Freedom from infection because of resistance. Inbreeding the breeding of closely related plants or animals. In plants, it is usually brought by self-pollination.

Imperfect flower.
A flower lacking either stamens or carpels.

Included phloem.
Secondary phloem included in the secondary xylem of certain dicotyledons. Term replaces intenylary phloem.

Incomplete flower.
A flower lacking one or more of the four kinds of floral organs: sepals, petals, stamens or pistils.

Indole-acetic acid (IAA).
A widely distributed plant growth hormone of the auxin type.

Inhibitor.
Beta-complex of phenolic and terpenoid inhibitors. Extracted by ether from plant tissues and which possess the retarding properties.

Inhibitors smethabolic.
Some antibiotics and its chemical analogues which depress a certain metabolic processes or reactions like protein synthesis, nucleic acid synthesis, coupling of respiration and phosphorilation.

Initial.
Cell in a meristem that by division gives rise to two cells one of which remains in the meristem, the other is added to the plant body. (2). Sometimes used to designate a cell in its earliest stage of specialization. More appropriate term for (2) primordium.

Intercalary meristem.
A meristematic region between two partly differentiated tissue regions. Meristematic tissue derived from the apical meristem and continuing meristematic activity some distance from that meristem; may be intercalated between tissues that are no longer meristematic.

Internode.
The region of the stem between any two nodes.

Karyokinesis.
Division of a nucleus as distinguished from the division of the cell, or cytokinesis.

Knot.
A portion of the base of a branch embedded in the wood.

Krebs cycle.
The long complicated series of reactions that results in the oxidation of pyruvic acid to hydrogen and carbon dioxide. The hydrogen, held on

hydrogen carrier molecules, then goes through the oxidative phosphorylation and terminal oxidation processes.

Lateral bud.

A bud in the axil of the leaf

Lateral meristem.

A meristem located parallel with the sides of the axis; refers to the vascular cambium and phellogen, or cork cambium.

Latex.

(L.,latex, juice) a white, yellowish or reddish thickened collodial sap, as in the spurges and milkweeds.

Leaching.

The process in which materials in or on the soil gradually dissolve and are carried by water seeping through the soil. It may result in the removal of valuable nutrients from the soil, or it may carry buried wastes into ground water, thereby contaminating it.

Leacliate.

The mixture of water and materials that are leaching

Leaf bud.

A bud that produces only leaves.

Leaf.

An organ of limited growth, arising laterally and from superficial tissues of a shoot apex. Usually dorsiventral in structure; may be simple or compound. Commonly consists of blade and petiole, but petiole may be absent (leaf sessile) or blade absent and petiole bladelike (phyllode).

Legume.

(L., legumen, a pulse) - a unicarpellate dry fruit which dehisces along both sutures; the fruit type of the Leguminosae and one of several common names used for the family.

Lignification.

The impregnation with lignin.

Lignin.

An organic substance or mixture of substances of high carbon content derived from phenylpropane and distinct from carbohydrates. Associated with cellulose in the walls of many cells.

Lipid.

A general term for the fatty substances, flits, and oils. Lipids have greatly reduced oxygen content as compared to carbohydrates.

Long-day plant.

A plant that flowers when days are longer when its critical light period.

Malt.

A cereal, usually barley, that has been allowed to sprout and then has been dried and ground. The product contains amylase and other enzymes.

Maltase.
An enzyme that hydrolyses maltose (malt sugar) to glucose.

Meiosis.
The two divisions during which the chromosomes are reduced from the diploid to the haploid number.

Meristem.
A region in which cell division continues, commonly for the life of the plant. Region of protoplasmic synthesis and tissue initiation.

Mesocotyl.
The internode between the scutenal node and the coleoptile in the embryo and seedling of Poaceae.

Mesophyll.
The photosynthetic parenchyma of a leaf blade located between the two-epidermal layers.

Metabolism.
The sum of all the chemical reactions that occur in an organism.

Micelia.
The threadlike feeding filaments of fungi.

Micropyle.
The opening in an ovule by which the pollen tube enters.

Microspore.
The spore that gives rise to the male gametophyte generation.

Microsporophyll.
A leaf usually modified, that bears one or more microsporangia.

Microtubules.
Nonmembranous tubules about 25 nanometers (250 angstroms) in diameter and of indefinite length. Located in the cytoplasm in a non-dividing eukaryotic cell, usually near the cell wall, and form the meiotic or mitotic spindle and the phragmoplast in a dividing cell. Sometimes called organelles.

Mitochondria.
Cytoplasmic organelles that contain the enzymes of the Krebs cycle and of oxidative phosphorylation.

Mitosis.
A process during which the chromosomes become doubled longitudinally, the daughter chromosomes then separating and forming two genetically identical daughter nuclei, Mitosis is usually accompanied by cell division.

Monocot.
A semi technical group name for those flowering plants which have a single cotyledon (hence the name), parallel venation, and flower pans in 3's or multiples thereof.

Monocotyledonous.
(Gk., one+ a vessel)-having one cotyledon.

Monoecious.
(Gk., one + a house)-having staminate and pistillate flowers on the same plant; note that the term is not synonymous with imperfect and should never be applied to individual flowers.

Morphogenesis.
The morphological and physiological events involved in the development of an organisms.

Mosaic.
Term designating the mottled or variegated appearance of a leaf resulting from the localized failure of chlorophyll formation; usually caused by virus infection.

Mutagenic.
Causing mutations.

Mutants.
Forms which appears after the effect of some chemical (mutagens) or physical factors (WV, X-rays) on genome.

Mutation.
A random change in one or more genes of an organism. Mutations may occur spontaneously in nature, but their number and degree are vastly increased by exposure to radiation and/or certain chemicals. Mutations generally result in a physical deformity and/or metabolic malfunction.

Mutualism.
Refers to a close relationship between two organisms in which both organisms benefit from the relationship.

Mycelium.
Collective term applied to a mass of hyphae.

Mycology.
The study of fungi.

Mycorrhiza.
The symbiotic association of fungi and roots of higher plants. The mycelia of certain fungi that grow symbiotically with the roots of some plants and provide for additional nutrient uptake.

Nastic movements.
A response to external stimuli in which the nature of the reaction is independent 1 of the direction from which the stimulus comes.

Natural phenolic inhibitors.
Inhibitors –some phenolic substances which forms and are accumulated in the tissues and depressed growth.

Natural selection.
The mechanism of evolution by which environmental factors favor the survival and reproduction of those variants of a population that are best adapted. Contrasted with artificial selection. Practiced by man the improvement of domestic plants and animals.

Nitrification.
The conversion of ammonia nitrogen to nitrate nitrogen by two specifics groups of bacteria, the nitrifying bacteria.

Nitrifying bacteria.
The bacteria that carry on nitrification.

Nitrogen bases.
Nitrogen-containing compounds (adenine, guanine, thymine, cytosine, and uracil) found in nucleotides.

Nitrogen-fixing bacteria.
Bacteria living in the soil, or in the roots of the legumes, that convert atmospheric nitrogen into nitrogen compounds in their own bodies.

Node.
(L.,nodus, a knot) -the point or region on a stem where one or more leaves are borne.

Nodules.
Enlargements on roots of plants, particularly in the Fabaceae, inhabited by nitrogen-fixing bacteria.

Nucellus.
Inner part of an ovule in which the embryo sac develops. Commonly considered to be equivalent to the megasporangium.

Nuclear membrane.
The outer, bounding membrane of the nucleus.

Nucleic acids.
A class of large acidic compounds composed of a series of nucleotides. There are of two kinds, DNA and RNA. Concerned with the storage and replication of heredity information and the synthesis of proteins.

Nucleic acids.
The class of natural organic macromolecules that function in the storage and transfer of genetic information.

Nucleolus.
Spherical body, composed mainly of RNA and protein, present in the eukaryotic cells, one or more to a nucleus; site of synthesis of ribosomes.

Nucleotides.
The structural units of DNA and RNA; compounds composed of an organic phosphate, a pentose sugar, and a nitrogen base. The different nucleotides vary in the nature of their nitrogen bases.

Nucleus.

The organelle in an eukaryotic cell bound by a double membrane and containing the chromosomes, nucleoli, and nucleoplasm.

Nut.

(L., nux, nut) a dry, hard, indehiscent, 1-seeded fruit derived from a syncarpous gynoecium.

Nutation.

The movement of growing parts, such as stem tips, leaves, or flowers, in which an irregularly circular path is traced in space as growth proceeds.

Nutrient.

Animal: Material such as protein, vitamins, and minerals required for growth, maintenance, and repair of the body and material such as carbohydrates required for energy Plant: An essential element in a particular ion or molecule that can be absorbed and used by the plant. For example, carbon, hydrogen, nitrogen, and phosphorus are essential elements; carbon dioxide, water, nitrate (NO_3^-) and phosphate ($-PO_4$) are the respective nutrients.

Nutrient cycle.

The repeated pathway of particular nutrients or elements from environment through one or more organisms back to the environment. Nutrient cycles include the carbon cycle, the nitrogen cycle, the phosphorus cycle and so on.

Ontogeny.

The development of all organism, organ, tissue, or cell from inception to maturity. The entire development of the individual, from the fertilized egg to the adult stage; the life history of an individual organism.

Open venation.

Leaf venation in which large veins end freely in the mesophyll instead of being connected by anastomoses with other veins.

Order.

A category of classification above the family and below the class; composed of one or more families. The suffix to the ordinal name is usually -ales.

Organelle.

Membrane-bound structure in an eukaryotic cell. Organelles partition the cell into regions which carry out different cellular functions. Mitochondria, the ER, and lysosomes are examples of organelles.

Organic compound.

A compound containing carbon and usually hydrogen, with or without other elements.

Organs.
A distinct and visibly differentiated part of a plant, such as root, stem, leaf, or part of a flower.

Osmosis.
The diffusion of water through a differentially permeable membrane from a region of higher diffusion pressure of water to one of lower diffusion pressure of water.

Ovary.
(L., ovum, an egg)-the lower swollen portion of the gynoecium which contains the ovules.

Ovulate.
(L., ovulatus, having eggs)- possessing ovules, as in certain conifer strobili.

Ovule.
At maturity, a structure composed of embryo sac, nucellus, integuments, and stalk. Following fertilization the ovule . An immature seed.

Oxidation.
The loss of electrons from an atom or a molecule; in biology, an energy-releasing process. Typically (1) the removal of hydrogen atoms, together with their electrons, or (2) the addition of oxygen to a compound.

Oxidative phosphorylation.
The production of ATP from ADP and P during the oxidation of $NADPH_2$ in respiration.

Ozone hole.
First discovered over the Antarctic, this is a region of stratospheric air that is severely depleted of its normal levels of ozone during the Antarctic spring because of CFCs from anthropogenic (human-made) sources

Ozone shield.
The layer of ozone gas (0_3) in the upper atmosphere that screens our harmful ultraviolet radiation from the sun.

Ozone.
A gas, O_3, that is a pollutant in the lower atmosphere but necessary to screen out ultraviolet radiation in the upper atmosphere May also he used for disinfecting water

Parallel.
(Gk., parallel) ertending in the same direction and equidistant, as in the vein pattern in most monocot leaves.

Parenchyma cell.
A cell with living contents, usually thin-walled, rhomboidal in shape, and relatively unspecialized.

Parenchyma.
A leaf tissue composed of columnar chloroplast-bearing cells with their long axes at right angles to the leaf surface.

Parenchyma.
An unspecialized, simple cell or tissue. The cells are isodiametric or sometimes elongated, usually thin-walled, living at maturity, and retaining a capacity for renewal of cell division.

Parthenocarpy.
The production of fruits in the absence of fertilization. Parthenocarpic fruits are usually seedless.

Pectin.
Noncellulosic polysaccharide which links to other pectins by demerized hydroxycinnamic acids such as diferulic acid. A complex organic compound present in the intercellular layer and primary wall of plant cells. The basis of fruit jellies.

Peptide bond.
A covalent bond between two amino acids,

Peptide.
Two or more amino acids linked by peptide bonds.

Perennial.
(L., perennis, lasting through the years) a plant which lives for three or more years, often flowering and fruiting repeatedly.

Pericycle.
A layer of cells (parenchyma) just outside the primary phloem and inside the endodermis. If the endodermis is lacking, as in many stems, the pericycle cannot be distinguished from the coflex.

Periderm.
Secondary protective tissue that replaces the epidermis in stems and roots, rarely in other an organs. Consists of phellem (cork), phellogen (cork cambium), and phelloderm.

Petal.
One of the units of the corolla of the flower.

pH.
A symbol denoting the negative logarithm of the concentration of the hydrogen ion in grams per liter. A measure of acidity or alkalinity of the solution.

Phenolic acid.
Metabolism in plants- initial phenylpropanoid metabolism and provides the biosynthesis steps of general precursors to lignin biosynthesis. Phenolic acids particularly hydroxycinnamates-low molecular compounds are present in plant cell walls, where they may act as molecular bridges. Phenolic acids involve in the biosynthesis of lignin precursors protect

plants from pathogen attack, providing signal molecules to regulate cell division or elongation and possibly playing role as natural antioxidants.

Phenotype.

The external, manifest, -or visible characters of an organism, as contrasted with its genetic constitution (the genotype).

Phloem elements.

Cells of the phloem tissue.

Phloem initial.

A cambial cell on the phloem side of the cambial zone that is the source of one or more cells arising by periclinal divisions and differentiating into phloem elements with or without additional divisions in various planes. Sometimes called phloem mother cell.

Phloem mother cell.

A cambial derivative that is the source of certain elements of the phloem tissue, such as, sieve element and its companion cells or phloem parenchyma cells forming a parenchyma strand. Used also in a wider sense synonymously with phloem initial.

Phloem parenchyma.

Parenchyma cells located in the phloem. In secondary phloem refers to axial parenchyma.

Phloem ray.

That part of a vascular ray which is located in the secondary phloem.

Phloem.

The principal food-conducting tissue of the vascular plant con~ose4 mainly of sieve elements, various fibers, and sclereids.

Phosphate.

A compound of phosphorus; in general, phosphoric acid.

Phosphorylation.

A phosphorylation reaction dependent upon the energy of light absorbed by the chlorophylls.

Photoperiod.

The optimum length or period of illumination required for the growth and maturation of a plant. The photoperiod is distinct from photosynthesis.

Photoperiodic response.

The response of plants to the relative length of the day and night.

Photoperiodism.

The reaction of plant to day length.

Phototropin.

Blue light absorbing chromoprotein, photoreceptor-mediating phototropism.

Phytochrome.
Plant photoreceptor: red - far-red reversible chromoprotein, mediating photomorphogenesis.

Placenta.
The region or area of the ovary to which one or more ovules (or seeds) are attached.

Plasmodesma.
Cytoplasmic threads that extend through openings in the cell walls, connecting the protoplasts of adjacent living cells.

Plastid.
Organelle with a double membrane in the cytoplasm of many eukaryotes. May be concerned with photosynthesis (chloroplast) or starch storage (amyloplast), or contain yellow or orange pigments (chromoplast)

Plastochron (or plastochrone).
The time interval between the inception of two successive repetitive events, as origin of leaf primordia, attainment of certain stage of development of a lea, etc. Variable in length as measured in time units.

Plate meristem.
A meristematic tissue consisting of parallel layers of cells dividing only anticlinally with reference to the wide surface of the tissue. Characteristic of ground meristem of plant parts that assumes a flat form, as a leaf

Plumule.
The young stem as it grows out of the seed.

Pollen grain.
A microspore in a seed plant included in an elaborately structured wall (one cell). Also a germinated microspore having formed a microgametophyte, immature (two cells) or mature (three cells).

Pollen sac.
A locute in the anther containing the pollen grains.

Pollen tube.
A tubular cell extension formed by the germinating pollen grain; carries the male gametes into the ovule.

Pollen.
A collective term for pollen grains.

Pollination.
The transfer of pollen from the anther to the receptive surface, stigma in angiosperms.

Polypeptide.
A long chain of amino acids.

Pome.
An accessory fruit with a leathery endocarp, as an apple.

Primary cell wall.

Version based on studies with the light microscope: cell wall, formed chiefly while the cell is increasing in size. Version based on studies with the electron microscope; cell wall in which the cellulose microfibrils show various orientations - from random to more or less parallel - that may change considerably during the increase in size of the cell. The two versions do not necessarily coincide in delimiting primary from secondary wall.

Primary endosperm nucleus.

Nucleus resulting from the fusion of the male gamete and two polar nuclei in the central cell of the embryo sac.

Primary growth.

The growth of successively formed roots and vegetative and reproductive shoots from the time of their initiation by the apical meristems and until the completion of their expansion. Has its inception in the apical meristems and continues in their derivative meristems, protoderm, ground meristem, and procambium, as well as in the partly differentiated primary tissues.

Primary phloem.

Phloem tissue differentiating from procambium during primary growth and differentiation of a vascular plant. Commonly divided into the earlier protophloem and the later primary cell wall.

Primary tissues.

Cells derived from the apical and subapical meristems of root and shoot; opposed to secondary tissues derived from a cambium or cork cambium.

Primary wall.

The wall layer on either side of the intercellular layer; (2) collective terms applied to two primary walls and the intercellular layer.

Procambium Primary.

Meristem or meristematic tissue giving which differentiates into the primary vascular tissue. Also called provascutar tissue.

Prop roots.

Roots that arise from the stem above soil level.

Proplastid.

A plastid in its earliest stages of development.

Protease.

An enzyme that digests protein.

Protein.

Complex organic compounds constructed from amino acids and composed of carbon, oxygen, and nitrogen. Many proteins also contain sulfur. Proteins are one of the three groups of plant foods, and the chief organic component of protoplasm.

Radicle.

The embryonic root. Forms the basal continuation of the hypocotyl in an embryo.

Ray.

A panel of tissue variable in height and width, formed by the ray initials ill the vascular cambium and extending radically in the secondary xylem and secondary phloem.

Receptacle.

The tip of a flower stalk, to which the floral parts are attached.

Redifferentiation.

A reversal in differentiation in a cell or tissue and subsequent differentiation into another type of cell or tissue.

Repression.

The inhibition of an operator gene by a regulator gene through the production of the protein repressor, The removal of this inhibition is known as depression.

Respiration.

An intracellular process in which food is oxidized with release of energy. The complete breakdown of sugar or other organic compounds to carbon dioxide and water is termed aerobic respiration, although the earlier steps are anaerobic.

Rhizobium-Legume.

Symbiosis-process, which begins with two fi living organisms and end with intimate cellular, co-exists Rhizobium bacteria recognize specific plants, provide development of a root nodule and invade the plant tissue. The bacteria brings fixed nitrogen to the plant receiving in turn plant biosynthetic products (sugars, amino acids).

Rhizoid.

(L.,rhizoideus, root-like)-a root-like structure in certain vascular plants which carries on the functions of the root, but which lacks it anatomical complexity.

Rhizome.

(L., rhizoma, a root)- an underground horizontal stem which bears reduced scaly leaves.

Root.

A portion of the plant axis; distinguished from rhizosomes and stolones by absence of nodes or internodes.

Root apex.

The meristematic tissue in the terminal pan of the root, that is, the root apical meristem, sometimes used loosely to include the root cap also.

Root hair.

A type of trichome on root epidermis that is a simple extension of an epidermal cell and is concerned with absorption of soil solution.

Root pressure.
The pressure developed in roots that causes guttation and exudation from cut stamps.

Rootcap.
A thimblelike mass of cells covering the apical meristem of the root.

Rosette.
(dim off., rose, a rose) a cluster of leaves in a radiating pattern, usually at the base of the 286 plant.

Scion.
A portion of a shoot used for grafting.

Scutellum.
The portion of the cotyledon that, in grasses, absorbs food from the endosperm at germination.

Secondary cell wall.
Version based on studies with the light microscope: cell wall deposited in some cells over the primary wall after the primary wall ceases to increase in surface.

Secondary endosperm nucleus.
Nucleus resulting from fusion of two polar nuclei in the central cell of the embryo sac of angiosperms.

Secondary growth.
In gymnosperms, most dicotiledons, and some monocotiledons, A type of growth characterized by an formation of secondary vascular tissues by the vascular cambium Commonly supplemented by activity of the cork cambium (phellogen) forming periderm.

Secondary phloem.
Phloem tissue formed by the vascular cambium during secondary growth in a vascular plant. Differentiated into axial and ray systems.

Secondary tissue.
Tissues produced by the cambium and cork cambium.

Secondary xylem.
Xylem tissue formed by the vascular cambium during secondary growth in a vascular plant. Differentiated into axial and ray systems.

Seed coat.
The outer coat of the seed derived from the integument or integuments. Also called testa.

Seed.
A structure formed by seed plants following fertilization. In conifers it consists of seed coat, embryo, and female gametophyte (n) storage tissue.

Some angiosperm seeds are composed only of seed coat and embryo; others also contain endosperm (3n) storage tissue.

Seminal adventitious root.

Root initiated in the embryo on the hypocotyl or higher on the axis.

Sepal.

(Gk., a covering) - one of the component parts of the calyx or a unit of the calyx.

Sex organs.

The archegonia and antheridia.

Shoot apex.

The meristematic tissue in the terminal part of the shoot; sometimes considered to consist only of the tissue above the youngest leaf primordia, but more commonly considered to extend below this essentially synonymous with apical meristem of shoot.

Shoot.

Collective term applied to the stem and leaves; any young growing branch or twig.

Short-day plants.

A plant that flowers when days are shorter than its critical light period.

Sieve tube.

A series of sieve elements (sieve tube members) arranged end to end and interconnected through sieve plates.

Simple fruit.

Fruit derived from a single pistil, simple or compound; ovary superior or inferior.

Simple leaf.

An undivided leaf; opposed to compound.

Species.

(L., species, a kind or son)- a kind of plant or animal, its distinctness seen in morphological, anatomical, cytological, and chemical discontinuities presumably brought about by reproductive isolation; thought by some to be entities with biological reality and by other to be convenient concepts which exists only in the mind of the taxonomist.

Species name.

The binomial consisting of the genus and the specific epithet.

Starch synthetase.

The primary enzyme concerned in starch synthesis.

Starch.

An insoluble carbohydrate, the chief food storage substance of plants, composed of anhydrous glucose residues of the formula C6H1005 into which it easily breaks down.

Stomata sign, stoma.
Microscopic pores in leaves, mostly on the undersurface, that allow the passage of carbon dioxide and oxygen into and out of the leaf and that also permit the loss of water vapor from the leaf. A pore in the surface of a leaf, surrounded by a pair of guard cells, through which carbon dioxide enters the leaf during photosynthesis.

Stratification.
The exposure of moist dormant seeds to low temperatures to break the dormant period.

Stroma.
The supporting framework of a plastid; in a chloroplast, the colourless portion in which the grana, which contain the chlorophyll, are embedded,

Style.
Extension of the top of the ovary, usually columnar, through which the pollen tube growth.

Substrate.
Substance acted upon by an enzyme.

Symbiosis.
The phenomenon of two organisms inhabiting one body for mutual benefit. The living together in close association of two or more dissimilar organisms. Includes parasitism, in which the association is harmful to one of the organisms; and mutualism, in which the association is advantageous to both.

Testa.
The tough outer coat of a seed.

Thorn.
A branch that has become hard, woody, and pointed.

Thylakoids.
Saclike membranous structures (cisternae) in a chloroplast combined into stacks (grana) and present singly in the stroma (stroma Thylakoids or frets) as interconnections between grana.

Tiller.
A branch from the axil of a lower leaf, as in grasses.

Tissue culture.
Growth of cells or tissues from plants or animals in sterile, synthetic media; an important research tool.

Totipotency.
The capacity of a cell or a group of cells to give rise to an entire organism.

Toxins.
Poisonous substances produced by living organisms. As generally employed, refers to is substances that stimulate the formation of antibodies.

Transgenic plants.
Plants with the insertion of active foreign (plant, bacteria, fungi) gene into native genome, resulted in the change of the morphological and physiological properties

Transpiration.
The evaporation of water from the leaves of a plant.

Tree.
A perennial woody plant with a single stem (trunk).

Trichrome.
(Gk., a growth of hair)- any hair-like outgrowth of the epidermis.

Trifoliate.
(L., trifoliatus, three-eaved)- a plant with three leaves, as in Trillium.

Trifoliolate.
(L., trifoliolatus, three leaflets)- having three leaflets.

Tropism.
Refers to movement or growth in response to an external stimulus the site of which determines the direction of the movement or growth.

Ultraviolet radiation.
Radiation similar to light but with wavelength slightly shorter that violet light and with more energy. The greater energy causes it to severely burn and otherwise damage biological tissues.

Vacuolation.
The development of vacuoles in a cell; in mature state the presence of vacuoles in a cell.

Vacuole.
Cavity within the cytoplasm filled with a watery fluid, the cell sap, and bound by a unit membrane, the tonoplast. Involved in uptake of water during germination and growth and maintenance of water in the cell. Also contains hydrolytic enzymes and has a lytic function.

Vascular bundle.
A strandlike pan of the vascular system composed of xylem and phloem.

Vascular cambium.
Lateral meristem that forms the secondary vascular tissues, secondary phloem and secondary xylem, in stem and root. Is located between those two off cells toward both tissues,

Vascular cylinder.
Vascular region of the axis. Term used synonymously with stele or central cylinder or in a more restricted sense excluding the pith.

Vascular meristem.
General term applicable to procambium and vascular cambium

Vascular system.
The total of the vascular tissues in their specific arrangement in a plant or plant organ.

Vegetable.
Botanically, any edible pan of a plant not formed from a mature ovary or from an ovary and associated parts.

Vegetative reproduction.
(1) In seed plants, reproduction by means other than by seeds; (2) in lower forms, reproduction by vegetative spores, fragmentation, or division of the plant body.

Vegetative.
Growth, tissues, pr process concerned with the maintenance of the plant body, as contacted with tissues or activities involved in sexual reproduction.

Venation.
The arrangement of veins in the leaf blade.

Vernalization.
The sensitizing of a plant to photoperiod by exposure to near-freezing temperatures at an earlier stage of development.

Vessel element.
A cell, usually lignified, of the wood in which the end walls are perforated.

Vessel.
A tubelike series of vessel members the common walls of which have perforation.

Vine.
(L., vinea, a vine) - herbaceous plants with elongate, flexible, non-self-supporting stems.

Virus.
A submicroscopic entity whose genetic material is composed either of DNA or RNA, and which is directly dependent upon a living host cell for its multiplication.

Vitamins.
Complex organic compounds constructed by green and some non green plants and necessary in minute amounts for normal growth.

Vittae.
(L., vitta, a fillet)- the oil tubes in the pericarp of most members of the Unbelliferae.

Wood fibers.
The supporting cells of the wood resembling tracheids but generally longer, thicker-walled, with more tapering ends and with reduced pits;

Wood.
Usually secondary xylem of gymnosperms and dicotyledons, but also applied of to any other xylem.

Woody plants.
Trees and shrubs in which increase in diameter of stems and roots continues from year to year.

Xanthophyll.
A yellow or orange pigment found in the chloroplasts of a plant, and in some chromoplasts.

Xanthophylls.
Yellow to orange carotenoid pigments associated with carotenes in the plastids of plant cells.

Xylary procambium.
The part of procambium that differentiates into primary xylem. Also called xylic or xyloic procambium.

Xylem ray.
That part of a vascular ray which is located in the secondary xylem.

Xylem initial.
A cambial cell on the xylem side of the cambial zone that is the source of one or more cells arising by periclinal divisions and differentiating into xylem elements either with or without additional divisions in various planes. Sometimes called xylem mother cell.

Xylem mother cell.
A cambial derivative that is the source of certain elements of the xylem, such as axial parenchyma cells forming a parenchyma strand. Used also in a wider sense synonymously with xylem initial.

Xylem.
The principal water-conducting tissue in vascular plants characterized by the presence of tracheary elements. The xylem may also serve as a supporting tissue, especially the secondary xylem (wood).

Zones of stress.
Regions where a species finds conditions tolerable but suboptimal. Where a species survives but under stress.

INDEX

A

B

C

D

E

F

G

H

I

K

L

P

Q

R

S

T

U

V

W

X

Z